# AC Corrosion of Pipelines

Y. Frank Cheng

AMPP™

**AMPP**
**Association for Materials Protection and Performance**


Printed in the United States of America

ISBN: 978-1-57590-400-9



AMPP
Association for Materials Protection and Performance
15835 Park Ten Place
Houston, TX 77084
ampp.org

# Contents

**CHAPTER 1: INTRODUCTION ........ 1**
1.1. Pipelines and pipeline integrity management ........ 1
1.2. Corrosion causing pipeline failure ........ 3
1.3. Pipeline failure under alternating current interference ........ 4
1.3.1. Pipelines in collocation with HVAC power lines ........ 5
1.3.2. Pipelines with adjacency to AC-powered rail transit systems ........ 6
1.4. Contents of the book ........ 7
References ........ 10

**CHAPTER 2: SOURCES AND PRINCIPLES OF AC INTERFERENCE ........ 13**
2.1. Sources of AC interference ........ 13
2.1.1. HVAC transmission power lines ........ 13
2.1.2. Railway electrification systems ........ 14
2.1.3. Other sources ........ 15
2.2. Types and principles of AC interference by HVAC power transmission lines ........ 15
2.2.1. Capacitive coupling effect ........ 15
2.2.2. Resistive coupling effect ........ 16
2.2.3. Inductive coupling effect ........ 18
2.3. AC interference and the associated risk ........ 20
2.3.1. Normal conditions of HVAC power lines ........ 20
2.3.2. Fault conditions of HVAC power lines ........ 21
2.4. AC interference from electrified rail systems ........ 22
References ........ 23

**CHAPTER 3: PIPELINE CORROSION IN THE PRESENCE OF AC INTERFERENCE ........ 25**
3.1. Recognition of AC corrosion of pipelines ........ 25

3.2. Basics of AC corrosion of pipelines ... 27
3.2.1. Characteristics of AC corrosion of pipelines ... 27
3.2.2. AC corrosion mechanisms ... 28
3.2.3. Parametric effects ... 34
3.3. Experimental setup enabling AC corrosion measurements ... 41
3.4. AC corrosion of pipelines: Active dissolution systems ... 43
3.4.1. Environments resulting in active dissolution of pipeline steels 43
3.4.2. AC accelerated pipeline corrosion ... 43
3.4.3. Mechanistic analysis ... 49
3.5. AC corrosion of pipelines: Passive systems ... 52
3.5.1. Environments resulting in passivation of pipeline steels ... 52
3.5.2. AC-induced passivity degradation ... 53
3.5.3. Mechanistic analysis ... 57
3.6. AC corrosion of pipelines at coating defects ... 60
3.6.1. Size of coating defects ... 60
3.6.2. Shape of coating defects ... 71
References ... 74

**CHAPTER 4: AC-INDUCED PITTING CORROSION OF PIPELINES ... 81**
4.1. Conditions for initiation of pitting corrosion on pipelines ... 81
4.1.1. Near-neutral pH bicarbonate environments ... 81
4.1.2. High pH carbonate–bicarbonate environments ... 85
4.2. AC-induced pitting corrosion ... 87
4.2.1. Near-neutral pH aqueous environments ... 87
4.2.2. High pH aqueous environments ... 90
4.2.3. Threshold conditions for initiation of pitting corrosion under AC interference ... 94
4.3. Mechanisms for AC-induced pitting corrosion on pipelines ... 94
References ... 97

**CHAPTER 5: COATING DEGRADATION AND FAILURE UNDER AC INTERFERENCE ... 99**
5.1. Essential coating properties for pipeline integrity maintenance ... 99
5.2. Effect of AC on the resistance of coatings to cathodic disbonding.. 100
5.2.1. Concept and principle of cathodic disbonding of pipeline coatings ... 100
5.2.2. Testing for resistance of coatings to cathodic disbonding under AC interference ... 101
5.2.3. Cathodic disbonding of pipeline coatings under AC interference ... 103
5.3. Effect of AC on water uptake by pipeline coatings ... 106
5.3.1. Coating properties associated with water resistance ... 106
5.3.2. Testing setup to measure water uptake by coatings under AC interference ... 108
5.3.3. Dependence of the water resistance of coatings on AC interference ... 108
5.4. Effect of AC on electrochemical impedance of the coatings ... 110

5.5. Coating structure and performance degradation under AC interference ........ 113
5.6. A conceptual model for coating degradation under AC interference ........ 117
References ........ 118

**CHAPTER 6: CATHODIC PROTECTION PERFORMANCE IN THE PRESENCE OF AC INTERFERENCE ........ 119**
6.1. Principle of impressed current CP on pipelines ........ 120
6.1.1. Impressed current cathodic protection ........ 120
6.1.2. Sacrificial anode cathodic protection ........ 121
6.1.3. Use of ICCP in pipeline corrosion protection ........ 121
6.2. CP shielding under disbonded coating in the presence of AC interference ........ 123
6.2.1. What is CP shielding? ........ 123
6.2.2. Effect of AC on CP permeation into the coating disbondment: Testing results ........ 124
6.2.3. Effect of AC on CP permeation into the coating disbondment: Mechanistic analysis ........ 129
6.3. Shift of CP potentials under AC interference ........ 131
6.3.1. At NACE-recommended CP potential ........ 132
6.3.2. At potentials more negative than the NACE-recommended value ........ 136
6.3.3. Mechanistic aspect of the effect of AC on CP potential shift ........ 140
6.4. CP performance under AC interference ........ 143
6.4.1. Potentiostatic current density and corrosion rate of CP-applied pipeline steels under AC interference ........ 144
6.4.2. Effect of AC on CP effectiveness for corrosion protection ........ 151
6.5. Recommendations of CP criteria for pipeline corrosion protection under AC interference ........ 153
References ........ 158

**CHAPTER 7: INSPECTION AND MONITORING TECHNIQUES FOR AC CORROSION OF PIPELINES ........ 163**
7.1. Conventional measurement techniques ........ 164
7.1.1. Buried coupon test station ........ 164
7.1.2. Ultrasonic-based coupon testing method ........ 168
7.1.3. AC voltage modulation technique ........ 169
7.2. An online AC corrosion monitoring and assessment technique ........ 172
7.3. A graphical user interface-based AC corrosion monitoring technique ........ 176
References ........ 177

**CHAPTER 8: MODELING FOR EVALUATION OF AC INTERFERENCE AND THE RISK OF AC CORROSION OF PIPELINES ........ 179**
8.1. Historical background ........ 179

8.2. Modeling for evaluation of AC interference .......... 181
8.2.1. Principles for modeling of AC interference .......... 181
8.2.2. Modeling of affecting factors and variables .......... 182
8.2.3. Modeling of AC harmonic interference .......... 190
8.3. Application of computational models for assessment of AC interference on pipelines .......... 192
8.3.1. Optimization of power line design for minimization of AC interference .......... 192
8.3.2. Modeling of AC voltages and AC current densities on pipelines .......... 192
8.3.3. Modeling of fault currents .......... 195
8.3.4. Modeling of the mutual coupling effect between a pipeline and grounding anodes .......... 197
8.4. Reliability analysis of the AC interference modeling .......... 198
References .......... 198

**CHAPTER 9: MITIGATION AND CONTROL OF AC CORROSION OF PIPELINES .......... 201**
9.1. General principles for mitigation and control of AC interference... 201
9.2. Surface and deep grounding .......... 203
9.2.1. Surface grounding .......... 204
9.2.2. Point drain grounding .......... 205
9.2.3. Deep grounding .......... 205
9.2.4. Grounding materials .......... 206
9.2.5. Backfill materials .......... 207
9.3. Overvoltage arresters .......... 208
9.4. Transformer use for AC interference mitigation .......... 210
9.5. DC decoupling technique .......... 211
9.6. Gradient control facilities .......... 212
9.7. Industry standards and practical recommendations .......... 213
9.7.1. NACE International criteria .......... 213
9.7.2. European Committee for Standardization .......... 214
9.8. Development of AC interference management program or pipelines .......... 214
References .......... 219

**CHAPTER 10: CORROSION OF PIPELINES UNDER DC INTERFERENCE...223**
10.1. Sources of DC interference .......... 224
10.1.1. DC device .......... 224
10.1.2. HVDC power lines .......... 226
10.1.3. DC transit systems .......... 228
10.1.4. Identification of DC interference .......... 230
10.2. Features and parametric effects on DC corrosion of pipelines...... 230
10.2.1. Recognition of DC corrosion of pipelines .......... 230
10.2.2. Parametric effects on DC corrosion of pipelines .......... 231
10.3. DC corrosion measurements .......... 232
10.3.1. Testing setup enabling DC corrosion measurements .......... 232

10.3.2. Effect of DC current density on steel potential ........ 233
10.3.3. Effect of DC current density on solution pH in the anodic and cathodic zones ........ 236
10.3.4. Corrosion rate and morphology of steel under various DC current densities ........ 237
10.4. Mechanism of DC corrosion ........ 239
10.5. Effect of DC on coating properties and performance ........ 241
10.5.1. Water permeability ........ 241
10.5.2. Electrochemical impedance spectroscopy ........ 243
10.5.3. Molecular structure of coatings ........ 246
10.5.4. Mechanistic aspect of the role of DC interference on coating degradation ........ 249
10.6. Adverse effect of DC interference on CP performance ........ 251
10.7. Management of DC corrosion of pipelines ........ 257
10.7.1. Detection of DC voltage and stray current for evaluation of DC interference ........ 257
10.7.2. Mitigation of DC corrosion ........ 259
10.7.3. Modeling and calculation of DC interference on pipelines .. 261
References ........ 262

Index ........ 267

# 1 Introduction

## 1.1. Pipelines and pipeline integrity management

Pipelines are regarded as energy lifelines, effectively and efficiently transporting crude oil, natural gas, and petrochemical products from their production sites—which are usually located in remote areas, ports of entry, and processing plants—to markets and consumers every day. There is a vast pipeline network across regions, countries, and even continents. The total length of oil and gas transmission pipelines around the world is estimated at 3,500,000 km.[1] The number can be multiplied many times if gathering and distribution pipelines are included. In Canada, there are 119,000 km of large-diameter transmission pipelines, while transmission, gathering, and distribution pipelines total approximately 825,000 km.[2] According to the U.S. Department of Energy's Energy Information Administration, world energy consumption is predicted to increase by 71% from 2003 to 2030, with fossil fuels continuing to supply much of the energy used worldwide.[3] It is thus expected that pipeline construction and operation activities will continue growing. Internationally, about 32,000 km of new pipelines are constructed each year.[1]

Compared with modes such as rail, trucks, and ships, pipelines are more economical and safer for oil and gas transportation.[4] On average, it costs about U.S. $5 per barrel to transport oil and gas by pipelines, compared with U.S. $10–15 a barrel for rail and U.S. $20 a barrel by truck.[5] According to the Canadian Energy Pipeline Association (CEPA), a train with 4,200 rail cars or 15,000 tanker trucks would be required to transport the 3 million barrels of crude oil transported by pipelines in Canada every day.[6] Moreover, pipelines require much less human capital than the other methods of transportation and are more efficient at moving products over large distances. In addition, compared with both train and truck transportation, pipelines are actually the safest method of transporting oil and gas. In the United

States, oil and gas transportation by pipelines resulted in 1.7 fatalities to operators, personnel, and the general public per year, while transportation by rail and truck resulted in 2.4 and 10.2 fatalities per year, respectively.[4] In a report released by the Fraser Institute, pipeline transportation of hydrocarbon products was shown to be 4.5 times safer than rail on a like-for-like basis from analysis of the North American data.[7]

In reality, a number of factors can adversely affect pipeline integrity, potentially causing pipeline failures. These factors include internal and external corrosion, stress corrosion cracking, mechanical fracture, materials and manufacturing faults, equipment and component failures, geotechnical factors, incorrect operation, and external interference such as excavation.[8, 9] Although occurring only occasionally, pipeline incidents usually attract wide attention from regulators, news media, and the public owing to the potential seriousness of their consequences and their potential impacts on the environment and ecology.[10] The most widely reported examples include (a) the rupture and release of Enbridge's oil pipeline in Marshall, Michigan, on July 25, 2010, resulting in the largest inland oil spill–and indeed one of the costliest spills–in U.S. history[11] and (b) Pacific Gas and Electric's natural gas transmission pipeline rupture and fire in San Bruno, California, on September 9, 2010, causing eight deaths, many injuries and massive evacuation from the area.[12]

Safety is recognized as the top priority to pipeline operators. The safe and reliable operation of pipeline systems is critical to both energy transportation and environmental conservation, as well as the national economy. A pipeline integrity management program has thus been developed and implemented by the pipeline industry to ensure the safety, reliability, and longevity of the system and to mitigate and prevent pipeline failure. The pipeline integrity management program is a process to develop, implement, measure, and manage the integrity of a pipeline through assessment, mitigation, and prevention of risks to ensure a safe, environmentally responsible, and reliable service.[13] A typical pipeline integrity management program consists of four essential steps: data gathering, feature assessment, risk assessment, and integrity planning.[14] In step one, pipeline operators collect substantial data from sampling and testing in the field, literature review, databases from relevant parties, and periodic inspections, particularly in-line inspections (ILIs). The data collection provides information on the fitness-for-service of the pipeline. Data from a single inspection are usually not sufficient to implement the pipeline integrity management program. Different inspections conducted on the same pipeline segment over a period of time are required. In step two, the collected data are analyzed, by either comparison with historical records or developed models or both, to assess the condition of the pipeline. Emphases are usually put on the critical features detected on the pipeline to determine the potential risk and, if necessary, schedule rehabilitation measures. Critical parameters, such as the failure pressure of the pipeline, the welding strength, the defect growth rate, and the remaining service life of the pipe, are calculated and determined. In the third step, the potential consequences and impacts resulting from pipeline failure are evaluated. These include not only energy loss and the direct financial effect on pipeline operators but also the impact on the environment, ecology, and public safety. The final step involves the consideration, design, and implementation of appropriate measures to address any issues identified on the pipeline. It also includes taking measures and actions to eliminate and control pipeline failures. This is the goal of the entire process.[15] Implementing the pipeline integrity management program is one of the core tasks of pipeline companies.

## 1.2. Corrosion causing pipeline failure

Corrosion has been recognized as one of the primary mechanisms causing pipeline failures.[9, 16, 17] According to statistics from Canadian Energy Pipeline Association members, the leading causes of pipeline failure include metal loss (i.e., reduction in the pipe wall thickness due to corrosion); material, manufacturing, or construction defects; and cracking.[9] Collectively, these accounted for more than 79% of the total incidents on Canadian pipelines over the period 2010–2014. Corrosion was responsible for 35.2% of all reported pipeline failures over the time period. This figure did not include the cracking relevant to corrosion. In a comparative analysis of pipeline performance issued by the National Energy Board, the primary cause of ruptures on National Energy Board–regulated pipelines between 1991 and 2009 was corrosion-related cracking (38%) and metal loss (27%).[16] The cracking included hydrogen-induced and mechanical damage delayed cracking, stress corrosion cracking, and corrosion fatigue. The metal loss included both internal and external corrosion. For liquid pipelines operated in the United States, data from the Pipeline and Hazardous Materials Safety Administration showed that corrosion was the second leading cause of reported oil spills from 2002 to 2012, representing 34% of cases.[17]

Corrosion can occur on pipelines both internally and externally. For long-distance oil and gas transmission pipelines, internal corrosion is not a big threat. The quality of the fluids carried (i.e., oil and natural gas) has been ensured through processing before they are transported by the pipelines. According to Federal Energy Regulatory Commission (FERC) regulations, the transported fluids are subject to a limitation on basic sediments and water of less than 0.5 vol%.[18] Pipelines do not corrode in crude oil unless the entrained water and wetted sands are separated from the oil phase and settle down on the pipe floor, making the pipe steel water-wetted over an extended time period.[19] Moreover, the pipeline operators conduct periodic pigging to clean out various deposits, which are usually composed of petroleum sludge, solid sands, water, chemicals, corrosion products, or microorganisms from the pipeline interior; this pigging removes the environment that might otherwise cause corrosion. ILIs allow operators to collect information about the geometry and location of pipeline flaws and analyze the fitness-for-service of pipelines and thereby enable proper measures to be taken.[9] The industry also depends on appropriate chemical treatment programs to add inhibitors and biocides to mitigate and control the corrosion and microbial corrosion of the pipelines.[20] Owing to the strictly enforced basic sediments and water limit in the pipelined fluids and to the periodic pigging (combined with the appropriate chemical treatment program), internal corrosion does not occur as frequently as external corrosion in transmission pipelines.

External corrosion of buried pipelines is protected against with coatings and cathodic protection (CP). The principle of the design is straightforward. The coating provides the first line of defense to protect the pipelines from environmental attack, and the CP serves as a backup. When the coatings have degraded and failed owing to either transportation and construction or long-term service, the CP can protect the pipeline at the coating failures. However, in reality, the CP can become shielded and lose its protective ability when the coating has failed. Three scenarios have been identified on pipelines where the CP loses its function while the coating degrades and fails.[21] First, the pipeline coatings (primarily polymeric coatings) contain pinholes, which are usually invisible to the eye but can be penetrated by water and corrosive species and gases to cause corrosion of pipeline steel at the bottom of the pinholes. Owing to the geometric limitation and deposit of corrosion products, the CP current is shielded from reaching the pinhole bottom for corrosion protection. As a result, the mouth of the pinhole is under CP, but its bottom suffers from corrosion.[22] Second, the coating is disbonded from the steel substrate at a holiday. Although CP is applied, the disbonding crevice is shielded from the CP current owing to the geometric limitation. Thus, corrosion occurs on the steel under the disbondment, while the holiday is under CP.[23, 24] Third, the CP current is shielded from defect-free coating membranes, which are usually composed of polyethylene (PE) components, such as PE tape, when the coatings are disbonded from the pipe steel. Owing to the intrinsic properties of PE molecules, there is low permeability of the coating membranes to water, gases, chemicals, and the CP current.[25]

Field experience has shown that the majority of pipeline failures and the resulting consequences can be attributed to the external corrosion of pipelines. For example, the rupture and release of Enbridge's oil pipeline in Marshall, Michigan, on July 25, 2010, was caused by external corrosion and the induced cracking at corrosion defects.[11] On November 22, 2013, the Donghuang II oil pipeline exploded in Qingdao, China, ripping roads and sidewalks apart, turning cars over and sending thick black smoke over the city. The blast killed 62 people and injured 136. The incident led to stoppages in electricity and water in nearby areas, and about 18,000 people were evacuated.[26] The direct cause of the explosion was the ignition of vapors produced from oil leaking from an underground pipeline owing to external corrosion under a disbonded coating.

## 1.3. Pipeline failure under alternating current interference

With increasing energy demand and the increased cost of right-of-way (ROW) suitable for pipelines and high-voltage alternating current (HVAC) power transmission lines, organizations are forced to construct their pipelines and transmission power lines to follow the same route (the so-called energy corridors). The collocation of pipelines and the HVAC power lines is mainly due to numerous restrictions, including private and governmental regulations for land use, minimization of environmental impact, high transportation efficiency, and lower construction costs.[27] These restrictions become even more severe when new construction approaches high-density urban areas. In North America, for example, the increased collocation of pipelines and alternating current (AC) power lines is due to the recent shale production.[28] As the shale oil and gas production is booming, the need to build infrastructure to move the products to the markets is urgent. At the same time, it is becoming increasingly difficult for pipeline and power line operators to secure new ROWs. Laying a pipeline adjacent to an existing HVAC power line thus becomes a quick and easy solution.[27]

Recently, a number of reports have been made of pipeline failures resulting from AC interference. In 2001, AC corrosion was determined to have caused defects on a section of a cathodically protected 412-km long, 10-inch ethylene pipeline in the U.K.[29] These defects caused local metal loss of up to 40%. In 2002, a natural gas leak occurred owing to a pinhole perforation leak near the center of a pit that was approximately 25 mm in diameter on a cathodically protected gas pipeline in the U.S.[30] The pipeline showed high levels of induced AC voltages by collocations with HVAC power lines, with the failure occurring at the location with peak AC voltages. The leak was finally attributed to AC-induced failure accompanied by the generation of corrosion pits.

It is generally acknowledged that pipelines will experience accelerated corrosion in the presence of AC interference, which is often called "AC corrosion."[31, 32, 33] It has also been found that AC is able to facilitate disbondment of coatings from the steel substrate, contributing to the generation of a corrosive environment under a disbonded coating.[34] Moreover, AC interference can adversely affect the effectiveness of CP on pipelines by shifting the applied CP potential from the design value.[35] All of these effects contribute to pipeline failure, as reported extensively across the world. Furthermore, AC interference is associated with personnel safety concerns. As AC current is induced on the pipelines, or in the event of a fault condition, a voltage gradient can occur around the aboveground structures. Touching or even standing near these structures could result in electrical shock or death.

### 1.3.1. Pipelines in collocation with HVAC power lines

In 1995, a routine ILI by an ultrasonic tool on a coal tar–coated gas pipeline, which was installed in 1972 by Union Gas, a Canadian natural gas company based in Chatham, Ontario, identified a number of irregular corrosion flaws, all of which were located on the section collocated with an AC power line (Gummow et al., 1996). Further analysis of the clustered flaws found that they had an average depth of 40% of the pipe wall thickness, with a maximum penetration up to 80% of the wall thickness. The field dig at the deepest defect identified that the adjacent soil was damp clay with a measured resistivity of 2,000 Ω·cm. A hard, tightly adhering tubercle, protruding about 50 mm, covered the defect. The coating was disbonded with a radius of 200 mm. The pH of the electrolyte at the defect was greater than 8.5. There was no evidence of bacterial-induced corrosion, and the local CP potential was –1.56 V (copper-sulfate electrode [CSE]). The induced AC voltage at this site was, on average, 33 V before the AC mitigation system was installed in 1994. The AC current density was calculated to be 84 A/m$^2$, which was almost identical to the AC current density at the maximum defect depth. A complete analysis concluded that the observed corrosion on the pipeline section was induced by AC interference.

A worldwide problem, AC corrosion of buried pipelines was documented in India on a crude oil pipeline operated by Indian Oil Corporation Limited.[44] The pipeline, which was commissioned in 1997, was coated with a three-layer PE in combination with a CP system. Intelligent pigging was conducted on the pipelines at fixed intervals to identify repair and maintenance requirements. During an ILI, a significant metal loss (78% of the pipe wall thickness) feature was reported, although the quarterly pipe-to-soil potential survey and close interval potential survey report did not indicate problems at the given location, which was just 1 m away from a 66 kV power transmission line running parallel to the underground pipeline. After excavation, the coating of the pipeline was found to be in relatively good condition except for pinhole-type damage on the exterior. Upon removal of the coating, a deep pit was observed, implying flow of a high current through the location.

A similar corrosion incident was also reported on a gas pipeline in Beijing, China.[36] The pipeline was protected from external corrosion by three-layer PE in conjunction with magnesium anode CP systems. The pipeline was parallel to six HVAC power lines, including four 220 kV and two 110 kV lines. The average AC voltage measured at the test stations was over 10 V, with the highest AC voltage being 27.3 V. The average AC current density exceeded 30 A/m$^2$, with the maximum value being 101.7 A/m$^2$. The testing results showed that the pipeline was under a high risk of AC interference.

The AC corrosion of buried pipelines collocated with HVAC electric lines over a certain distance has been reported frequently in Europe. In 1993, corrosion was detected on a gas pipeline that had been installed in 1992, with 3 km of parallelism with a 400 kV electric power line.[37] Field tests showed that the pipe-to-remote-earth AC voltage was 13 V. Although the protective potential indicated effective CP, the corrosion depths of the pipe were up to 0.8 mm. The pH was 13 under the disbonded coating and near the corrosion products, which were mainly composed of iron oxide (i.e., magnetite). Field and laboratory tests both concluded that a high level of AC current density, combined with the small-sized coating defects and the low soil resistivity, caused AC corrosion of the pipeline. AC-induced pipeline failures have also been reported in France, Germany, the U.K., and other European countries.[38]

### 1.3.2. Pipelines with adjacency to AC-powered rail transit systems

In addition to HVAC power lines, AC-powered rail transit systems are another source of interference that can result in corrosion and failures of adjacent pipelines. The first documented AC corrosion case in Europe occurred in Germany in 1986, when two corrosion perforations were observed on a PE-coated gas pipeline that had been installed in 1980 parallel to an AC-powered (16Ð Hz) rail transit system.[39] A subsequent investigation attributed the pipeline corrosion to AC interference from the transit operation.[40]

Corrosion failure was detected on a natural gas pipeline in Switzerland in 1987.[41] The pipeline was coated with PE and was combined with impressed current CP, with protective current densities below 5 $mA/m^2$. The pipeline was parallel to a 16Ð Hz, 15 kV single phase traction system at varying distances. The potential mapping indicated that the CP system was effective. The soil resistivity was measured to be 2,200 Ω·m. Field tests showed that the pipe-to-remote-earth AC voltage ranged from 36 V to 60 V near the corrosion sections. The testing found the corrosive solution to have an alkaline pH. Cations and anions were both contained in the corrosion products, indicating that both the anodic and cathodic reactions occurred simultaneously. The applied CP was under common protection criteria. It was finally concluded that the pipeline corrosion was caused by AC.

A corrosion incident was reported on an oil pipeline in Hebei, China.[42] The pipeline was buried near an electric railway, and the railway track crossed the pipeline at several locations. The pipeline was made of 16Mn steel and protected from external corrosion by bitumen coating in conjunction with an impressed current CP system. The pipelines were attacked by AC current, resulting in the generation of several hemispherical pits of 10 mm diameter, with an average depth of 3 mm. The applied CP was effective for corrosion protection but was insufficient where the stray AC current was active. Corrosion occurred on the pipeline segments where the coating failed.

Compared with conventional rail systems, high-speed electrified rails require more power to operate. It is expected that severer AC interference would be induced by the high-speed rail system on the adjacent pipelines. It has been reported that, when a high-speed train passed, the instant stray AC current was able to break the external coating at some locations to flow into the pipeline steel, shifting the CP potential from the applied value and causing corrosion of the adjacent pipelines.[43]

## 1.4. Contents of the book

Investigations of accelerated corrosion of steels by AC can be tracked back as early as the 1900s.[32] In 1986, the corrosion failure of a PE-coated gas pipeline caused by AC interference from an AC-powered rail transit system was first reported in Europe.[39] Since then, pipeline failures in the presence of AC interference have been reported worldwide. Nowadays, there is increasing exposure of oil and gas pipelines to AC interference from HVAC power lines or AC-powered rail systems or both. This presents an essential threat to the integrity and safety of the pipeline system. At the same time, with the huge efforts made in the past several decades, our understanding of AC corrosion of pipelines has been evolving to a stage where a comprehensive and advanced summary to this phenomenon can be given. Effective techniques for assessment, inspection, mitigation, and prevention of the problem have been developed. All of them facilitated the development of this book.

The book starts, in Chapter 2, with an introduction to the wide variety of sources generating AC interference and the main principles of the interference. The generation of AC interference under steady-state operation and the fault conditions are detailed. In particular, the modes and mechanisms of AC interference (i.e., capacitive coupling, resistive coupling, inductive coupling, etc.) are reviewed systematically for educational and practical purposes.

Pipeline corrosion in the presence of AC interference is covered in Chapter 3. First, the specific features associated with AC corrosion are listed to distinguish this particular corrosion problem from other types of corrosion and to help with recognizing the occurrence of the problem on pipelines. The fundamentals of AC corrosion, including the parametric effects, are imparted to aid understanding of the problem. To assist in investigation and measurement of AC corrosion in the laboratory, the typical experimental setup is described under which the obtained data are sufficiently accurate and reproducible. Depending on the type and failure mode of pipeline coatings, as well as their permeability to CP current, the corrosive environments generated on the steel surface can be either anaerobic near-neutral pH or aerobic high pH. As a result, the pipeline steel will be in an active dissolution or a passive state, respectively. The AC corrosion of pipelines in both environments would have different mechanisms and processes, which are given in the chapter and based on the author's first-hand research results. Finally, AC corrosion occurs at coating defects where the AC current flows into the pipeline steel. The essential effect of the defect's geometry on AC corrosion is also included.

In addition to accelerated uniform corrosion, AC can induce localized pitting corrosion on pipelines. In Chapter 4, the conditions critical to initiating corrosion pits on pipelines under AC interference are reviewed. On the basis of the research work conducted in both near-neutral pH and high pH environments, the threshold AC current densities for pitting corrosion of pipelines are determined. More importantly, the mechanistic aspects of pit initiation and growth are discussed in terms of the effect of AC on the double-charge layer at the steel–solution interface. Since pitting corrosion, compared with uniform corrosion, presents a more dangerous mechanism that can result in pipeline leaking, this chapter develops recommendations to pipeline operators to improve the integrity of their pipelines.

Coatings provide the first line of defense to protect pipelines from external environments. Chapter 5 starts with an introduction of the essential properties of coating candidates required to maintain pipeline integrity. Four typical coating properties and features—cathodic disbonding resistance, water permeability, electrochemical impedance and molecular structure—are characterized in circumstances in which AC is applied. Detailed discussions are included on the effect of AC interference on coating properties and performance. Conceptual models are developed to illustrate the coating degradation due to AC interference. It is expected that the industry understands, both qualitatively and quantitatively, coating performance and degradation under AC interference, a problem not fully realized today.

As all pipelines are protected by both coatings and CP, the effect of AC on CP performance is included in Chapter 6. After introduction of the CP principle for pipeline corrosion protection, one of the most common modes of coating failure in the field (i.e., coating disbondment from the steel substrate) is described mechanistically, and the role of AC interference in coating disbondment is discussed. The shift in CP potential caused by AC interference is quantified in terms of the applied CP potentials, and the true cathodic potential exerted on pipelines is derived. In combination with corrosion rate measurements, CP performance is evaluated and a new CP design is recommended for pipelines in the presence of AC interference.

In Chapters 7, 8, and 9, the techniques and methods used to inspect, evaluate, and mitigate or control AC corrosion of pipelines are given. In Chapter 7, conventional methods for AC corrosion inspection, including buried coupons, ultrasonic-based coupon testing, and alternating voltage modulation, are introduced in detail in terms of their principles, affecting factors, advantages, and disadvantages. A novel online AC corrosion monitoring and assessment technique is also included. It is expected that the technique will enable convenient and accurate measurement of pipeline AC corrosion in the field. In Chapter 8, commercial software packages for modeling the AC interference and the induced AC corrosion of pipelines are introduced, along with the modeling principles and analysis of the factors affecting the modeling accuracy. Examples of modeling and determining AC interference as a function of various conditions from HVAC power lines, soils, and pipelines are given. Practical applications of computational modeling of AC interference are described in relation to optimized power line design, determination of the induced AC voltage and AC current density on the pipelines, modeling of the fault current, and so on. Finally, various methods and techniques used in the mitigation, control, and management of pipeline AC corrosion in practice are covered in Chapter 9. The general principle for mitigation of AC interference is analyzed, and various methods and techniques used today are covered in detail. Moreover, practical recommendations for AC corrosion control included in various industry standards developed in legal territories are reviewed and compared. Finally, an AC interference management program for pipelines is introduced.

Pipeline operation in the presence of direct current (DC) interference is covered in Chapter 10. In contrast to AC interference, DC interference occurs when underground pipelines are collocated with high-voltage direct current (HVDC) power lines, DC-powered transit systems, and other DC devices. The sources of DC interference are summarized and commented on, and the mechanisms and features of DC corrosion of pipelines are analyzed. Moreover, coating properties and CP performance under DC interference are investigated and determined on the basis of a specific experimental setup designed for DC corrosion testing. The effect of DC on the double-charge layer at the steel–solution interface is discussed. Solutions effective at controlling the DC corrosion of pipelines in practice are recommended.

## References

1. Hopkins, P., "Pipelines: Past, present, and future," 5th Asian Pacific IIW International Congress, Sydney; International Institute of Welding, Genoa, Italy, 2007.
2. Natural Resources Canada, *Pipelines across Canada*, Ottawa, 2014.
3. Department of Energy, *International Energy Outlook*, Energy Information Administration Office of Integrated Analysis and Forecasting, Washington, DC, 2006.
4. Hansen, M.E., Dursteler, E., *Pipelines, Rail & Trucks: Economic, Environmental, and Safety Impacts of Transporting Oil and Gas in the U.S.*, Strata, Logan, UT, 2017.
5. Frittelli, J., Andrews, A., Parfomak, P.W., Pirog, R., Ramseur, J.L., Ratner, M. (2014), *U.S. Rail Transportation of Crude Oil: Background and Issues for Congress*, Congressional Research Service 7-5700, Washington, DC.
6. CEPA (Canadian Energy Pipeline Association), *Let Us Talk Pipelines*, Calgary, Alberta, 2016.
7. Green, K.P., Jackson, T., "Canadian analysis shows pipelines safer than rail," *Pipeline Coatings* 11: 18–20, 2015.
8. Godin, M., *Innovation Roadmap for Transmission Pipeline Transportation of Petroleum Products*, Alberta Innovates–Energy and Environment Solutions; Canadian Energy Pipeline Association, Calgary, Alberta, 2014.
9. CEPA, *2015 Pipeline Industry Performance Report*, Calgary, Alberta, 2015.
10. Cheng, Y.F., "Environmental hazard: Monitor safety of aged fuel pipelines," *Nature* 529: 156, 2016.
11. National Transportation Safety Board, (NTSB), *Pipeline Accident Report: Enbridge Incorporated Hazardous Liquid Pipeline Rupture and Release, Marshall, Michigan, July 25, 2010*, NTSB/PAR-12/01, PB2012-916501, Washington, DC, 2012.
12. NTSB, *Pipeline Accident Report: Pacific Gas and Electric Company Natural Gas Transmission Pipeline Rupture and Fire, San Bruno, California, September 9, 2010*, NTSB/PAR-11/01, PB2011-916501, Washington, DC, 2011.
13. Nelson, B.R., "Pipeline integrity: Program development, risk assessment and data management," 11th Annual GIS for Oil & Gas Conference, Houston, TX; Geospatial Information and Technology Association, Aurora, CO, 2002.
14. ASME (American Society of Mechanical Engineering), *Managing System Integrity of Gas Pipelines*, ASME B31.8S, New York, 2014.
15. Focke, J., "The future of pipeline integrity management," *Pipelines International* 3: 28–29, 2015.
16. National Energy Board, *Focus on Safety and Environment: A Comparative Analysis of Pipeline Performance 2000–2009*, Calgary, Alberta, 2011.
17. Keener, B., *Hazardous Liquids Accident Data with Emphasis on Crude Oil and Internal Corrosion*, U.S. Department of Transportation, Pipeline and Hazardous Materials Safety Administration, Washington, DC, 2012.
18. FERC (Federal Energy Regulatory Commission), *TransCanada Keystone Pipeline, LP, Containing Rules and Regulations Applying to the Transportation of Petroleum*, FERC No. 5.1.0, Washington, DC, 2011.

19. National Academy of Sciences, *Effects of Diluted Bitumen on Crude Oil Transmission Pipelines*, Transportation Research Board Special Report 311, Washington, DC, 2013.
20. Banff Pipeline Workshop, *Internal Corrosion in Transmission Pipelines*, Banff, Alberta, 2015.
21. Cheng, Y.F., Norsworthy, R., *Pipeline Coatings*, NACE, Houston, TX, 2017.
22. Xu, L.Y., Cheng, Y.F., "Experimental and numerical studies of effectiveness of cathodic protection at corrosion defects on pipelines," *Corrosion Science* 78: 162–171, 2014.
23. Kuang, D., Cheng, Y.F., "Study of cathodic protection shielding under coating disbondment on pipelines," *Corrosion Science* 99: 249–257, 2015.
24. Kuang, D., Cheng, Y.F. (2015b), "Probing potential and solution pH under disbanded coating on pipelines," *MP* 54: 40–45.
25. Fu, A.Q., Cheng, Y.F., "Characterization of the permeability of a high performance composite coating to cathodic protection and its implications on pipeline integrity," *Progress in Organic Coatings* 72: 423–428, 2011.
26. Agence France Presse, "Chinese oil giant Sinopec to pay big over pipeline blast that killed more than 60 people," Business Insider, January 13, 2014.
27. Saied, M.M. (2004), "The capacitive coupling between EHV lines and nearby pipelines," *IEEE Transactions on Power Delivery* 19: 1225–1231.
28. Brelsford, C., "Mitigating AC corrosion on cathodically protected pipelines," *Pipeline & Gas Journal* 242, 10: 1–3, 2015.
29. Ellis, R., *AC Induced Corrosion on Onshore Pipelines: A Case History*, Shell UK, London, 2001.
30. Wakelin, R.G., Sheldon, C., "Investigation and mitigation of AC corrosion on a 300 MM natural gas pipeline," CORROSION/2004, paper no. 04205, NACE, Houston, TX, 2004.
31. Pourbaix, A., "Detection and assessment of alternating current corrosion," *MP* 39: 34–37, 2000.
32. Yunovich, M., Thompson, N.G., "AC corrosion: Corrosion rate and mitigation requirements," CORROSION/2004, paper no. 04206, NACE, Houston, TX, 2004.
33. Fu, A.Q., Cheng, Y.F., "Effects of alternating current on corrosion of a coated pipeline steel in a chloride-containing carbonate/bicarbonate solution," *Corrosion Science* 52: 612–619, 2010.
34. Kuang, D., Cheng, Y.F., "AC corrosion at coating defect on pipelines," *Corrosion* 71, 3: 267–276, 2015.
35. Xu, L.Y., Cheng, Y.F., "Effect of alternating current on cathodic protection on pipelines," *Corrosion Science* 66: 263–268, 2013.
36. Li, X., Liang, Y., Du, Y.X., Tang, D.Z., "Case study of AC interference on an urban gas pipeline: Field test, AC mitigation design and assessment," CORROSION/2017, paper no. 8914, NACE, Houston, TX, 2017.
37. Ragault, I., "AC corrosion induced by V.H.V. electrical lines on polyethylene coated steel gas line," CORROSION/98, paper no. 557, NACE, Houston, TX, 1998.
38. Kim, D.K., Bae, J.H., Scantlebury, D., "AC induced corrosion: Summary of AC corrosion cases," *Journal of Corrosion Science and Engineering* 4: 1–15, 2004.

39. Gummow, R.A., Wakelin, R.G., Segall, S.M., "AC corrosion: A new threat to pipeline integrity?" Proc. 1st International Pipeline Conference, Vol. 1, ASME, New York, 1996.
40. Prinz, W., "AC induced corrosion on cathodically protected pipelines," UK Corrosion '92, Sandown Park, United Kingdom; Institute of Corrosion, Northampton, U.K., 1992.
41. Stalder, F., "Pipeline failures," *Materials Science Forum* 247: 139–146, 1997.
42. Zhu, Q., Cao, A., Wang Z., Song, J., Chen S., "Stray current corrosion in buried pipeline," *Anti-Corrosion Methods and Materials* 58: 234–237, 2011.
43. Liu, B., Zhao, Y., Zhao, S., Wang, S., Chen, L., "Analysis of interference of high-speed rail current with cathodic protection of pipeline and countermeasures," *Corrosion Science and Protection Technology* 30: 94–98 (in Chinese), 2018.
44. Yadav, B.D. (2016), "Pipeline corrosion due to interference from high voltage transmission line," *Pipeline Technology Journal* 1: 30–34.

**Other Reference**

Hopkins, P., "Pipelines: Past, present, and future," 5th Asian Pacific IIW International Congress, Sydney; International Institute of Welding, Genoa, Italy, 2007.

# 2 Sources and Principles of AC Interference

## 2.1. Sources of AC interference

### 2.1.1. HVAC transmission power lines

Generally, electrical interference requires a source of disturbance, a coupling mechanism, and a receptor. Pipelines that share, parallel, or cross the ROW of HVAC power transmission lines may be subject to electrical interference, where the power lines and pipelines are the disturbance source and receptor, respectively. HVAC power lines are used to transmit electric power over long distances or used as electric power transmission from one central station to another for load sharing. HVAC power lines, which usually carry voltages of 115 kV or above, can interconnect with one another, forming a high-voltage transmission network. The transmission lines mostly use a three-phase AC system, with one- or double-circuit structures, each of which is carried on separate line held by pylons or towers along the ROW. Aluminum alloys reinforced with steel strands are usually used as the conductor materials owing to their advantages in weight and cost compared with copper. In the transmission network, individual substations decrease the voltage of incoming electricity by means of voltage transformers, allowing the connection of HVAC power lines and the low-voltage distribution network. At the user end, the energy is transformed to lower voltages in accordance with government and customer requirements.

The voltage sequence of the transmitted power is sinusoidal AC power phases, each 120° out of phase with the other two phases. If each phase is equal, the sum of the AC currents in the three phases and the sum of the magnetic fields resulting from the AC in each phase should equal zero in the balanced three-phase system.

When a phase imbalance occurs owing to, for example, the different distances between the buried pipelines and each phase in the transmission power line, AC interference would be induced on the pipelines.[1] Recently, collocation of underground oil and gas pipelines with HVAC power transmission lines in joint utility corridors has been increasing. Owing to established infrastructure, convenient maintenance, difficulty in finding proper pipeline routes, and reduced impact to the environment, it is expected that joint corridors of pipelines and HVAC power lines will be even more common in the near future.[2]

## 2.1.2. Railway electrification systems

In addition to HVAC power transmission lines, AC-powered transit systems with electrified rails present another source for the induction of AC interference on adjacent buried pipelines, as shown in Figure 2-1. There is a wide variety of electrical traction systems around the world in terms of the types of railway, the location, and the available technology at the time of installation. Electrified railway networks can use either AC or DC. The AC-powered rail systems, which always use overhead wires, are used mainly for long distance, while the DC systems are the preferred option for shorter lines, urban systems, and tramways. The 25 kV AC railway electrification system is commonly used in railways worldwide, especially for high-speed rail. The choice of 25 kV is based on the efficiency of power transmission as a function of voltage and cost.[3]

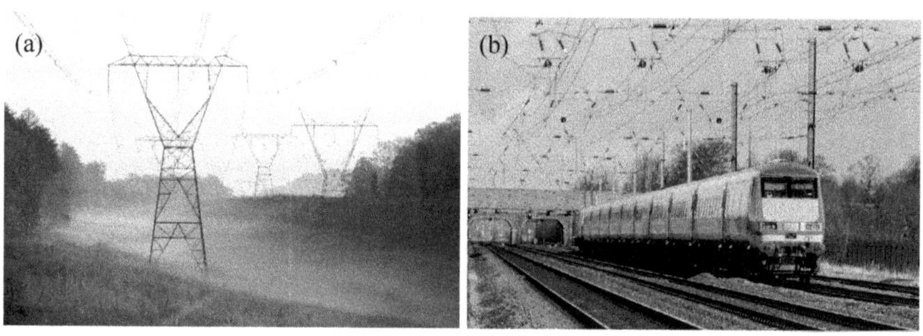


**Figure 2-1.** Examples of (a) HVAC power transmission lines and (b) an AC traction rail system, which can be sources for the induction of AC interference on collocated pipelines.

Interference from AC-powered rail systems can cause electric shock and pose a threat to collocated pipelines during both fault and load conditions. A pipeline buried in the vicinity of AC-powered rail lines can become energized by the electric and magnetic fields originating from the rail lines. Moreover, AC current can flow into the pipeline when electric charges are injected into the soil under both steady-state and fault conditions of the rail lines, resulting in AC corrosion of the pipeline.[4]

### 2.1.3. Other sources

Other possible sources generating AC interference include geomagnetically induced current and low-frequency communication systems. AC interference can occur on pipelines during construction and operation. While in construction, the pipeline is installed above ground and is electrically isolated from the ground before burial. High voltages can thus be induced on the pipe. After being buried in the soil, AC stray currents flow through the soil and along the pipeline because the external coating electrically isolates the pipeline from the soil. In addition to the pipelines, AC voltages can also be produced on other equipment and vehicles working near the interference sources, thereby causing hazards to those resources and their personnel.

## 2.2. Types and principles of AC interference by HVAC power transmission lines

When the electrical interference occurs on pipelines collocated with HVAC transmission lines, the interference mechanism can be a capacitive, resistive, or inductive coupling effect. The first two types of coupling effect primarily introduce personnel safety hazards, while the third (inductive coupling) contributes to AC interference by continuously coupling AC power to the adjacent pipelines, resulting in AC corrosion of the pipelines.[5]

### 2.2.1. Capacitive coupling effect

Theoretically, when different voltages are applied to two conductors separated by a dielectric medium (i.e., an insulator), the attraction of charges is defined as capacitance. The tendency of the electrical charges to combine with each other is described as the energy stored in the capacitor. The stored energy can move electrons from the higher potential to the lower potential when a path exists for the electrons to flow. Generally, the capacitive coupling effect refers to the mutual influence of two or more circuits on one another, through a dielectric medium, by means of the electric fields acting between them.[6]

For any ungrounded or poorly grounded facilities, such as coated pipelines, the collocated HVAC power lines produce electric fields that can induce AC voltage on the pipelines. The capacitive coupling occurs in the immediate vicinity of the overhead power lines when the pipeline is laid on a foundation that is well insulated from the ground. This usually occurs during the construction and installation of pipelines. The pipeline picks up a voltage relative to the soil, with the amplitude proportional to the voltage carried in the power line, as shown schematically in Figure 2-2.[2] In some cases, the voltage can be above the maximum safe voltage limitations for a pipe. However, contacting the pipe will only cause a slight electric shock, and the pipe voltage will be immediately reduced to zero.[7] The resulting current is a function of the size of the structure, the distance to power conductors, the voltage of power conductors, and their geometrical arrangement. Normally, the stray current is not significant. As a result, it does not play a role in corrosion of the pipeline. The electrical circuit can be closed if the pipeline coating is damaged: the current can flow out and return to its source through a conductor such as a human body. Since the capacitance between the pipeline and earth is negligible, the electric field normally does not induce an AC voltage on buried pipelines, especially when dielectric coatings are used.

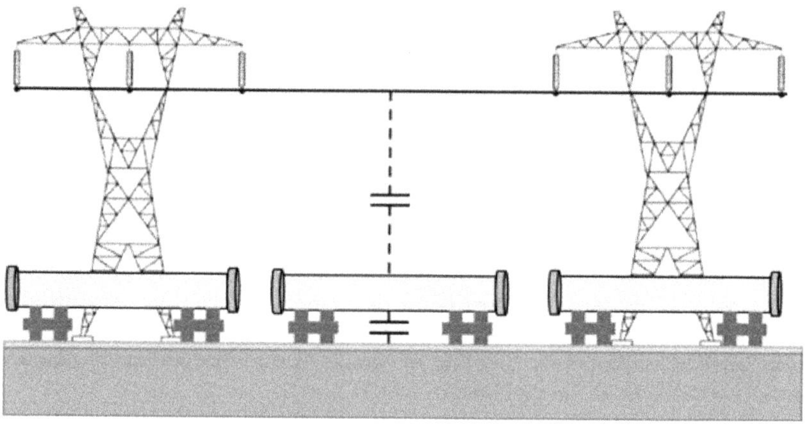

**Figure 2-2.** Schematic diagram of the capacitive coupling effect between HVAC power lines and the adjacent aboveground pipelines.

Welded pipe segments near HVAC power lines must be grounded when the nominal voltage in the power lines exceeds 115 kV and the length of the welded section exceeds a hundred or so meters. Proper grounding of the pipelines can eliminate the capacitive coupling effect.

## 2.2.2. Resistive coupling effect

The resistive coupling effect is due to the mutual influence of two or more circuits by means of conductive paths (e.g., metallic, semiconductive, electrolytic) between the circuits.[6] Soils, especially those with an appreciable moisture content, are good conductors. The HVAC power lines can become electrically coupled with the adjacent buried infrastructure, such as pipelines, through the supporting towers that are grounded to the soil, as shown in Figure 2-3.

When lightning strikes a power transmission structure and causes electrical faults to occur, or the AC system is not perfectly phase balanced, the current flows through the cable and supporting towers into the soil. Part of the current can flow into the pipelines. As a result, AC voltages generate on the pipelines to a level that results in hazards to the working personnel. In particular, if the AC voltages result from the HVAC system during fault conditions for reasons such as breaks in conductors, failure of insulators to cause short circuits, system imbalances, improper system maintenance, or lighting strikes, a large part of the current in the HVAC power lines would flow to the earth through foundations and grounding systems such as towers or substations.[2] The current flow raises the electric potential of the earth near the pipelines, often to thousands of volts with respect to remote earth, resulting in a considerable AC voltage across the coating of the buried pipeline. Since the impedance in the ground fault circuit can be significantly lower than that in the normal circuit, the magnitude of the fault current would be much greater than the normal balanced phase current.[7] The potential difference between the earth and the pipeline can represent an electric shock hazard. Under some circumstances, the electric potential of the pipelines may be sufficiently high to transfer hazardous voltages over considerable distances. The resistive coupling effect depends on a number of factors, including the total short-circuit current; the geometrical size of the grounding system of poles, towers, and substations through which the current flows; the soil resistivity; the distance between the HVAC power lines and the buried pipelines; and the properties and conditions of the pipeline coating.[6] Generally, protective measures for people are required if the contact voltage exceeds 65 V for long-term interference or 1,000 V for short-term interference.

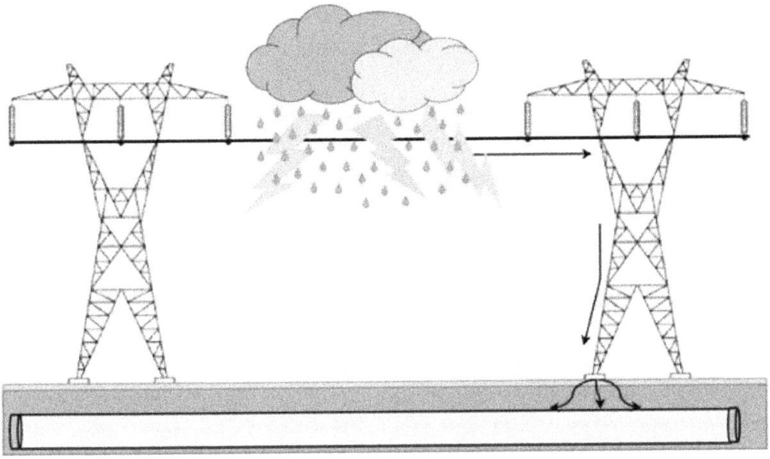

**Figure 2-3.** Schematic diagram of the resistive coupling effect between HVAC power lines and the adjacent buried pipelines.

Lightning strikes to HVAC power lines can initiate fault current conditions. Generally, lightning strikes to a structure or to earth in the vicinity of the structure can produce electrical effects that are similar to those caused by AC fault currents. If a lightning strike occurred between the support tower of the HVAC power lines and an overhead cloud, the potential of the tower could be raised to an extremely high voltage. A current could then flow from the tower to the earth and spread uniformly in all radial directions if the soil resistivity is relatively homogenous. Therefore, the soil around the pipelines is at a relatively high potential with respect to the pipeline potential, which can result in coating damage due to the heating effect. If the fault potential gradient is of sufficient magnitude, it could puncture a hole in the coating. If the potential gradient occurred at an existing coating defect (e.g., a pinhole, a holiday), a heating effect could evaporate the moisture in the defect to steam and enlarge the defect by detaching the previously adherent coating surrounding it.[7] Moreover, the pipe surface could burn and cause mechanical fracture. Even if the mechanical failure did not occur instantly, it might impact the integrity of the pipeline and its coating or allow corrosion to occur at an accelerated rate. The fault current could also enhance the coating disbondment and induce further corrosion under the disbonded coating.

Hazards to personnel also exist in the vicinity of ground fault currents. If a person touches an aboveground pipeline appurtenance (e.g., a main line valve, a hydrant), a large potential difference could generate between the person's hands and feet. While the potential of the person's hands would be of the value of remote earth owing to the coated pipeline, the feet resting on earth would undergo the influence of the fault current. As a result, the person might be in a voltage gradient of many thousands of volts.[7] However, if the pipeline coating is in poor condition with low resistance (e.g., if pinholes or holidays exist in the coating), the pipeline can collect a significant amount of current from the surrounding soil and rise in potential. The ground potential in the vicinity of the pipeline then decreases. Consequently, the potential difference between the pipeline and the ground can be reduced significantly.

### 2.2.3. Inductive coupling effect

Inductive coupling is also called electromagnetic coupling, where the current flowing in HVAC power lines produces a strong electromagnetic field, which can–through the inductive coupling effect–generate AC voltages on pipelines buried within the field, as shown mechanistically in Figure 2-4. Inductive coupling is the main coupling mechanism of HVAC power lines on adjacent pipelines during normal or steady-state operation, where the power lines and the pipelines behave as two coils of an autotransformer. The current flows in the conductor create an electromagnetic field, which expands away from the conductor and toward the buried pipelines at a rate that is a function of the system frequency.[7] Phase conductor currents can range from several hundred amperes to 2,000 A, and the resulting current on the pipelines is the net magnetic field produced by all the phase conductors. If a conductor is located equally distant from each phase and if each phase transmits the same current, there will be no net magnetic field. As a result, there is no net current induced in this conductor. In practice, it is impossible to locate a pipeline with an identical distance to each phase. Moreover, the phase currents in each circuit are not identical (i.e., there is phase imbalance). Therefore, a steady-state AC current is inevitably induced on a pipeline that is collocated with a power line owing to the influence of the electromagnetic field produced by the individual phase currents.

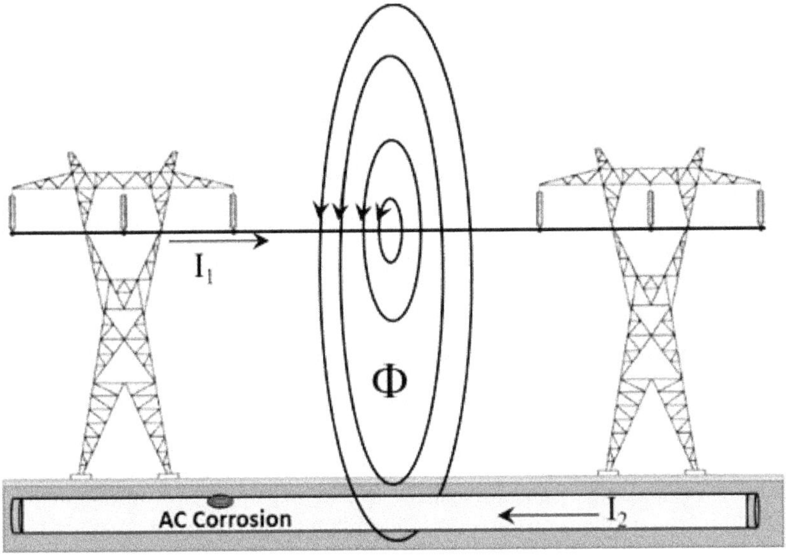


**Figure 2-4. Schematic diagram of the inductive coupling effect between HVAC power lines and the adjacent buried pipelines.**

Inductive coupling is one of the major reasons that AC corrosion occurs. For buried pipelines, external coatings act as an ideal insulator to break the conduction of electricity. When defects such as pinholes and holidays are contained in the coatings, the current finds a path by which to enter and escape through the defects, causing the so-called tubercles.[5] Generally, the AC current induced by the inductive coupling effect produces a longitudinally induced voltage along the length of the pipeline. The magnitude of the voltage depends on the factors shown below.[6]

- The overall separation distance between the buried pipelines and the HVAC power lines, and the length of their mutual parallelism
- The current flowing in the power lines
- The type of conductor used to shield the power lines from lightning
- The properties and performance condition of the coating
- The soil resistivity as a function of the burial depth
- The grounding of the pipelines

## 2.3. AC interference and the associated risk

### 2.3.1. Normal conditions of HVAC power lines

Of the three types of coupling effects for AC interference stated above, electromagnetic inductive coupling is the primary effect of HVAC transmission power lines on buried pipelines during normal operation, where the pipeline is in a stationary electric field.[8] Under normal operation, the individual phase conductors of a power line circuit carry a similar load current, but with different phase angles attributed to them (i.e., 120° of difference in three-phase systems). Each phase conductor generates an electromagnetic field, and the summation of the fields generated by all current-carrying conductors is the net electromagnetic field imposed on the affected pipeline, inducing AC voltages along its length.[9] The magnitude of the AC voltage and the location of the voltage peaks depend on the parallel ROW configuration, pipeline conditions (coatings, CP), power line parameters, and soil properties (e.g., resistivity, moisture content). For a well-coated pipeline collocated with an HVAC power line operating at a high current capacity, the induced voltages can be up to hundreds of volts. The inducing AC interference can impact the pipelines in three ways: a shock hazard to personnel working on the pipelines or standing nearby, accelerated corrosion degrading the pipeline integrity, and depolarization of the pipelines under applied CP.[10]

If the electrical interference is significant enough during normal operation of the transmission lines, a potential risk of electrical shock exists for the people who touch an exposed part of the pipeline system, such as valves, CP test stations, or aboveground appurtenances of the pipeline. The voltage is the main concern to people when touching or working near the pipeline. Two important concepts related to the shock hazard should be distinguished. The "touch potential" refers to the voltage between an exposed feature of the pipeline, such as a CP test station, and the surrounding soil or a nearby isolated metal object, such as a fence, that can be touched at the same time. The "step potential" is the voltage across a person's two feet and is defined as the difference of the earth surface potentials between two spots 1 m apart.[8] Generally, the touch potential is a concern during both the normal inductive and fault conductive conditions of an HVAC power line. The step potential is an important indicator of the risk during the fault conductive condition only, owing to high currents and voltage gradients in the soil.

During normal operation of power transmission lines, AC current density at a coating defect (i.e., a pinhole, a holiday) may cause accelerated corrosion to the pipeline when the local current is above a certain threshold in a specific environment. Nowadays, various criteria based on AC current density or the ratio of AC current density to CP current density are used to evaluate the risk of AC corrosion of buried pipelines.[11, 12]

In addition to shock hazard to personnel and accelerated pipeline corrosion, the AC interference generated during normal conditions can depolarize a coated pipeline that is under CP.[13] Thus, the presence of AC may result in misleading results on CP potential measurements. This is especially important when a close interval potential survey is conducted as the trailing wire from the reading voltmeter to the test station is subject to induced AC interference from the adjacent power lines.

### 2.3.2. Fault conditions of HVAC power lines

Nonstationary AC interference takes place when the electric field the pipelines are exposed to is time dependent. Particularly, power line faults usually occur during harsh weather such as winds, ice storms, or lightning. A phase-to-ground fault on a power line results in conduction of the phase current from one or more AC power line phase conductors via the metallic tower to ground, or directly to ground as a result of conductor lines falling to the ground.[10] Resistive coupling effects are primarily a concern when a fault occurs in an area where the pipeline is close to the power line.[10]

If a pipeline is close to the foundation (e.g., support towers) or the grounding points of a faulted power line, an arc could develop between the power line grounding and the adjacent pipeline through the resistive coupling effect, which could damage the steel pipe and the external coating. Even when the pipeline is sufficiently far away from the powerline structure that arcing is prevented, hazards still exist for the personnel working on the pipeline and for the coating integrity. When the fault current flows to the ground through the tower foundation, the potential of the ground could be up to several thousand volts, resulting in a large potential differential between the pipeline and the ground, with most of the potential differential across the coating.[10] While some AC fault current transfers through the coating, the majority of the current tends to flow at the coating defects, where the local resistance is much lower than at the coating film. As a result, serious localized corrosion (i.e., pitting corrosion) can happen on the pipelines. The fault current is typically more than an order of magnitude higher than the steady-state phase current, and the induced voltage can reach thousands of volts for the duration of the power line fault. For example, it was found that, for six cycles of a duration fault, perforation of a steel pipe along the wall thicknesses was between 5.6 mm and 9.54 mm, resulting from fault currents ranging from 15.1 kA to 44 kA.[14] Field fault tests were conducted on a coated steel pipe near a 230 kV power line, with a maximum testing fault current of 7.8 kA in 100 Ω·m clay.[15] It was found that both the depth and the diameter of the corroded area on the pipeline linearly increased with the fault current.

In a number of reported cases, pipeline failures were caused by power line faults, where substantial damage occurred on external coatings, pipeline steel, and the associated facilities in the whole pipeline system.[16, 17] Moreover, the fault current and voltage generated during arcing can transmit along the pipeline away from the fault location and produce a hazardous voltage to anyone touching a pipeline appurtenance. There is also a risk of the arc crossing foreign pipelines, potentially damaging multiple pipelines and the associated equipment.

Even if a fault condition does not result in perforation of the pipe wall, it can still cause damage to the external coating, depending on the fault voltage across the coating and its dielectric strength. For example, during the pulse period–typically varying from about 50 ms to 1 s–the fault voltage levels that would break fusion-bonded epoxy (FBE), coal tar epoxy, coal tar, and coal tar enamel (CTE) coatings with a thickness of 25 microns are 1,000 V, 3,500 V, 4,500 V, and 5,000 V, respectively.[18]

The probability of personnel being killed by a fault current is usually small because the fault duration is short, normally less than 10 cycles (less than 1 s in time). The likelihood of people touching the pipeline during poor weather such as lightning is also small. However, since a fault current can be transmitted along the pipeline over some distance from the faulted location, the safety of pipeline personnel should be an important consideration.

## 2.4. AC interference from electrified rail systems

Generally, three types of AC interference from electrified rails can affect the pipelines nearby: capacitive coupling, resistive coupling, and inductive coupling interference.[19] Since the capacitive coupling effect exists for aboveground pipelines that are ungrounded with the soil, it can thus be ignored for buried pipelines.

Rails tracks are the key components of the traction power supply system for electrified rails. As the steel tracks are not fully insulated from the ground, a part of the current used for rail traction can flow into the soil through the tracks. Thus, the resistive coupling effect is generated when the current flows into the adjacent pipelines. Figure 2-5 shows schematically the principle behind the generation of the resistive coupling effect on the buried pipelines from the electrified rails.[19] When the rail power system is in fault, several hundred, and even thousands, of amperes of current can flow into the soil to generate a strong electric field, breaking the external coatings, damaging the steel pipelines, and even destroying the CP equipment.[20, 21]

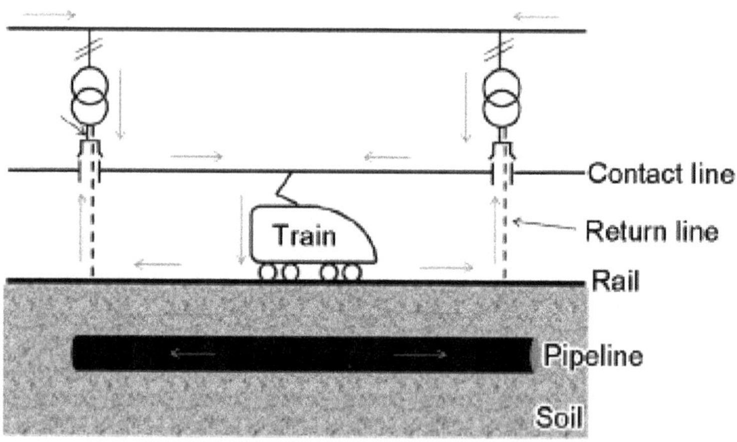


**Figure 2-5.** Schematic diagram of the principle behind the generation of the resistive coupling effect on buried pipelines from electrified rails.[18]

For the generation of inductive coupling interference, an alternating electromagnetic field is produced by the AC current flow between buried pipelines and the electrified rails that are parallel or in proximity. An inductive voltage and an inductive current are generated on the pipelines under the electromagnetic field, as shown in Figure 2-6.

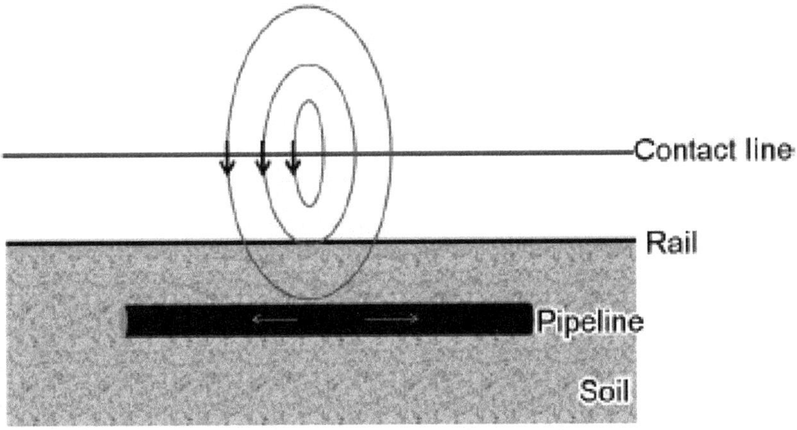


**Figure 2-6.** Schematic diagram of the principle behind generation of the inductive coupling effect on buried pipelines from electrified rails.[18]

Although resistive and inductive coupling effects both present the main types of AC interference from electrified rail systems, their relative importance and affecting factors have not been fully established.[22]

## References


1. Shwehdi, M.H., Johar, U.M., "Transmission line EMF interference with buried pipeline: Essential & cautions," International Conference on Non-Ionizing Radiation at UNITEN, Kuala Lumpur, 2003.
2. Kroon, D.H., "Alternating current interference and alternating current corrosion of pipelines—Part 1," *MP* 53: 30–33, 2014.
3. Yin, K., Tang, M., Hsiong, C., "Corrosion of buried pipeline under effect of electrical field with industrial frequency," *International Petroleum Exhibition and Technical Symposium*, Society of Petroleum Engineers, Richardson, TX, 1982.
4. Sun, P.Q., "AC interference induced by electrified railway on buried metal pipelines and its protective measures," *Urban Gas* 4: 7–11, 2011.
5. Brelsford, C., "Mitigating AC corrosion on cathodically protected pipelines," *Pipeline & Gas Journal* 242, 10: 1–3, 2015.
6. NACE, *Mitigation of Alternating Current and Lightning Effects on Metallic Structures and Corrosion Control Systems*, NACE Standard Practice SP0177, Houston, TX, 2007.

7. Bonds, R.W., *The Effect of Overhead AC Power Lines Paralleling Ductile Iron Pipelines*, Ductile Iron Pipe Research Association, Golden, CO, 2017.
8. Garrity, K.C., "Managing interference," *World Pipelines* 14, 9: 36–40, 2014.
9. Fieltsch, W., Silva, D., "AC interference risk ranking," CORROSION/2016, paper no. 7393, NACE, Houston, TX, 2016.
10. CEPA (Canadian Energy Pipeline Association), *A/C Interference Guideline Final Report*, Calgary, Alberta, 2014.
11. Baeckmann, W.V., Schwenk, W., Prinz, W. (1997), *Handbook of Cathodic Corrosion Protection*, 3rd ed., Gulf Publishing House, Houston, TX.
12. British Standards Institution, *Evaluation of AC Corrosion Likelihood of Buried Pipelines Applicable to Cathodically Protected Pipelines*, BS EN 15280, London, 2013.
13. Kajiyama, F., Nakamura, Y., "Effect of induced alternating current voltage on cathodically protected pipelines paralleling electric power transmission lines," *Corrosion* 55: 200–205, 1999.
14. Webster, C., Zawadzki, J.A., Stefanski, T., *Powerline Ground Fault Effects on Pipelines*, Report 239T817, Canadian Electricity Association, Ottawa, 1994.
15. Drakos, J.E., Akhtar, A., *Problems Associated with Pipelines Occupying Joint-Use Corridors with AC Transmission Lines*, Report 75-02, Canadian Electricity Association, Ottawa, 1979.
16. Gleekman, L.W., "Corrosion from lightning and power fault current," *MP* 12: 24–27, 1973.
17. Akhtar, A., Hunter, M. "Damage of a gas pipeline resulting from a high impedance fault on a 25 kV AC distribution power line," *MP* 24: 21–24, 1985.
18. Gummow, R.A., *Cathodic Protection Considerations for Pipelines with AC Mitigation Facilities*, Report PR-262-9809, Corrosion Supervisory Committee, Pipeline Research Council International, Chantilly, VA, 1999.
19. Li, W., Du Y., Jiang, Z., Lu, M., "Research progress on AC interference of electrified railway on buried pipeline," *Journal of Chinese Society for Corrosion and Protection* 36: 381–388, 2016.
20. Chinese Ministry of Railways, *Engineering Design Guidelines for Prevention of Electromagnetic Interference on Communication Lines and Facilities*, Engineering Design Center, China Railway Publishing, Beijing, 2009.
21. Zhang, X.Y., "Calculation and protective measures of electromagnetic interference induced by electrified railway on the oil and gas pipelines," *Oil Depot and Gas Station* 19: 29–35 2010.
22. Braunstein, R., Schmautzer, E., Propst, G., "Comparison and discussion on potential mitigating measures regarding inductive interference of metallic pipelines," 2010.

**Other References:**

International Conference on Electrical Railway Systems for Aircraft, Railway and Ship Propulsion, Bologna, Italy.

Fickert, L., Schmautzer, E., Braunstein, R., Lindinger, M., "Reduction of the electrical potential of interfered pipelines due to currents of high voltage power lines and electric railways," *Elektronische Information* 127: 362–366, 2010.

# Pipeline Corrosion in the Presence of AC Interference

## 3.1. Recognition of AC corrosion of pipelines

Accelerated corrosion of metals in the presence of AC interference was investigated as early as the 1900s, when researchers from the National Institute of Science and Technology in the U.S. studied the effect of the frequency of AC current on corrosion in electrolytes.[1] In 1906, Hayden investigated whether and, if so, to what extent, AC current flowing between a metallic conductor and the ground would produce AC corrosion.[2] But until the mid-1980s, corrosion engineers had not considered AC-enhanced corrosion of metallic structures an important problem. The reason was that the research results obtained by that time showed that AC current might cause corrosion at a rate that was only about 1% of that of a similar electric quantity of DC,[3, 4] and the AC corrosion rate was less than 1% of the equivalent DC corrosion rate at the same current density.[5] Moreover, the corrosion of steels under AC interference at 60 Hz was only about 1% of that induced by an equivalent amount of DC.[6] It was thus believed that, even if AC corrosion occurred, it could be controlled to a negligible level with CP applied at industry standards. Another reason that AC corrosion was not paid proper attention for such a long time was that the instruments normally used to measure the electric parameters in DC could not correctly detect the presence of AC with frequencies between 50 and 100 Hz.[7]

The situation did not change until 1986, when an investigation conducted on a natural gas pipeline in Germany indicated that the pipeline failure was solely due to AC corrosion.[8] The pipeline accident was reported in a number of publications in the early 1990s.[9, 10, 11] The failure occurred on a PE-coated pipeline that was parallel to a 16.6 Hz AC-powered railway. The pipeline was protected by CP at −1,000 mV (saturated calomel electrode [SCE]), typical of the European industry CP standard at that time, through an impressed current system. An important finding from the failure and investigations was that, in the presence of AC interference, a standard level of CP would not be able to control the AC corrosion of pipelines.[12] Since then, numerous cases have been reported on pipeline failure resulting from AC corrosion.[13, 14] With the growing number of HVAC transmission lines collocated with buried pipelines, nowadays AC corrosion is recognized as a serious threat to the integrity and safety of pipelines.

There are big scatters in the reported pipeline AC corrosion rates obtained in both field tests and laboratory measurements. It is relatively easy to misdiagnose the damage caused by AC corrosion on pipelines since it is normally characterized by localized pitting corrosion that can be attributed to a number of other factors.[15] It can take several years for a correct diagnosis to emerge and for final conclusions and recommendations to be developed. To date, there has been much evidence collected of AC corrosion–induced pipe wall loss and even failures of pipelines.

For example, an AC corrosion case was reported on a natural gas pipeline located under an HVAC utility corridor near a power plant in Rainier, OR, U.S., in 2001.[16] Four holes were found in the pipeline after only five months of use, with a measured AC voltage of 90 V. The corrosion rate was calculated to be over 400 mpy (mils per year), in other words exceeding 10 mm/y. Researchers have reported that, in field testing and studies, corrosion rates ranged from 0.4 to 1.4 mm/y for AC current densities between 75 and 200 A/m$^2$.[17] In soils with an electrolyte pH between 6 and 7, it has been estimated that the corrosion rate of steels was below 0.07 mm/y at an AC current density of 160 A/m$^2$, and between 0.25 and 0.50 mm/y at 780 A/m$^2$.[18] A corrosion rate of 2.4 mm/y (i.e., 60 mpy) was found on pipelines with high-quality dielectric coatings in the field.[19] The statistical results of corrosion rate data obtained on cathodically protected steels in the presence of AC interference showed that the corrosion rate could reach up to 2.5 mm/y under an AC current density of about 300 A/m$^2$.[17] In recent laboratory tests, it was found that, in the absence of CP application, the corrosion rate of X65 pipeline steel in a deoxygenated, near-neutral pH bicarbonate solution was 1.25 mm/y at the AC current density of 400 A/m$^2$.[20] When the steel was under CP of –850 mV (CSE), the corrosion rates were 0.1 and 0.6 mm/y at the AC current densities of 10 A/m$^2$ and 400 A/m$^2$, respectively. With a further negative shift of the CP potential to –1,000 mV (CSE), the corrosion rate was smaller than 0.1 mm/y, which was acceptable to industry for the corrosion of carbon steels in practice.[21] Obviously, the AC corrosion rate of metals depends heavily on the specific conditions and parameters that are relevant to the service of the pipelines. Moreover, an accurate model that is able to predict the long-term progression of AC corrosion of pipelines has not been developed yet.

# 3.2. Basics of AC corrosion of pipelines

## 3.2.1. Characteristics of AC corrosion of pipelines

An investigation was conducted on a bitumen-coated natural gas transmission pipeline that had failed owing to AC corrosion induced by a railway (16.7 Hz system) that was parallel to the pipeline for approximately 10 km.[22] It was found that the leaking site on the pipeline was covered by a large cap–approximately 200 mm in diameter–of hard, agglomerated soil (Figure 3-1a). It was further observed that the bituminous coating became soft and sticky in the area surrounding the leak, which was probably caused by the AC heating effect. After cleaning, localized corrosion with deep cavities was found (Figure 3-1b). The corrosion products sampled from the corroded area were analyzed, and a mixture of materials from different sources was detected. The mixture contained black bitumen with iron oxide (magnetite) as the predominant corrosion product; a gray soil material of silicate incorporating a minor amount of sodium, potassium, calcium, and magnesium ions; and a white mixture of $NaHCO_3$ and $Na_2CO_3$. Localized pitting corrosion had occurred on the pipeline in the presence of AC interference. Moreover, the pH of the trapped electrolyte collected from the corroded area had increased compared with the pH value of the groundwater.

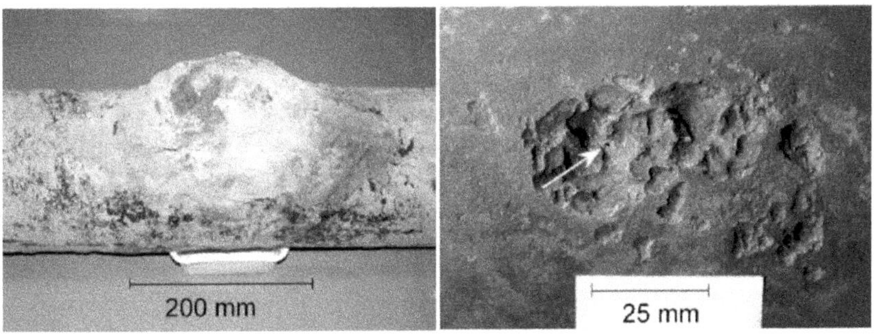


**Figure 3-1.** Optical views of the morphology of the leak site on the failed gas transmission pipeline due to AC corrosion (a) before and (b) after cleaning of deposit, where the arrow indicates the leak.[22]

There is much evidence to further support the morphological features of AC-induced corrosion on buried pipelines. Similar tubercles of stone-hard soil comprising a mixture of corrosion products and soil were also observed at coating defects on a pipeline where AC corrosion occurred.[23] The size of the tubercles was considerably bigger than that of the original coating defects. It was stated that this type of distinct tubercle was frequently observed on pipelines experiencing AC corrosion. The resistivity of the tubercles was lower than that of the surrounding soil. It was also reported that a field dig found a nodule of 20 mm in diameter and 20 mm in height at the leak site on a gas pipeline paralleled by four single-circuit

345 kV power lines for 5.3 km and a double-circuit 115 kV power line for 9.4 km in the U.S.[13]. The pH inside the nodule was measured as 13.1. A corrosion pit was found beneath the nodule and a coating holiday, with a round, smooth, and dish-shaped geometry (30 mm in diameter and 1.35 mm in depth).

In investigating AC corrosion cases on Canadian pipelines, Wakelin et al. summarized the morphological characteristics of AC corrosion on pipelines (i.e., corrosion pits with a rounded bottom and a hard dome over the pits formed by corrosion products).[13, 17] Thick and nonadhesive corrosion products were also observed in the laboratory when environments causing pipeline AC corrosion were simulated.[24]

In terms of the compositional characteristics of the corrosion products generated during pipeline AC corrosion, it was found that the corrosion products mainly consisted of magnetite mixed with soil,[25] which was identical to the finding in a later work.[22] Actually, magnetite being the primary corrosion product in the AC corrosion of pipeline steels was reported as early as the 1960s.[4]

### 3.2.2. AC corrosion mechanisms

Historically, a consensus of the mechanisms for AC corrosion has been unavailable. There were at least two reasons for this situation: (a) the mutual interaction between AC and DC currents that affects the electrochemical corrosion processes was too complicated to be fully understood and (b) the instrumentation used to measure the electric parameters in DC and AC with frequencies between 50 and 100 Hz was not available.[2] In the past several decades, great efforts have been made worldwide to understand fundamentally the phenomenon of AC corrosion occurring on pipelines, and various theories and models have been proposed to explain the pipeline corrosion processes and parametric effects under AC interference. Some important and extensively cited theories and models developed to date are described and compared herein, as a summary of the essential steps made to advance our understanding of this important phenomenon.

Gummow and others reviewed the theories for AC corrosion proposed up to the 1980s in their well-cited review article published in 1996.[12] As early as 1916 some researchers have speculated that AC corrosion was due to the irreversibility of the corrosion reaction of metals occurring during the anodic half cycle of the AC, while the corroded part did not "self-recover" back to the metal during the negative half cycle.[1] Others have thought that, in addition to the dissolution reaction, the AC corrosion may have partially resulted from a "thermal activation."[18] However, a well-defined correlation of AC corrosion rates with temperature has not been established.[5] In terms of the AC corrosion rate being low compared with DC corrosion, it was believed that the majority of AC current with a sine function and high-frequency harmonics was partially used by the double-charge layer capacitance and did not contribute to the metallic oxidation at the metal–solution interface.[26] This effect became more obvious when the corrosion process was under diffusion-controlled conditions (e.g., dissolved oxygen participating in the cathodic reaction). Jones also found that the accelerated corrosion by AC was greater in de-aerated environments than in aerated ones. [27] Jones thus proposed that the AC could cause depolarization of the anodic reaction, which probably resulted from anion desorption or reduction of the surface film during the cathodic half cycle. For metals that were passivated in the environments, the AC caused breakdown of the passivity. The effect increased with increasing AC current density.[28] However, Hamlin believed that "AC does not have any significant effect on the polarization or depolarization of cathodically protected steels."[29]

Despite the divergence and even controversial results and conclusions regarding the AC corrosion phenomenon, the early-stage researchers and their great efforts laid important and valuable foundations to help advance understanding of the fundamentals of the phenomenon. Nielsen and Cohn proposed an electrical equivalent circuit model to illustrate the occurrence and growth of AC corrosion.[23] According to the model, when a DC source (which could be from a CP system) and an AC source are applied individually on the buried pipeline at a coating defect, the resistance of the soil electrolyte affects the AC current flow through the soil to the coating defect and should be included in the electrochemical impedance circuit. The soil resistance is connected in series with the parallel connection of the electric circuits for anodic and cathodic partial reactions. Each circuit representing the anodic–cathodic partial reaction includes a parallel connection of a double-charge layer capacitance with a charge-transfer resistance, while the latter is connected in series with a diffusion Warburg impedance to restrict the reaction rate where the reactions are rate limited by the diffusion processes of reactants. The key element of the corrosion process is the charge-transfer resistance, which is associated with metal dissolution. AC corrosion occurs if the electrical charge during the anodic half cycle exceeds the cathodic charge. The model attributed the AC corrosion occurrence to the anodic electrical charge passing to the pipeline steel and used the net charge (i.e., the anodic charge subtracted by the absolute value of the cathodic charge) to quantify the amount of pipeline steel that would be expected to corrode.

Drawing on the theory initially proposed by McCollum and Ahlborn[1] and modified by Nielsen and Cohn,[23] Yunovich and Thompson[30] analyzed the reversibility of cathodic and anodic processes occurring during AC corrosion and determined that the process occurring during the anodic half cycle of the AC signals was not completely reversed during the cathodic half cycle. As a result, the AC corrosion of metals was attributed to the charges that cause metal dissolution during the anodic cycle being greater than those during the cathodic cycle used for metal deposition. Furthermore, an AC corrosion mechanism based on the electrode reactions occurring in the positive and negative half cycles of the applied AC signals was developed, as illustrated in Figure 3-2.[31] During the positive half cycle, the metal is oxidized and an oxide film forms on the metal surface. During the negative half cycle, the oxide film is reduced to iron hydroxide. A new oxide film grows in the following anodic cycle. The amount of iron hydroxide increases owing to the further reduction of the oxide film in the negative half cycle. As a result, each AC cycle results in a net oxidation of the metal and, thus, a metal loss in the presence of AC interference. It was proposed that the reactions occurring in the anodic and cathodic processes were iron oxidation and the reduction of iron oxide, respectively:

$$3Fe + 4H_2O \rightarrow Fe_3O_4 + 8H^+ + 8e \quad (3\text{-}1)$$

$$Fe_3O_4 + 4H_2O + 2e \rightarrow 3Fe(OH)_2 + 2OH^- \quad (3\text{-}2)$$

Although the model described briefly and clearly the AC corrosion process, it did not consider the condition of CP application on the metal. The influences of AC on the interfacial electrode steps, both thermodynamically and kinetically, were not included either.

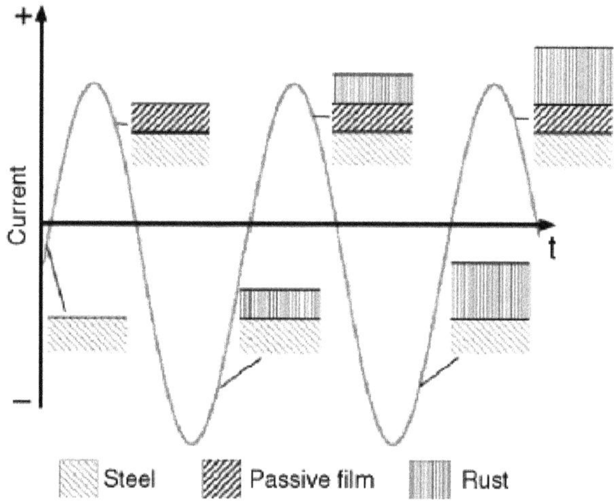


***Figure 3-2.*** **Schematic diagram illustrating the AC corrosion process.**[31]

With substantial evidence to support electrolyte alkalization during the AC corrosion of pipelines, another mechanism considering the environmental changes resulting from AC interference was proposed in 2005, in which the combined action of high pH and potential oscillations caused by AC interference led to corrosion attack on the pipeline steel.[32] As shown in Figure 3-3, the hydroxide ions ($OH^-$) produced by CP-driven cathodic reactions, which were the electrochemical reduction of either dissolved oxygen in aerated environments or water molecules in deoxygenated ones, accumulated near the coating defects where the AC current flew in. An incubation period, which was defined as the time taken to reach a critical pH value of the electrolyte in contact with the metal, was required for $OH^-$ accumulation. Potential fluctuations between the passivity, immunity, and high pH corrosion regions in the Pourbaix diagram (Figure 3-4), caused by the AC interference could induce corrosion of the metal owing to different time constants being associated with the iron dissolution (fast) and the subsequent formation of passive film (slow). At a very alkaline pH, the formation of dissolved $HFeO_2^-$ may cause metal corrosion at a high rate. This model considered an AC corrosion pathway that was different from previously proposed ones, where AC corrosion occurred at a negative potential in an alkaline electrolyte, and the steel corrosion resulted in the generation of oxide films and $HFeO_2^-$ ions. While this could happen at "big" coating defects, where full CP potential was applied to cause the electrolyte alkalization, the model did not consider "small" coating defects, such as pinholes where the CP was shielded but AC current flew in to cause AC corrosion locally. Moreover, in reality, pipeline steels are not always passivated to form oxide films in all electrolytes. For example, it has been well established that pipeline steels are in an active dissolution condition in the near-neutral pH bicarbonate solution trapped under disbonded PE-type coatings.[33]

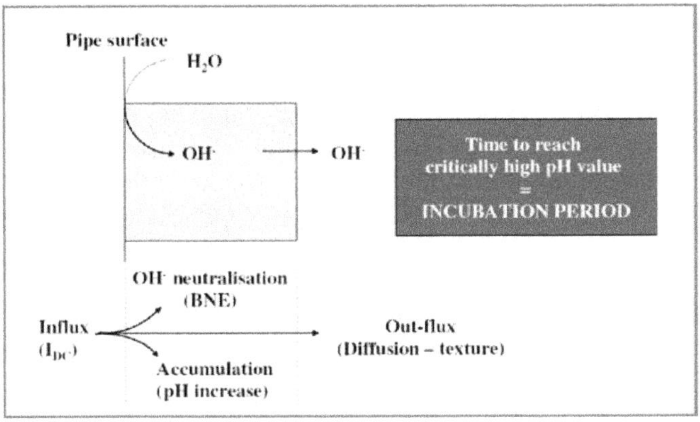


**Figure 3-3.** Schematic diagram illustrating the electrolyte alkalization mechanism during AC corrosion.[32]

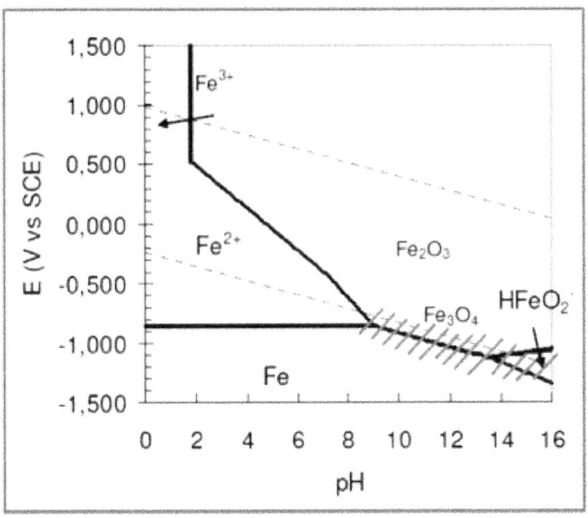


**Figure 3-4.** Pourbaix diagram where the red-marked area indicates the critical AC corrosion zone.[32]

A similar mechanism was proposed later, in which AC corrosion was a consequence of the oscillations of events occurring at the metal–electrolyte interface, depending on the electrolyte pH,[34] making the steel alternately in the corrosion–immunity region and the passivity–immunity region. It was thought that the AC corrosion was due to both the irreversibility of the iron dissolution reaction in acidic, neutral, and slightly alkaline electrolytes and the impossibility of the formation of a passive film in electrolytes with a pH $< 14$. Under excessively alkaline electrolytes with a pH $> 14$, corrosion resulted from the reduction of a thin oxide film previously formed on the metal surface.

A modified two-step AC corrosion concept based on the alkalization mechanism was proposed in 2013.[35] In step one, AC causes the breakdown of the passive film formed on the steel surface owing to electromechanical stresses present in the film. The presence of a high AC field within the passive film could cause the local breakdown of the film. In step two, corrosion occurs only if the pH of the environment at the steel–electrolyte interface is up to 14. The high pH can be reached owing to the high CP current supplied to the metal. As stated by the authors, further tests are required to confirm the model. Moreover, AC corrosion can happen in electrolytes with a pH lower than 14. There have been no sufficient experimental results to demonstrate whether the pH at the steel–electrolyte interface reaches 4 or to quantify the electrostriction stress needed to break the passive film.

Regarding the AC corrosion reaction mechanism and kinetics, the reversibility of the electrochemical reactions in response to the frequency of the interfacial signals was regarded as having great importance.[36] If the frequency was not high enough and the chemical reaction was not reversed, the metal could degrade by the dissolution and migration of metal ions into the solution. At the same time, the AC corrosion kinetics were modeled by a modified Butler–Volmer equation, in which, for activation-controlled corrosion kinetics, a mean current density (i.e., a rectification current) appeared in response to the asymmetrical, sinusoidal potential to affect the ratio of anodic and cathodic Tafel slopes ($r = b_a/b_c$).[37, 38]

For example, in the presence of AC-induced voltage on metals, the corrosion potential would shift negatively or positively, or would remain unchanged, when $r < 1$, $r > 1$, or $r = 1$, respectively. Thus, both reversibility and faradic rectification are critical to the AC corrosion mechanism and kinetics. Goidanich et al. performed galvanostatic polarization tests on carbon steel in iron sulfate solutions under various AC current densities.[39] It was found that the corrosion process occurring during the anodic half cycle would not completely be reversed during the cathodic half cycle. As a result, the double-charge layer at the steel–solution interface was modified, resulting in a significant effect on the corrosion kinetics. It was further confirmed that models based on a pure faradic effect were not accurate in describing the AC corrosion because the interfacial processes were not completely reversible.

Xu et al. proposed an equivalent circuit for electrochemical corrosion processes of pipeline steels in the presence of AC, as schematically shown in Figure 3-5.[40] The total input AC potential, $\dot{U}$, consists of three components: (a) AC potential in the solution, $\dot{U}_S$, (b) AC potential in the corrosion product film, $\dot{U}_f$, and (c) AC potential in the double-charge layer, $\dot{U}_{dl}$. They are related quantitatively by:

$$\dot{U} = \dot{U}_S + \dot{U}_f + \dot{U}_{dl} = i_{AC}(R_S + Z_f + Z_p) \tag{3-3}$$

where $i_{AC}$ is AC current density, $R_s$ is solution resistance, and $Z_f$ and $Z_p$ are the impedance of the corrosion product film and the double-charge layer, respectively. Furthermore, the measured real-time AC potential between the reference electrode and the steel, $\dot{U}_x$, is written as:

$$\dot{U}_x = E_{DC} + \dot{U}_f + \dot{U}_{dl} = E_{DC} + i_{AC}(Z_f + Z_p) = E_{DC} + 0.5V_{pp}\sin(\omega t) \tag{3-4}$$

where $E_{DC}$ is the DC potential of the steel, $\dot{U}_f + \dot{U}_{dl}$ represent the sum of the measured AC potential of the corrosion product film and the double-charge layer, $V_{pp}$ is peak-to-peak voltage of AC signal $\omega$ is the angular frequency, and $t$ is time. The dependence of $E_{DC}$ on $i_{AC}$ can be affected by the formation of a compact corrosion product on the steel surface, which usually occurs under high AC current densities. This leads to a time-dependent increase of the surface electric resistance for AC current flowing through the steel surface. It is usually easy to measure the AC potential, but very difficult to measure the AC current density directly owing to a lack of information about the coating defect and the AC current flowing through it. Using measured information, such as the soil resistivity and the distance between the buried pipe and the ground, it is feasible to measure AC potential components at coating defects and then derive the AC current density.

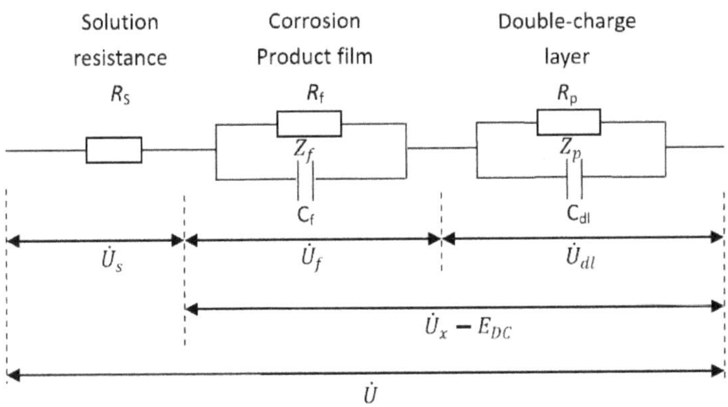


**Figure 3-5. Schematic diagram of the equivalent circuit for electrochemical corrosion of steels in the presence of AC application.**[40]

In summary, the AC corrosion mechanisms developed to date consider three main aspects: (a) the reversibility of corrosion reactions during positive and negative half cycles of AC signals, (b) the alkalization of environmental electrolyte, and (c) the change of the double-charge layer at the steel–solution interface under AC interference. All three apply to AC corrosion processes on pipeline steels, and each has its own limitations in explaining the AC corrosion phenomenon. Therefore, a complete understanding of the fundamentals of pipeline corrosion under AC interference requires consideration of all relevant conditions and changes.

### 3.2.3. Parametric effects

Both field testing and laboratory work have demonstrated that a great number of factors affect the occurrence, growth, and severity of AC corrosion of pipelines, and the affecting factors interact with one another, making the problem even more complicated. For example, the AC corrosion rate is related to AC current density, which depends on the area of the coating defect and the electrolyte resistivity under the induced AC voltage ($V_{AC}$). When the AC interference is primarily generated by the electromagnetic field from an adjacent HVAC power line, the induced voltage on the pipeline is a function of the HVAC transmission parameters and the mutual impedance between phase conductors of the power line and the pipeline.[41]:

$$V_{AC} = \sum_{n=1}^{3} Z_n I_n \tag{3-5}$$

where $I_n$ is the current load at the $n$th phase conductor of an HVAC power line, and $Z_n$ is the impedance (a complex number) with earth between the $n$th phase conductor of the HVAC power line and the pipeline, which can be calculated by:

$$Z_n = \left(\mu_0 \times \frac{\omega}{8}\right) + j\left[\left(\mu_0 \times \frac{\omega}{2\pi}\right) \ln \frac{1.85}{D_n\sqrt{\frac{\omega\mu_0}{\rho soil}}}\right] \tag{3-6}$$

where $\mu_0$ is the space permeability, $\omega$ is the angular frequency, $D_n$ is the distance between the $n$th phase conductor and the pipeline, and $\rho_{soil}$ is the soil resistivity. Assuming a circular holiday present on the external coating, the AC current density at the defect can be written as:

$$i_{AC} = \frac{8V_{AC}}{\rho_{soil}\pi d} \tag{3-7}$$

where $d$ is the diameter of a circular coating defect. This formula is applicable to cases when the defect size is greater than the thickness of the coating.[14] It can thus be seen that the AC corrosion rate increases with increased current load carried in HVAC power lines, reduced distance between the pipeline and the power line, decreased soil resistivity, and decreased area of the coating defect.

Equations (3-5) to (3-7) apply to the specific pipeline segments, rather than the whole length of the pipeline. Segmentation of pipelines is a must for the determination of the AC corrosion rate and the associated risk, because the geographical closeness and other parameters, such as soil resistivity, are not constant over the whole interfering distance.

**Collocation of buried pipelines with HVAC power lines.** The presence of adjacent HVAC power transmission systems is a key indicator that AC coupling could adversely affect a pipeline. The following conditions are generally associated with a high likelihood of induced AC corrosion on pipelines:

- Power transmission systems that operate at 69 kV or greater
- Pipelines and AC power systems that are collocated for 1.6 km or more
- Pipelines and AC power systems with a lateral separation of 150 m or less

The orientation of the buried pipelines relative to HVAC power lines is critical to the magnitude and distribution of induced AC voltage on the pipelines. Generally, the magnitude of the inductive coupling effect decreases when the crossing angle between the pipelines and the electric power lines increases, which is attributed to a reduced section of the pipelines being exposed to the electromagnetic field. For a perpendicular crossing, where the pipelines cross at or near 90° to the power lines, the effective parallel length between them is minimized. As a result, the inductive coupling effect on the pipeline is minimal. A numerical modeling to simulate a 152.4 mm (6 in.) pipeline coated with FBE in 10 Ω·m soil running under a 345 kV AC transmission line showed that a maximum induced voltage occurred at crossing angles less than 15°.[42] With a current load of up to 1,000 A, a crossing angle greater than 45° would induce about 2 V of AC voltage, and a crossing angle greater than 60° would induce fewer potentials. Industry experience has suggested that crossing angles greater than 60° between buried pipelines and HVAC power lines are not considered serious in terms of AC corrosion risk.[43] However, at high AC current, such as 3,000 A or 5,000 A, even a crossing angle of 80° may be considered significant in accelerating AC corrosion.[42] Thus, the AC corrosion risk to the affected pipelines depends on a complete consideration of both the carried current and the orientation between the pipelines and the HVAC power lines.

The level of AC interference for a given collocation or crossing with buried pipelines also depends on the distance between the power lines and the pipelines. The proximity of the phase wires limits the strength of the longitudinal electric field to which the pipelines are exposed. It was found that, as the distance between the pipelines and the power lines increased, the AC induction on the pipelines decreased.[44] Furthermore, in addition to the carrying load of the power lines, including AC voltages and current loads, the phase arrangement is critical to the induced AC interference. Under identical conditions, a double-circuit power line would cause a smaller effect on adjacent pipelines than would a single-circuit power line. This is attributed to a double-circuit tower (i.e., a full-roll arrangement) being conducive to canceling the effect that minimizes the magnetic field induced on the pipelines.[13]

**AC frequency.** It is generally accepted that the AC corrosion rate of pipelines decreases with increasing frequency because, with the increase of the AC frequency, the interval between successive anodic and cathodic half cycles becomes shorter.[45, 46] Consequently, the metallic cations generated in the anodic cycles are available for immediate redeposition in the cathodic cycles. In addition, at high frequencies, hydrogen atoms produced during the cathodic half cycles by electrochemical reductive reactions do not have sufficient time to combine and form hydrogen molecules. During the sequent anodic half cycle, hydrogen atoms cover the metal surface to prevent the metal from further dissolution reaction.[47] Moreover, it was found that the AC frequency affected the morphology and density of corrosion pits.[48] Generally, the passive current density of steels in electrolytes decreases with increasing AC frequency, indicating that the passivity of the steels under higher AC frequencies is more stable than that under lower AC frequencies.

**AC voltage.** The AC voltages affect the corrosion rate of pipelines. It is generally believed that an elevated AC voltage can cause a higher corrosion rate of pipeline steels, in addition to raised electric shock hazards. British Standard 15280 suggests that AC voltage be used as an important parameter to evaluate the AC corrosion risk.[49] The specification recommends that to reduce AC corrosion, the AC voltage induced on pipelines should not at any time exceed 10 V or 4 V if the local soil resistivity is greater than 25 Ω·m or lower than 25 Ω·m, respectively. The NACE SP0177 states that a steady-state AC voltage of 15 V or more with respect to earth could result in a shock hazard to personnel.[50] However, it is generally believed that assessing AC corrosion based on AC voltage is not reliable and even generates misleading information.[51] Particularly, AC current density, rather than AC voltage, should be used to characterize AC corrosion and give precise assessments.

**AC current density.** AC current density is the most critical factor in the prediction of AC corrosion risk to pipelines. Field experience and lab testing results have both indicated that AC corrosion of pipelines occurs when the AC current density is above a threshold value in a certain environment.[52] At high AC current densities, corrosion can occur even in the presence of applied CP on the pipelines.[53] Funk et al. found that AC current densities greater than 30 $A/m^2$ caused corrosion rates greater than 0.1 mm/y at a constant CP current density of 2 $A/m^2$.[56] It was also reported that corrosion rates could be up to 1.3 mm/y at AC current densities of 100 $A/m^2$ to 200 $A/m^2$ and that AC corrosion could not be ignored at AC current densities starting at 15 $A/m^2$.[12]

For threshold AC current densities determined in certain environments, below which AC corrosion is not a concern, there have been so far no universal standards in terms of the magnitude of the value. The relevant AC current density criteria obtained in early investigations included the following:[54]

- AC corrosion did not occur at AC current densities less than 20 $A/m^2$.
- AC corrosion was unpredictable for AC current densities between 20 $A/m^2$ and 100 $A/m^2$.
- AC corrosion occurred at AC current densities greater than 100 $A/m^2$.

Both German Standard DIN 50925 and British Standard BS EN 15280 suggest a critical AC current density of 30 A/m$^2$, above which the AC is detrimental to buried steel structures.[49, 55] Relevant work also indicated that, generally, AC corrosion could only occur if the AC current density exceeded 30 A/m$^2$.[10, 56] However, it was also reported that higher current densities did not necessarily lead to corrosion attack.[31] Baeckmann et al. used 20 A/m$^2$ as the threshold AC current density for AC corrosion occurrence.[57] However, experimental testing has shown that a theoretical "safe" AC density may not exist.[58] It was also found by Italian researchers that the AC current density of 10 A/m$^2$ may be hazardous to pipelines.[59] Their results confirmed that the corrosion rate of pipeline steels at this AC current density was increased two-fold in a simulated soil solution compared with that under the AC-free conditions.

In addition to enhanced general corrosion, localized corrosion can be caused on pipelines under high AC current densities. Extensive corrosion pits were observed on the surface of a pipeline steel in a neutral pH bicarbonate solution when the AC current density was increased to 800 A/m$^2$.[60] Kuang and Cheng demonstrated that, at sufficiently high AC current densities, an X65 pipeline steel suffered from pitting corrosion in both the high pH carbonate–bicarbonate solution and the soil-extracted, neutral pH bicarbonate solution, and that the steel was in passive and active dissolution state, respectively.[61] Their study determined that the threshold values of AC current density below which pitting would not occur on the steel were 300 A/m$^2$ and 200 A/m$^2$ for high pH and neutral pH bicarbonate solutions, respectively. Field inspections at selected coating holidays on a gas pipeline detected corrosion pits of 1 mm in depth after 2.5 years of operation in a sandy soil of 20,000 Ω·cm at AC current densities of 74 A/m$^2$ to 165 A/m$^2$.[9]

It is obvious that pipeline operators should consider AC current density in any AC interference analysis and mitigation program. All factors that contribute to AC current density, such as AC pipe-to-soil potentials, soil resistance, size of coating defects, pH of electrolyte on the pipeline surface, and changes in pipeline or AC power system alignment, should be considered in AC corrosion risk evaluation.

Nowadays, the AC current density is regarded as a main parameter for evaluation of the AC corrosion risk.[25] As reviewed above, both field and laboratory experience have given certain ranges and threshold values of AC current density for ranking the risk of AC corrosion of pipelines. However, Ghanbari et al. stated that "AC current density is unreliable to be considered as a parameter for AC corrosion threshold because AC current consists of both Faradaic current and non-Faradaic current (i.e., capacitive current), while only the Faradaic current contributes to corrosion."[62] Thus, a quantitative relationship between the AC current density and the resulting corrosion cannot be established. Instead, it was reported that, in AC corrosion testing on X65 pipeline steel in 0.1 M NaCl solution, the average AC current, as measured by the sum of the anodic and cathodic half cycles of the sine wave, correlated well with the current density measured by weight loss analysis.[63] Average AC current was thus suggested as a good indicator of the AC corrosion rate.

**Coatings and coating defects.** Coating systems available today for pipeline protection possess exceptional dielectric and mechanical characteristics, as well as high coating resistance, especially in soils with high resistivity.[64] The coating resistance to soil environments is critical in controlling the level of AC voltage built on the buried pipelines. Generally, the coating resistance depends on coating type, performance condition, coating thickness, soil chemistry, and soil resistivity, among other factors, all of which can vary along the pathway collocated with the power lines. When the coating resistance reduces owing to degradation (such as water permeation) during service, multiple paths exist to dissipate the built AC voltage. This would reduce the buildup of induction, resulting in a low AC voltage and a low AC current density discharging at individual holidays (or defects).[44] For coatings with a high dielectric strength and excellent performance condition, the resistance of the coating is high, allowing a great induction to build up over a long distance. This can generate elevated AC potentials, which are hazardous to the safety of working personnel. Moreover, the risk of AC corrosion increases as the AC current density discharging from a limited number of holidays (defects) can be quite high.

It is generally accepted that FBE-coated pipelines are most susceptible to AC corrosion because of the high dielectric strength of FBE coatings combined with their propensity to have small pinholes or holidays. This results in high induced AC voltages combined with high localized AC current discharge at the small holidays. It has been reported that many AC corrosion incidents have occurred on FBE-coated pipelines.[19] AC corrosion can also occur on coal tar, tape wrap, extruded PE, or asphalt coated pipelines with holidays.

Pipeline coatings are polymeric materials, such as those made of epoxies and PE. Defects in varied geometries can be introduced into the coatings during manufacturing, application, transportation, and construction.[65] Yet the key parameter in AC corrosion of pipelines is perhaps the spread resistance of the coating defect,[66] which relates the protection potential and induced AC voltage on a pipeline to the DC and AC current densities. The spread resistance depends on both the soil resistivity and the geometry of the coating defects, including their size, depth, and shape. In Jiang et al. the coating defect directly influenced the AC corrosion of pipelines when AC current discharged locally from these defects.[67] It was found that the smaller the defects, the greater the AC current density, and thus the AC corrosion rate.[41, 68] Moreover, the applied CP current usually has more difficulty penetrating a smaller defect than a larger defect to reach the defect bottom for corrosion protection owing to the shielding effect.[69, 70] Field investigations have indicated that most AC corrosion occurs at coating holidays with an area of 1 $cm^2$ or smaller.[71]

**Soil properties and chemistry.** The soil resistivity along the collocation between buried pipelines and the HVAC power lines is a critical parameter in the evaluation of AC inductive and resistive coupling effects, the calculation of AC current density, the assessment of AC corrosion risk, and the design of AC mitigation systems. The local soil resistivity is usually controlled by the type and amount of soluble salts and the water content, which in turn affect the electrochemical corrosion processes occurring on the steel surface. In reality, the local environment at the coating defect exposed to the soil is quite complex, with multiple factors affecting the environmental evolution with time. The electrical resistance of the soil can either increase or decrease with time when the pH of the electrolyte is affected by the applied CP, which drives the electrochemical reduction of hydrogen ions or dissolved oxygen in acidic or neutral to alkaline environments, respectively, or by dissociation of water under anaerobic conditions. The earth-alkaline ions such as $Ca^{2+}$ and $Mg^{2+}$ contained in the soil react with $OH^-$ to form hydroxides with a relatively low solubility, which can significantly increase the resistance at the coating defect.[72] However, alkaline cations such as $Na^+$, $K^+$, or $Li^+$ can form highly soluble hygroscopic hydroxides, decreasing the electrical resistance at the coating defect. Therefore, the AC current density at the coating defect is highly dependent on the soil composition–specifically, the ratio of alkali to earth-alkaline cations–and the concentration of hydroxyl ions generated by the CP-facilitated reductive reactions. All of them would affect the soil resistivity. It has been found that, when the CP potential was less negative than –1.19 V (CSE), the AC corrosion rate of an X70 pipeline steel in a simulated soil solution concentrated with $Ca^{2+}$ and $Mg^{2+}$ ions was smaller than that in the soil solution containing a high content of $Na^+$ ions at identical AC current densities.[73]

The spread resistance of the coating defect, which critically determines the local AC current density flowing at the defect, is a highly dynamic parameter under AC interference, depending on the chemical environment of the soil, moisture, and buildup of calcareous deposits.[72] Generally, soils with a low resistivity yield a low spread resistance owing to the amount of dissolved species in the electrolyte.[66] The hydroxyl ions produced by CP-driven cathodic reactions lower the resistivity, but the increased pH alters the solubility of different species in the soil. The calcareous or hydroxide deposits would increase the spread resistance and thus limit AC corrosion.[74]

Furthermore, it was found that soils with high resistivity or calcareous deposits may lower the AC current density threshold for corrosion. However, the presence of calcareous deposits in the soil would increase the DC current density threshold. In either case, the environment will limit both AC and DC current densities that cause very high corrosion rates. Moreover, chloride may have a reducing effect on high AC corrosion rates, which is related to the mechanism that AC corrosion propagates by the continuous oxidation and reduction of a passive film to a porous layer.[31] Chlorides can destabilize the passive film on steels, limiting the formation of the passive film. Sulfate ions may adsorb to the steel surface and inhibit the destabilizing effect of chloride ions on the passive film.[75] Sulfates could increase the corrosivity of aqueous solutions and decrease the charge-transfer resistance of interfacial reactions.[76] Thus, sulfates may increase the AC corrosion rate. The phosphate buffer and the bicarbonate have little effect on AC corrosion. At high polarization, their presence can enhance the steel corrosion.[72]

In addition, the interfacial capacitance of the pipeline steel, which is also controlled by the soil chemistry, is important in determining the AC corrosion rate.[62] Particularly, the weakly dissolvable scale formed at the coating defect would change the local interfacial capacitance. Thus, pipelines in soils that are prone to forming alkaline metal scale have a remarkable difference in AC corrosion rate from those where the scale is not formed under the equivalent AC current and pipe-to-soil potential. Generally, the formed scale provides some protection to the steel, with a reduced corrosion rate under AC interference.[62] The role of scale in the corrosion of steels under AC interference was investigated in terms of its effect on the corrosion rate of the unscaled portion of the steel.[77] (It was found that the iron carbonate scale itself did not result in a higher or lower corrosion rate on an X65 pipeline steel exposed at pinholes. Thus, while the presence of the scale reduced the area of steel exposed to the solution, it did not affect the corrosion rate of the scaled steel compared with the corrosion of bare steel in an identical solution.

Soil resistivity varies widely owing to soil composition, chemistry, and moisture content. For example, in areas with a uniform soil composition and little or no change in moisture content, the seasonal variability in soil resistivity is low. However, in many areas, pipelines are buried in soils of mixed and variable compositions that are subject to significant seasonal changes in soil moisture at pipe burial depths. The variability of the soil resistivity would be remarkable.[78] For evaluation of AC interference and AC corrosion of pipelines, the soil resistivity should be measured frequently. Measurements conducted in relatively small test areas are often used to extrapolate the soil resistivity in wide areas with similar soil composition and moisture content.

In evaluation and control of AC interference, it is necessary to consider both the bulk and local soil resistivity. The bulk resistivity at the pipeline burial depth is one of the controlled factors in the analysis of AC voltage. The bulk resistivity was determined from the average soil resistance in a half hemisphere to the depth of the pipe, while the specific local resistivity of the soil layer referred to that directly adjacent to the pipe surface.[44] The local soil resistivity is one of the primary factors affecting the corrosion of pipeline steel at a coating holiday. The bulk soil resistivity, combined with the coating resistance, affects the level of induced AC voltage along the pipeline.[44] It has been shown that the induced AC voltage increases logarithmically with increasing soil resistivity. Furthermore, the soil resistivity has a great influence on AC current density. While an increase in soil resistivity can result in an increase in the magnitude of induced AC voltage, the AC current density and the associated AC corrosion risk decrease linearly with increased resistivity of the soil.[44]

**CP potential or CP current density.** It was demonstrated that the operation parameters of the CP system strongly influence the AC corrosion rate of pipelines, which were confirmed by extensive investigations conducted in artificial soil solutions,[30, 58] in soils,[79] and in sodium hydroxide solutions.[80] Generally, AC corrosion can be mitigated by a negative shift of CP potential.[81] At low AC current densities such as 20 $A/m^2$, a CP potential of –950 mV (CSE) can provide full protection to an X65 pipeline steel from corrosion in carbonate-bicarbonate solutions. When the AC current density is increased to 100 A/m$^2$, the CP potential of −1,050 mV (CSE) is required for full corrosion protection of the steel.[53] Similarly, it was found that an increase in the CP current density from 2 A/m$^2$ to 5 A/m$^2$ decreased the AC corrosion rate by at least one half at an AC current density of 50 A/m$^2$.[56]

**Metallurgical microstructure of steels.** It was reported that the difference in the metallurgical microstructure of X70 pipeline steel is associated with different corrosion rates and pitting corrosion behavior under AC interference in a simulated acidic soil solution.[82] It was found that a hot-rolled steel experienced uniform AC corrosion, with few corrosion pits. Accelerated corrosion and pitting corrosion were observed on the normalized and quenched steels. The annealed steel exhibited nonuniform corrosion with some pits. The normalized steel had the largest corrosion rate, and the hot-rolled steel possessed the smallest corrosion rate; the quenched steel was in between. The normalized microstructure composed of polygonal ferrite and pearlite was the most susceptible to AC corrosion. The exact mechanism linking the steel metallurgy with AC corrosion has not yet been established.

## 3.3. Experimental setup enabling AC corrosion measurements

The design and preparation of an experimental setup enabling measurements of the AC corrosion of a metal specimen are usually more complicated and difficult than expected. The difficulty mainly results from the mutual interference of the AC circuit and the electrochemical corrosion measurement circuit, which are applied on the tested specimen at the same time. The interference exhibits in three ways: (a) deviation of the AC signal on the specimen from the applied value, (b) deviation of the electrochemical potential or current signals recorded on the specimen from the actual values (i.e., the recorded values include the interference signal from the AC circuit), and (c) the adverse effect or even damage to the electrochemical potentiostat caused by the applied AC. When the test specimen is under CP, the situation becomes further complicated, as the CP is applied on the specimen through an additional circuit. Thus, study of AC corrosion is heavily dependent on a reasonable experimental setup.

Figure 3-6 shows the schematic diagram of an experimental setup for AC corrosion measurements, where WE, CE, and RE refer to working electrode (i.e., the tested steel specimen), counter electrode, and reference electrode, respectively.[40] The AC current is applied on the working electrode through a capacitor (C) and a resistor (R). The AC current density is adjusted by a slide rheostat, and its value is obtained by measuring the voltage on the resistor and dividing it by the resistance and the area of the specimen. The capacitor is used to block DC signals, and an inductor is used to prevent AC current from flowing into the electrochemical potentiostat. In addition, a data acquisition (DAQ) system, with a high internal resistance, is installed to measure the potential between the working electrode and the reference electrode. Moreover, the ohmic drop for potential measurements is kept as low as possible by either increasing the solution conductivity or positioning the reference electrode sufficiently close to the working electrode. An electric switch (K) is used to separate the AC circuit from the electrochemical measurement circuit. This provides a reliable setup for AC corrosion measurements while avoiding both signal deviation and adverse impact on the potentiostat. A similar circuit setup was followed by Guo et al.,[46] but the DAQ system was not included.

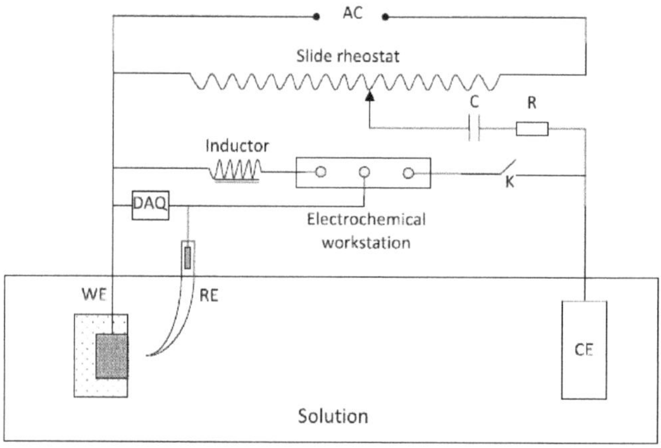


**Figure 3-6. Schematic diagram of the experimental setup for AC corrosion measurements in an aqueous solution.**[40]

Experimental arrangements that are different from the setup shown in Figure 3-6 have also been used in AC corrosion investigations. The methods suffer from various problems, as analyzed above. For example, most work has ignored the adverse impact of the applied AC signals on the electrochemical potentiostat and has not used the inductor in the circuit;[83, 84] This can also cause the measured potential of the target metal to deviate from the actual value. Other common mistakes include directly connecting the AC signal generator with the potentiostat and not considering the interference of DC signals or the deviation of the measurement signals.[77] Moreover, it is not uncommon for the measurement circuit used in the AC corrosion study not to be provided in open publications,[8, 31, 85] making it impossible to evaluate whether a reasonable setup was used.

# 3.4. AC corrosion of pipelines: Active dissolution systems

## 3.4.1. Environments resulting in active dissolution of pipeline steels

External coatings prevent buried pipelines from coming into direct contact with the soil environment. When the coatings stay intact (i.e., the coatings are adhered to the substrate steel and do not contain defects such as pinholes or holidays), corrosion should not occur on the pipelines. Pipelines corrode only when the coatings degrade in various ways, as described previously, so that a corrosive environment generates on the pipe steel surface and, at the same time, the applied CP is shielded from reaching the steel for corrosion protection. As is well acknowledged, the generation of corrosive environments and their properties depends on the property of the coatings and their failure modes.[33]

When coatings that are impermeable to CP current, such as PE-based coatings, are disbonded from the pipe steel for reasons such as poor surface preparation, inappropriate coating application, or significant soil stress being applied on the coated pipe, a corrosive environment can generate under the disbondment upon entry of groundwater from the crevice at the tape overlaps. Frequently, $CO_2$ generated from the decay of organic matter in the soil can dissolve in the trapped water under the disbonded coating. Since PE is impermeable to oxygen ingress (the oxygen amount is usually quite low at the pipeline burial depth), the trapped electrolyte is usually anaerobic. In addition, as CP current is blocked by the coating from reaching the pipe steel, the potential of the steel in the electrolyte is at its open-circuit potential. It has been established that pipeline steels, from the early low-grade (e.g., X46) to the high-grade (i.e., X80 and X100) steels, are in an active dissolution state in this corrosive environment.[33, 64, 86] Extensive field experience, through analysis of the sampled electrolytes trapped under disbonded PE-based coatings, has demonstrated that the electrolyte shows a pH ranging from 5.5 to 7.0 (i.e., the so-called near-neutral pH electrolytes).[87, 88] As a result, the pipeline corrosion occurring in the electrolytes under disbonded coatings that are impermeable to CP current is also called "near-neutral pH corrosion," where the pipe steel is in the active dissolution state. In addition, the groundwater and extracted soil solutions are usually composed of diluted bicarbonate with a near-neutral pH range.

## 3.4.2. AC accelerated pipeline corrosion

To quantitatively determine the AC accelerated corrosion of pipeline steels, weight loss testing by measuring the corrosion rate of the steels in the corrosive solution in the absence and presence of AC interference is regarded as the most accurate and reliable method. Figure 3-7 shows the corrosion rate of a 16Mn steel coupon (wt%: C 0.16, Mn 1.40, S 0.025, P 0.009, Si 0.36, Cr < 0.1, Mo < 0.05, Al < 0.03, V < 0.03, Ti < 0.03, Cu < 0.055, Fe balance) as a function of AC current density determined by weight loss measurements after 48 h of immersion in a simulated soil solution, where the steel is in an active dissolution state.[40] It can be seen that the steel corrosion is accelerated by AC interference. When $i_{AC}$ is smaller than 100 A/m$^2$, the corrosion rate increases slightly with the AC current density. When $i_{AC}$ exceeds 100 A/m$^2$, the corrosion rate increases rapidly and reaches a maximum value at 400 A/m$^2$. With a further increase in the $i_{AC}$ up to 600 A/m$^2$ and 800 A/m$^2$, the corrosion rate of the steel decreases.

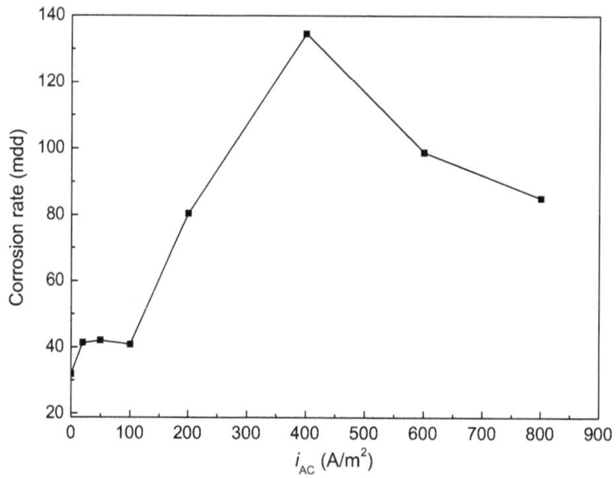


**Figure 3-7. Weight loss measurements of a 16Mn steel coupon under various AC densities after 48 h of immersion in a simulated soil solution, where mdd is the unit of corrosion rate of mg/dm$^2$day.[40]**

To be consistent with the corrosion rate results, the corrosion morphology of the steel coupon under various AC current densities is shown in Figure 3-8.[40] In the absence of AC, most of the coupon is shining morphologically, with rust present at isolated locations (Figure 3-8a). Upon application of 50 A/m$^2$ of AC current density, there are dark brown, porous corrosion products on the whole coupon surface (Figure 3-8b), indicating the increased corrosion of the steel. With the further increase of $i_{AC}$ to 100 A/m$^2$, 200 A/m$^2$, and 400 A/m$^2$, the corrosion rust layer becomes denser and more complete, covering the whole surface of the coupon. This corresponds to the stage of the rapidly increasing corrosion rate in Figure 3-7 with the increase of $i_{AC}$. When $i_{AC}$ is increased to 800 A/m$^2$, the coupon surface is covered by a compact corrosion product film (as seen in Figure 3-8f), which provides some protection against further corrosion of the steel. Thus, the corrosion rate decreases, as shown in Figure 3-7.

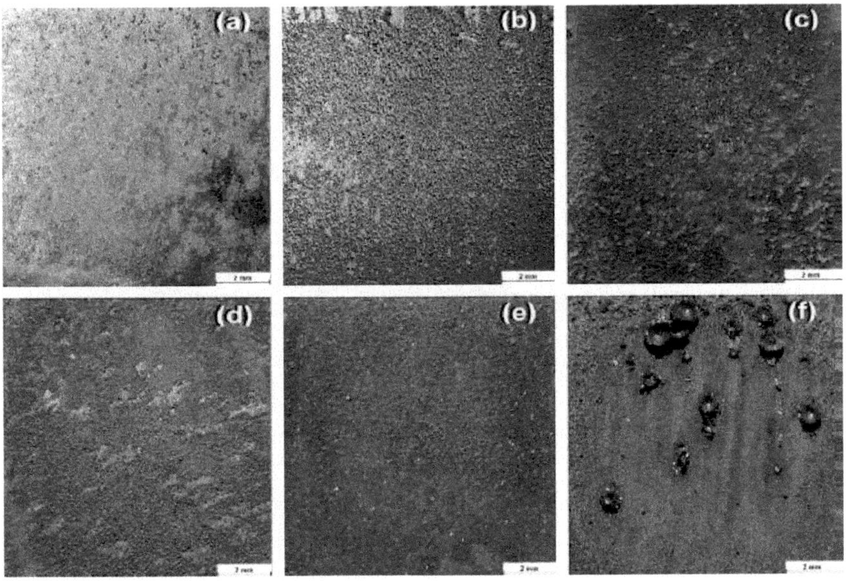


**Figure 3-8.** Optical views of the morphology of the steel coupon in the simulated soil solution under various AC densities: (a) 0 A/m$^2$, (b) 50 A/m$^2$, (c) 100 A/m$^2$, (d) 200 A/m$^2$, (e) 400 A/m$^2$, (f) 800 A/m$^2$.[40]

In addition to the increased corrosion reaction kinetics, the AC also affects the electrochemical thermodynamic activity of pipeline steels in corrosive environments. Figure 3-9 shows the real-time DC potentials that are separated from the mixed potential signals recorded on the 16Mn steel electrode at various AC current densities in the simulated soil solution.[40] Three potential regions are observed: (i) from 0 s to 1,000 s of testing, no AC is applied on the steel electrode, and the recorded DC potential is the corrosion potential of the steel; (ii) from 1,000 s to 2,000 s, the AC-ON potential is recorded and $i_{AC}$ is applied; and (iii) after 2,000 s, AC application is stopped and the AC-OFF potential is recorded. In the first region, DC potentials decrease rapidly and reach a steady-state value of about -0.75 V (SCE). When the AC current is applied at 1,000 s, the DC potentials drop immediately. Moreover, the higher the $i_{AC}$, the more negative the DC potential. When $i_{AC}$ is smaller than 400 A/m$^2$, a steady-state potential is achieved. When $i_{AC}$ is up to 600 A/m$^2$ and 800 A/m$^2$, the DC potentials increase to −0.63 V (SCE) and -0.60 V (SCE), respectively, which are much less negative than the original value of -0.75 V (SCE) before the AC application. When AC is stopped at 2,000 s, the DC potentials recover immediately back to a steady-state value at AC densities below 400 A/m$^2$. The steady-state potential becomes more negative with increasing $i_{AC}$. However, for $i_{AC}$ of 600 A/m$^2$ and 800 A/m$^2$, the DC potentials jump instantly to higher values and then decrease gradually to steady-state DC potentials, which are more negative than those recorded at $i_{AC}$ below 400 A/m$^2$.

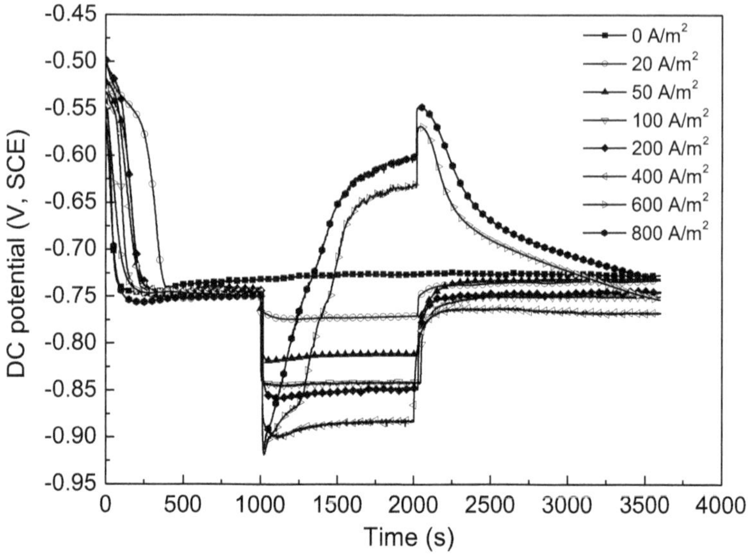


**Figure 3-9.** DC potential separated from the mixed potential signals recorded on the 16Mn steel electrode at various AC current densities in the simulated soil solution: (a) the corrosion potential before 1,000 s, (b) the AC-ON potential from 1,000 s to 2,000 s, (c) the AC-OFF potential after 2,000 s.[40]

The effect of AC current density on the shift of DC potential, which is defined as corrosion potential at stage (i) subtracted from the steady-state DC potential at stage (ii) in Figure 3-9, is shown in Figure 3-10. It can be seen that, when the AC current density is less than 400 $A/m^2$, the DC potential shifts negatively, indicating that the DC potential at stage (ii) is more negative than the corrosion potential of the steel and decreases with increasing $i_{AC}$. However, when the AC current density increases to 600 $A/m^2$ and 800 $A/m^2$, the shift of the DC potential becomes positive. Generally, the DC potential of a steel indicates its electrochemical activity in aqueous solutions (i.e., the more negative the DC potential, the more electrochemically active the steel). Thus, as the applied AC current density increases, the steel becomes more active electrochemically. However, at high AC current densities, such as 600 $A/m^2$ and 800 $A/m^2$, the electrochemical activity of the steel reduces. The change in the activity of the steel is associated with the corrosion behavior in the solution at different AC current densities. The reduced electrochemical activity is due to generation of a dense, compact, and complete corrosion product layer on the steel surface, as shown in Figure 3-8.

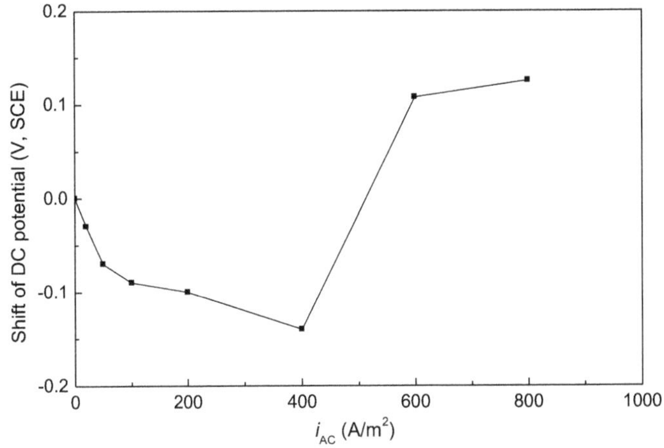


**Figure 3-10. Shift of DC potential of the steel under various AC current densities.**[40]

Figure 3-11 shows the potentiodynamic polarization curves of the 16Mn steel electrode measured under various AC current densities in the simulated soil solution.40 It can be seen that, although a limited potential range is associated with a possible "passivity" of the steel in the absence of AC, the AC application makes the "passive" behavior disappear. The anodic current density increases, and the corrosion potential ($E_{corr}$) is shifted negatively with the increase of $i_{AC}$ from 0 A/m$^2$ to 400 A/m$^2$, as shown in Figure 3-11a. The cathodic current density also increases, but not as apparently as the anodic one. With the further increase of $i_{AC}$ up to 600 A/m$^2$ and 800 A/m$^2$, the anodic current density decreases, and the corrosion potential becomes less negative (Figure 3-11b). The electrochemical corrosion parameters are fitted from the measured polarization curves, and the results are shown in Table 3-1. It can be seen that the corrosion current density ($i_{corr}$) increases with increasing $i_{AC}$, and a maximum value is observed at the AC current density of 400 A/m$^2$. With the further increase of $i_{AC}$, the corrosion current density decreases. The results are consistent with the results obtained from the weight loss testing. Furthermore, there are apparent variations of cathodic and anodic Tafel slopes ($b_c$ and $b_a$) with increasing AC current density, indicating that the applied AC alters the cathodic and anodic reaction kinetics. This is not surprising considering the enhanced oxygen and hydrogen evolutions and the film formation under AC, especially at high AC current densities. An increase in either AC peak voltage or AC current density can increase the corrosion rate of pipeline steels in active corrosion systems, but this occurs in the Tafel region only on the measured potentiodynamic polarization curves. The AC effect on the steel corrosion becomes diminished at higher potentials, where the effect of DC potential on the steel corrosion predominates.[63] It is thus suggested that AC has a more pronounced effect near the open-circuit potential of the steel in corrosive environments.

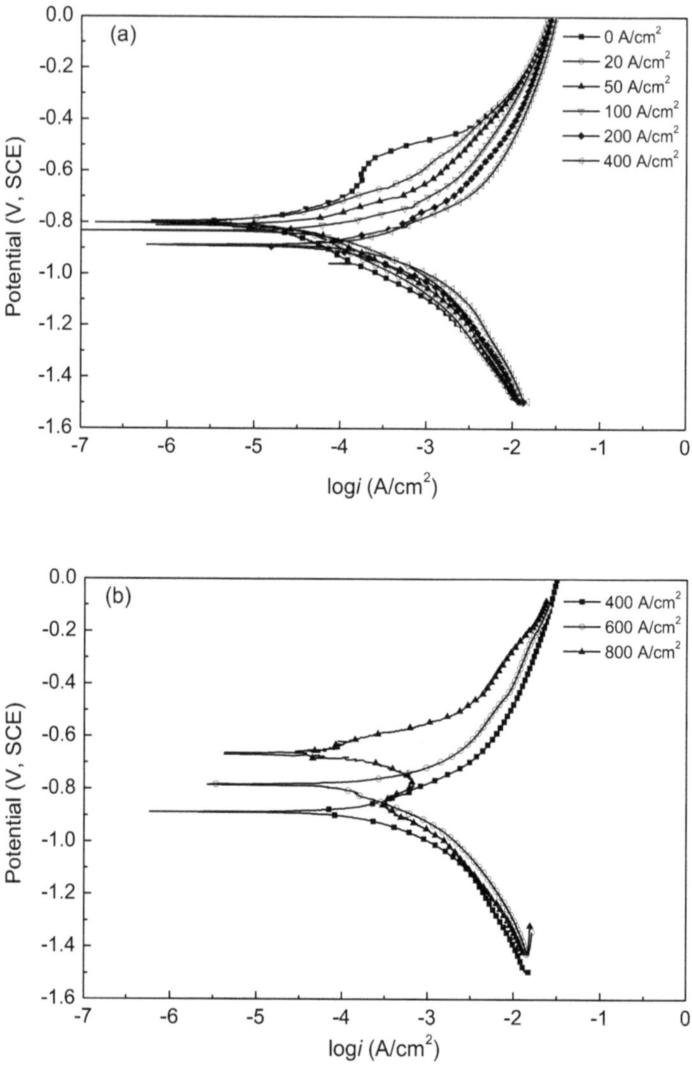


**Figure 3-11.** Potentiodynamic polarization curves of 16Mn steel measured at various AC current densities in the simulated soil solution: (a) $i_{AC}$ lower than 400 A/m$^2$; (b) $i_{AC}$ higher than 400 A/m$^2$.[40]

| $i_{AC}$ (A/m$^2$) | $E_{corr}$ (V, SCE) | $i_{corr}$ (A/cm$^2$) | $\beta_a$ (mV/decade) | $\beta_c$ (mV/decade) | $\beta_a/\beta_c$ | $\Delta i_{corr}/\Delta i_{AC}$ (%) |
|---|---|---|---|---|---|---|
| 0 | -0.803 | $1.49\times10^{-5}$ | 130.23 | 151.11 | 0.86 | 0 |
| 20 | -0.798 | $2.00\times10^{-5}$ | 110.76 | 142.16 | 0.78 | 0.25 |
| 50 | -0.811 | $4.84\times10^{-5}$ | 117.81 | 182.23 | 0.65 | 0.67 |
| 100 | -0.834 | $9.05\times10^{-5}$ | 122.47 | 161.13 | 0.76 | 0.76 |
| 200 | -0.889 | $1.34\times10^{-4}$ | 149.79 | 158.53 | 0.94 | 0.60 |
| 400 | -0.887 | $2.66\times10^{-4}$ | 181.21 | 182.12 | 1.00 | 0.63 |
| 600 | -0.784 | $2.09\times10^{-4}$ | 116.83 | 237.02 | 0.49 | 0.32 |
| 800 | -0.659 | $4.07\times10^{-5}$ | 82.27 | 123.25 | 0.67 | 0.03 |

**Table 3-1.** Electrochemical corrosion parameters fitted from the measured polarization curves in Figure 3-11.[40]

### 3.4.3. Mechanistic analysis

In electrochemical active dissolution systems, the anodic and cathodic reactions during the corrosion of pipeline steels in near-neutral pH soil solution include steel oxidation and the reduction of dissolved oxygen or water in aerobic and anaerobic environments, respectively. Upon application of AC current density, the most positive and negative potentials of the steel could be up to 2.25 V (SCE) and -3.46 V (SCE), respectively, under the AC current density of 800 A/m$^2$.[40] As a result, the oxygen and hydrogen evolution reactions occur, accompanying corrosion of the steel under AC interference:

$$\text{At AC positive potentials: } 2H_2O - 4e \rightarrow O_2 + 4H^+ \quad (3\text{-}8)$$

$$\text{At AC negative potentials: } 2H_2O + 2e \rightarrow H_2 + 2OH^- \quad (3\text{-}9)$$

Moreover, upon application of AC, the steel is under periodic anodic and cathodic polarization. Dissolution of the steel can be accelerated by the anodic polarization and mitigated temporarily by the cathodic polarization. A layer of corrosion products is formed and deposited on the steel surface through the following processes:[89]

$$Fe^{2+} + 2OH^- \rightarrow Fe(OH)_2 \text{ (anaerobic)} \quad (3\text{-}10)$$

$$6Fe(OH)_2 + O_2 \rightarrow 2Fe_3O_4 + 6H_2O \text{ (aerobic)} \quad (3\text{-}11)$$

$$4Fe(OH)_2 + O_2 \rightarrow 4FeOOH + 2H_2O \text{ (aerobic)} \quad (3\text{-}12)$$

The DC potential is shifted negatively by the AC application when $i_{AC}$ is smaller than 400 A/m², indicating increasing corrosion activity of the steel due to AC superimposition. Moreover, a new, steady-state DC potential is established with AC-ON, showing that the steel corrodes at a new state with a high corrosion rate, as confirmed by the weight loss measurements in Figure 3-7. With the further increase of $i_{AC}$ to 600 A/m² and 800 A/m², the DC potential drops immediately and then increases gradually with time. For example, the steady-state DC potential is 140 mV more negative than the corrosion potential at an $i_{AC}$ of 400 A/m², while it is 100 mV less negative than the corrosion potential at an $i_{AC}$ of 600 A/m². Thus, a sufficiently high AC current density, such as 600 A/m², would decrease the corrosion activity of the steel. This is attributed to the steel "passivity" occurring at a sufficiently high AC current density, where the oxygen evolution in reaction (3-8) could result in the formation of a compact and complete surface scale, reducing corrosion of the steel.

It can also be seen from Figure 3-9 that, when AC is removed, the DC potential reaches a new steady-state value that is more negative than the corrosion potential at stage (i). This indicates that the AC-induced activation of the steel will sustain even after the AC application is stopped. Thus, the effect of AC on the steel in enhanced corrosion activity is irreversible. In addition, the DC potentials at AC current densities of 600 A/m² and 800 A/m² shift negatively to corrosion potential after 1,500 s, indicating that the effect can only be maintained in the presence of applied AC, where positive potentials are achieved during AC to cause the oxygen evolution. When AC is removed, the decreased corrosion rate at high AC current densities would not be able to sustain owing to activation of the steel by aggressive ions, particularly chloride ions.

According to Lalvani and Lin,[37] the effect of AC on corrosion current density and corrosion potential are expressed by:

$$\bar{i}_a = i_{corr,DC} \exp\left(\frac{E_{DC} - E_{corr,DC}}{\beta_a}\right) \sum_{K=0}^{\infty} \frac{1}{(K!)^2} \left(\frac{E_p}{2\beta_a}\right)^{2K} \tag{3-13}$$

$$\bar{i}_c = i_{corr,DC} \exp\left(\frac{-E_{DC} + E_{corr,DC}}{\beta_c}\right) \sum_{K=0}^{\infty} \frac{1}{(K!)^2} \left(\frac{E_p}{2\beta_c}\right)^{2K} \tag{3-14}$$

where $\bar{i}_a$ and $\bar{i}_c$ are time-averaged anodic and cathodic current densities under AC interference, respectively; $E_{DC}$ is the DC potential of the steel when AC is applied; $i_{corr,DC}$ and $E_{corr,DC}$ are corrosion current density and corrosion potential in the absence of AC, respectively; $E_p$ is the amplitude of the applied AC voltage; $\beta_a$ and $\beta_c$ are anodic and cathodic Tafel slopes, respectively; and $K$ is a real number. Since $E_{DC}$ is equal to the corrosion potential of the steel in the presence of AC ($E_{corr,AC}$), the shift of DC potential, $\alpha$, under applied AC is:

$$\alpha = E_{corr,DC} - E_{corr,AC} = \frac{\beta_a \beta_c}{\beta_a + \beta_c} \ln \left[ \frac{\sum_{K=0}^{\infty} \frac{1}{(K!)^2} \left( \frac{E_p}{2\beta_a} \right)^{2K}}{\sum_{K=0}^{\infty} \frac{1}{(K!)^2} \left( \frac{E_p}{2\beta_c} \right)^{2K}} \right] \tag{3-15}$$

It can be seen that when the ratio of Tafel slopes $\beta_a$ to $\beta_c$ is smaller than 1 (i.e., $\beta_a/\beta_c < 1$), the increment of anodic current density is larger than that of the cathodic current density, and the corrosion potential is shifted negatively with the increase of $i_{AC}$, which is consistent with the results obtained when $i_{AC}$ does not exceed 400 A/m$^2$, as shown in Figures 3-10 and 3-11a, as well as Table 3-1. However, this model does not work at high AC current densities, such as 600 A/m$^2$ and 800 A/m$^2$, where the corrosion current density decreases and the corrosion potential shifts positively with increasing $i_{AC}$, as confirmed by both weight loss testing and polarization curve measurements in Figures 3-7 and 3-11b. This is primarily attributed to the formation of a corrosion scale on the steel surface.

The ratio of the increment of corrosion current density ($\Delta i_{corr}$) to the increment of AC density ($\Delta i_{AC}$) is also shown in Table 3-1. It can be seen that the AC-enhanced corrosion rate is extremely small in comparison with the increment of AC current density. It indicates that only a small percentage of AC current participates in the steel dissolution, acting as faradic current, while the majority of AC either participates in the charging–discharging process, acting as non-faradic current, in the double-charge layer or gets involved in the redox reactions of water in reactions (3-8) and (3-9).[90, 91] A mechanistic model has been developed to illustrate the accelerated corrosion of pipeline steels by the "small percentage" of applied AC based on a typical double-charge layer and corrosion reaction model, as shown in Figure 3-12 and as mentioned above.

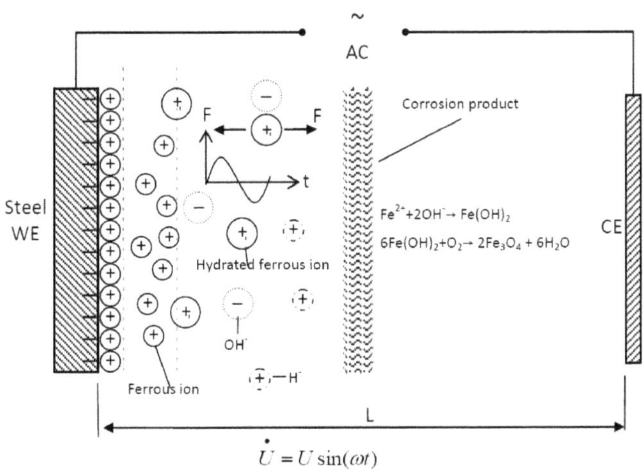


**Figure 3-12.** Schematic diagram of the double-charge layer at the steel–solution interface and the AC corrosion model.[40]

According to the proposed model, mechanistically, factors that affect the AC corrosion of steels include:

- **Double-charge layer effect.** During corrosion, charge-transfer reactions occur primarily in the double-charge layer owing to the extremely high strength of the electric field established locally. However, the electric field can be disturbed by AC. This effect does not contribute significantly to the corrosion of the steel. The high frequency of AC current (60 Hz is used in most work, including the results obtained in most studies mentioned so far) makes reductive and oxidative reactions difficult in such a short time period. It has been confirmed that a decreasing AC frequency can increase the corrosion rate notably.[48, 90] Moreover, the majority of AC current flowing through the double-charge layer acts as the non-faradic charging–discharging current, which is usually much higher than the faradic current for reductive or oxidative reaction.[92]
- **Vibrating effect.** AC application could generate a "vibrating" effect owing to the low-frequency alternating electric field force exerting on the reactants in the solution. The mobility rate of solvated ions in the solution (e.g., $Fe^{2+}$, $OH^-$, $H^+$) is generally small and is sensitive to the low-frequency electric field force. Upon AC application, the ions become accelerated by the electric field force, and consequently the "free path" for the ionic collision is shortened. This would increase the kinetic energy of reactants and the opportunity of reactions among them. Corrosion is thus enhanced.
- **Evolution of hydrogen and oxygen.** The effect of the hydrogen and oxygen evolutions by applied AC on the steel corrosion is not negligible. It has been demonstrated that hydrogen atoms generated during corrosion reaction, upon penetration into steels, can enhance the corrosion activity of the steels.[93] The influence of the oxygen evolution on corrosion of the steel supplies more cathodic reactants (i.e., dissolved oxygen) for the reductive reaction.

## 3.5. AC corrosion of pipelines: Passive systems

### 3.5.1. Environments resulting in passivation of pipeline steels

For pipelines coated with coatings that are permeable to CP current, such as epoxy-based coatings (e.g., FBE), the corrosive environment generated under the disbonded coatings would be different from that developed under an impermeable coating. The permeable coatings permit penetration of water, moisture, chemicals, and so forth, into the coating, especially after a long-term service in the field. Moreover, one of the most important characteristics the permeable coatings possess is their compatibility with CP (i.e., the ability of the coatings to allow CP current to penetrate). Thus, when a permeable coating degrades, the pipeline is still protected from corrosion as the CP current can pass through the coating. The permeable coatings are thus named "fail-safe" coatings.

Although coatings such as FBE possess an excellent adhesion to the steel substrate, disbondment can occur for a number of reasons, such as poor coating application. When water, gases such as $CO_2$, or chemicals penetrate through the coating, a corrosive electrolyte generates under it. Since FBE is permeable to oxygen, the electrolyte is aerobic, while the amount of dissolved oxygen is probably low. In addition, the CP current permeates through the coatings to reach the pipe steel. The potential of the steel in the electrolyte is at a cathodic polarization potential. Owing to the CP-driven cathodic reactions (e.g., the electrochemical reduction of dissolved oxygen or water), a large number of hydroxyl ions are produced to elevate the solution pH, resulting in alkaline electrolyte trapped under a permeable coating, with a pH range of 9–10.5.[33] With the dissolution of $CO_2$, the electrolyte becomes a concentrated carbonate–bicarbonate solution. It has been established that pipeline steels can be "passivated" in this solution, and the steel surface will become covered with a passive film.[33, 64] Thus, pipeline corrosion occurring in the electrolytes under disbonded coatings that are permeable to CP current is called "high pH corrosion," where the pipeline steel is in a passivation state.[87, 88]

### 3.5.2. AC-induced passivity degradation

The corrosion rates of test coupons made of an X65 pipeline steel (wt%: C 0.04, Si 0.2, Mn 1.5, P 0.011, S 0.003, Mo 0.02, and Fe balance), determined by weight loss testing in a concentrated carbonate–bicarbonate solution (pH 9.6) under various AC current densities, are shown in Figure 3-13. It can be seen that, without AC, the corrosion rate is negligible as the steel is passivated in the solution. Upon AC application, the corrosion rate of the steel increases with the AC current density.

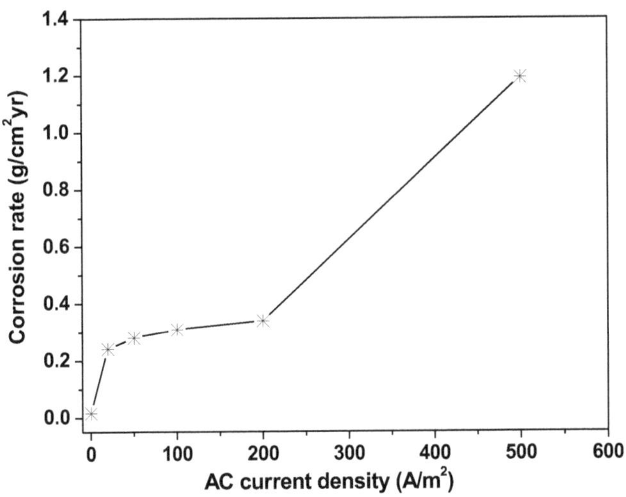


**Figure 3-13.** Corrosion rates of X65 steel coupons in a concentrated carbonate-bicarbonate solution at various AC current densities.[94]

According to Faraday's law, if the weight loss of a steel caused by DC current is $m_{DC}$, and the weight loss caused by an applied AC current is $m_{AC}$, the ratio of $m_{AC}$ to $m_{DC}$ is used as the percentage of the equivalent DC current density to the AC current density. The calculated results are shown in Table 3-2. Generally, AC causes less than 1% of the corrosion of the equivalent DC. A majority of AC flows through the double-charge layer capacitor causing charging and discharging as faradic current, and only a minority of AC results in charge-transfer corrosion reactions as faradic current, as discussed previously. Thus, the corrosion rate caused by AC is less than 1% of that caused by DC, except at low AC current densities such as 20 $A/m^2$.

| AC current density ($A/m^2$) | 0 | 20 | 50 | 100 | 200 | 500 |
|---|---|---|---|---|---|---|
| % of the equivalent DC current density | N/A | 1.32 | 0.46 | 0.34 | 0.19 | 0.26 |

**Table 3-2.** AC current densities expressed as the percentage of the equivalent DC current densities for the identical amount of corrosion of the steel.[94]

Figures 3-14 and 3-15 show the optical and scanning electron microscope (SEM) views of corrosion morphology of the steel coupons under various AC current densities in the concentrated carbonate–bicarbonate solution. It can be seen that, in the absence of AC, the steel surface is smooth after testing, without visible corrosion products (Figures 3-14a and 3-15a). The steel experiences uniform corrosion at the AC current densities of 20 $A/m^2$ (Figures 3-14b and 3-15b) and 50 $A/m^2$ (Figures 3-14c and 3-15c). When the applied AC density is up to 100 $A/m^2$, localized corrosion occurs, as shown by some corrosion pits in zone A in Figure 3-15d. The number of pits further increases when the AC current density increases to 200 $A/m^2$ (Figures 3-14e and 3-15e). When the AC current density is as high as 500 $A/m^2$, the steel experiences extensive pitting corrosion, as shown in Figures 3-14f and 3-15f.

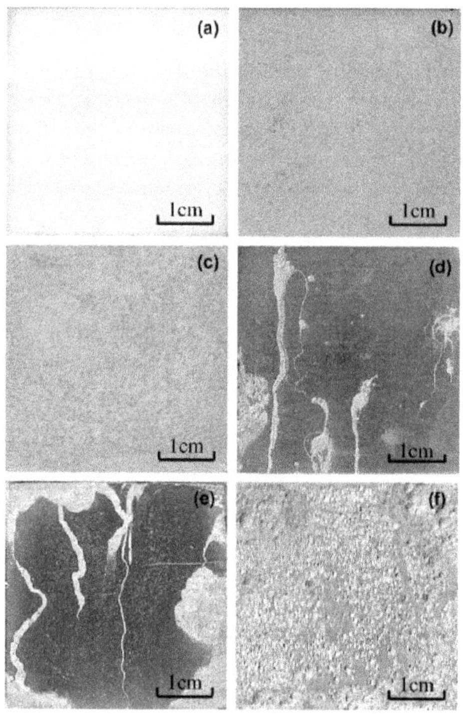


**Figure 3-14.** Optical views of X65 steel electrode in concentrated carbonate–bicarbonate solution at various AC current densities after two days of testing: (a) 0 $A/m^2$, (b) 20 $A/m^2$, (c) 50 $A/m^2$, (d) 100 $A/m^2$, (e) 200 $A/m^2$, (f) 500 $A/m^2$.[94]

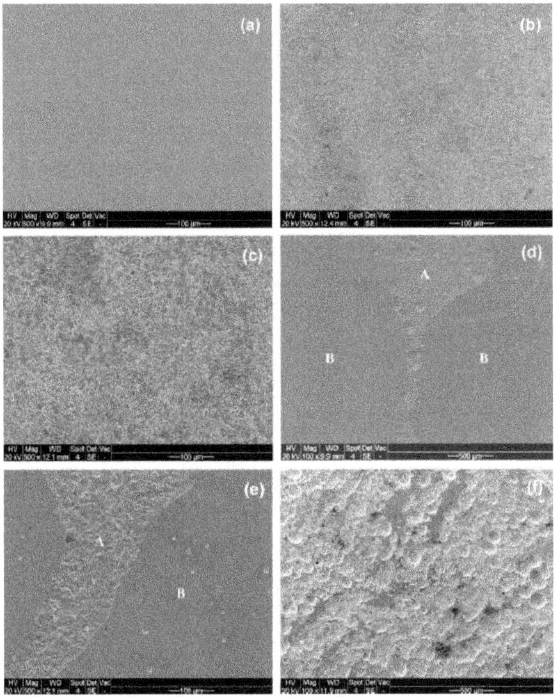


**Figure 3-15.** SEM views of X65 steel electrode in concentrated carbonate-bicarbonate solution at various AC current densities after two days of testing: (a) 0 $A/m^2$, (b) 20 $A/m^2$, (c) 50 $A/m^2$, (d) 100 $A/m^2$, (e) 200 $A/m^2$, (f) 500 $A/m^2$ .[94]

Figure 3-16 shows the time dependence of the corrosion potential of the steel electrode in the concentrated carbonate-bicarbonate solution at various AC current densities.[94] It can be seen that the corrosion potential shifts negatively with increasing AC current density. As mentioned previously, a negative shift of corrosion potential indicates increasing electrochemical activity of the steel. Thus, the corrosion activity of the steel is enhanced by AC in the solution, which is identical to the behavior in diluted bicarbonate solutions, such as the simulated soil solution in Figure 3-9.

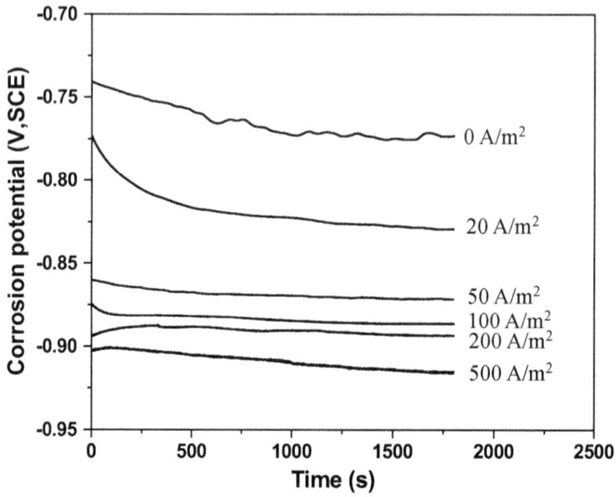


**Figure 3-16.** Time dependence of the corrosion potential of steel electrode in concentrated carbonate-bicarbonate solution at various AC current densities.[94]

Figure 3-17 shows the potentiodynamic polarization curves of X65 steel electrode in concentrated carbonate-bicarbonate solution at various AC current densities. It can be seen that in the absence of AC (Figure 3-17a), the polarization curve exhibits its distinct active, active-passive transition, passive, and transpassive regions in the anodic curve. Upon application of AC, current oscillations are observed, and the oscillating amplitude increases with the AC current density. Moreover, the passive region becomes less apparent. When the AC current density is up to 500 $A/m^2$, the passivity is not observed, as shown in Figure 3-17f. Furthermore, the AC current density has little effect on pitting potential, which is defined as the potential in which the anodic current density increases rapidly from the passive to the transpassive regions. In this work, the pitting potential is approximately -0.15 V (SCE).

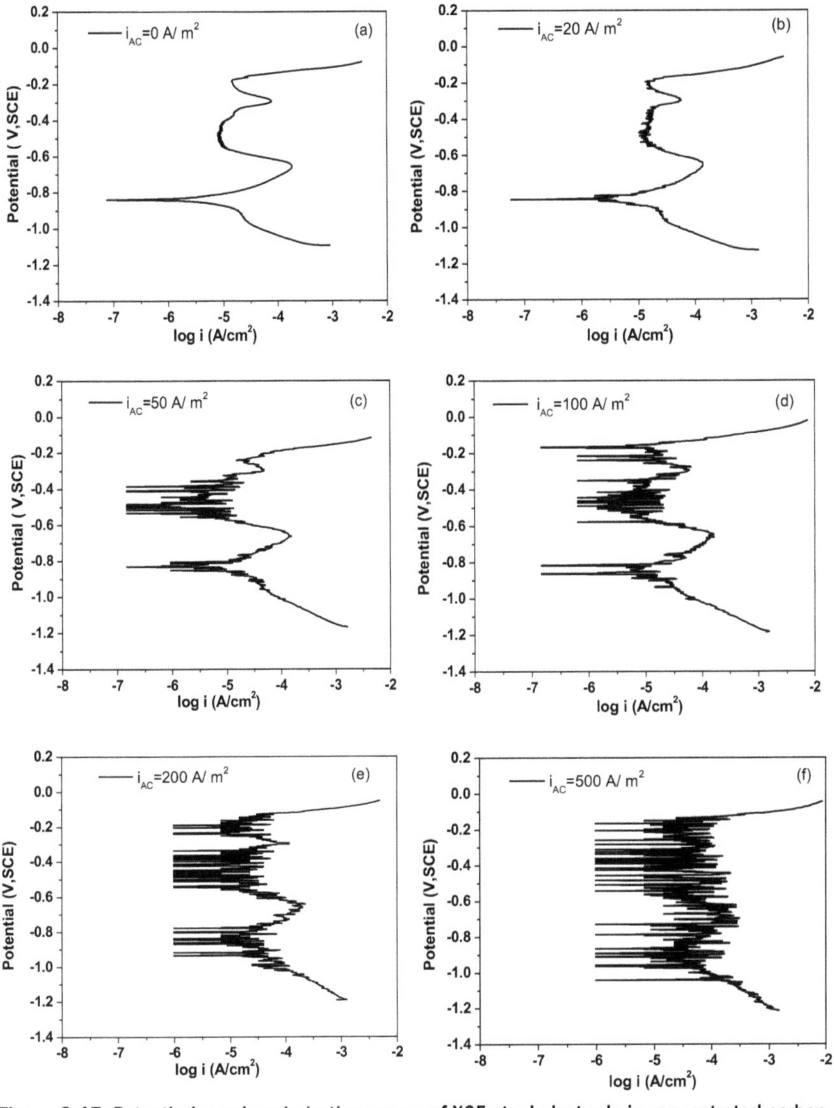


**Figure 3-17.** Potentiodynamic polarization curves of X65 steel electrode in concentrated carbonate-bicarbonate solution at various AC current densities.[94]

## 3.5.3. Mechanistic analysis

Pipeline steels such as X65 steel are passivated in concentrated carbonate-bicarbonate solutions, as demonstrated in the measured polarization curves in Figure 3-17, in which the anodic current peak at about –0.7 V (SCE) is attributed to the formation of a $FeCO_3$ and $Fe(OH)_2$ deposit layer on the steel surface through the following reaction:[95, 96]

$$Fe + HCO_3^- \rightarrow FeCO_3 + H^+ + 2e \qquad (3\text{-}16)$$

The current peak observed at about −0.3 V (SCE) is due to the oxidation of ferrous species to $Fe_2O_3$ and $Fe_3O_4$:[97, 98, 99]

$$4FeCO_3 + O_2 + 4H_2O \rightarrow 2Fe_2O_3 + 4HCO_3^- + 4H^+ \qquad (3\text{-}17)$$

$$6FeCO_3 + O_2 + 6H_2O \rightarrow 2Fe_3O_4 + 6HCO_3^- + 6H^+ \qquad (3\text{-}18)$$

Owing to the presence of chloride ions in the solution, pitting corrosion occurs, resulting in a rapid increase of anodic current density at about −0.2 V (SCE).

A mechanistic model is proposed to illustrate the effect of AC on steel corrosion in the passive system. According to the model,

$$\mathrm{E}_{\mathrm{corr},AC} = E_{corr} - \alpha \qquad (3\text{-}19)$$

$$\alpha = \left(\frac{b_a}{\frac{b_a}{b_c} - 1}\right) \ln \left[\frac{\sum_{k=1}^{\infty} \frac{1}{(k!)^2} \left(\frac{E_p}{2b_c}\right)^{2k} + 1}{\sum_{k=1}^{\infty} \frac{1}{(k!)^2} \left(\frac{E_p}{2b_a}\right)^{2k} + 1}\right] \qquad (3\text{-}20)$$

where $E_{\mathrm{corr,AC}}$ and $E_{\mathrm{corr}}$ are corrosion potentials of the steel in the presence and absence of applied AC, respectively; $b_a$ and $b_c$ are the anodic and cathodic Tafel slopes, respectively; $E_p$ is the amplitude of the applied AC voltage; and $k$ is an integral number. It can be seen that the variation of corrosion potential caused by AC is a function of the ratio of anodic Tafel slope to cathodic Tafel slope (i.e., $r = b_a/b_c$) and $E_p$. The corrosion potential shifts negatively as the $E_p$ increases. The measured corrosion potentials shift negatively with increasing AC current density since $E_p$ is proportional to the AC current density for a steel electrode with a fixed surface area in the solution. Therefore, AC results in increasing electrochemical corrosion activity of the steel.

An important observation of the polarization curves measured in the concentrated carbonate–bicarbonate solution is that the application of AC on the steel causes passive current oscillations and the degradation of the steel passivity. Thus, a superimposed AC would degrade the stability of the passive film formed on the steel. At low AC, such as 20 A/m² and 50 A/m², although the passive film is degraded by AC application, it still provides some protection of the steel. The steel electrode experiences a uniform corrosion only, as seen in Figures 3-14b, 3-14c, 3-15b, and 3-15c. With the increase of the AC, the passivity of the steel is further degraded. Corrosion pits generate locally, as indicated in zone A in Figures 3-15d and 3-15e. When the AC is as high as 500 A/m$^2$, passivity cannot be maintained, and pitting corrosion occurs extensively on the steel surface (Figure 3-15f). Moreover, AC causes an increase of the passive current density and a decrease of the transpassive potential if the AC current density is higher than 50 A/m$^2$. When the AC current density is increased to 100 A/m$^2$, the steel cannot be passivated (i.e., the anodic curve does not show a passive range). Similar results were observed by Ormellese et al., who confirmed that AC-induced corrosion was localized on a passivated carbon steel and that the penetration rate was up to 1 mm/year.[100] They also found that for a stainless steel passivated in a neutral solution, AC had harmful effects on the passive condition: reducing anodic overpotential, causing passive film breakdown over a critical AC interference level, and initiating corrosion pits.[84]

Although AC current density is suggested as a primary indicator evaluating the AC corrosion risk, in fact, the ratio of AC to DC current densities (i.e., $r_{AC/DC}$) is more accurate than the level of AC current density in the evaluation of AC corrosion.[94] The magnitude of current oscillations in the measured polarization curves is a direct reflection of $r_{AC/DC}$. Generally, uniform corrosion occurs at a low $r_{AC/DC}$, and severe pitting corrosion takes place at high $r_{AC/DC}$. At an intermediate $r_{AC/DC}$, localized corrosion starts to occur on the steel. Tan and Chin suggested that AC would play a similar role to chloride ions in causing pitting corrosion.[101] Although the consequence may be identical, the mechanistic aspects are completely different.

**Microbial corrosion of buried pipelines under AC interference.** Although the exact mechanism of the microbial culture and growth under AC interference has not been fully understood, it was claimed that a synergism existed between AC and nitrate-reducing bacteria growth to facilitate corrosion and stress corrosion cracking of an X80 steel in a soil solution.[102] It was reported that, depending on the AC current density, AC could favor the growth of microorganisms or kill microorganisms. Moreover, the AC affected the adhesion condition and the structure of the microbial membrane. When the cycling condition of the AC electric field and the life cycle of the microbes were consistent, AC could open the cell membrane and ionic channels, which were regulated by enzymatic activity, making it difficult for microbes to breed under the action of AC.[103] Moreover, the number of live cells increased with a negative shift of the CP potential at a suitable DC current density.[104] When the electric field intensity increased to a certain level, the cell permeability could be destroyed, and the microorganism was killed. It was found in Wan et al. that an AC voltage of 2 V and an AC current density of 30 A/m$^2$ did not obviously affect *B. cereus* survival.[102] Furthermore, AC could increase the organic material transfer and accelerate the steel corrosion. The bacteria *B. cereus* reduced nitrate to $NH_4^+$ under the corrosion product film or biofilm. It was finally concluded that AC and nitrate-reducing bacteria synergistically enhanced the corrosion of X80 steel in the soil solution.

The AC corrosion of pipelines in soil-extracted solutions containing sulfate-reducing bacteria was investigated by another research group in China, and it was found that a 10 mA/cm$^2$ AC current density inhibited the growth and metabolism of planktonic and sessile sulfate-reducing bacteria cells.[105] The mechanism of AC-influenced corrosion in the presence of sulfate-reducing bacteria was controlled by both the active dissolution of iron and the film degradation.

In recent years, there have been limited field reports regarding the microbial corrosion of pipelines that are relevant to AC interference. The exact role of AC in bacterial growth and survival and the activity of microorganisms has not been understood at a sufficiently advanced level. The occasional reports about the relationship between the microbial corrosion of pipelines and AC interference need further supportive evidence and data to validate them to enable conclusive comments to be made.

## 3.6. AC corrosion of pipelines at coating defects

Buried pipelines are coated infrastructure. Pipeline coatings are generally polymeric materials possessing a high resistance to water permeation and a sufficient dielectric strength.[64] When the polymeric coatings are intact (i.e., contain no defect), even if AC current flows from the nearby HVAC power lines or rails, it will be blocked from reaching the pipe steel by the coatings. Thus, AC corrosion should not occur on well-coated pipelines. While FBE coatings have been the primary coating of choice on North American pipelines and 3-layer PE coatings are widely used in Europe and China, both coatings have raised induced AC voltages on pipelines.[2]

Coating performance plays a critical role in the AC corrosion of pipelines. Pipeline coatings are not perfect forever. Natural and third-party-induced degradation of the coatings can occur during service. AC corrosion of pipelines usually occurs at coating defects, where the AC stray current reaches the steel to affect electrochemical charge-transfer reactions in the corrosive environments. The defects in the coatings can be introduced during various stages, such as coating manufacturing and transportation, pipeline construction, and pipeline operation. The geometry of coating defects, such as their size and shape, significantly affects the AC corrosion processes.[67]

### 3.6.1. Size of coating defects

Defects contained in pipeline coatings are usually assumed to be circularly shaped. Figure 3-18 shows the DC potential of an X65 steel electrode coated with high-performance composite coating (HPCC) containing a 20 mm diameter coating defect under various AC currents in a diluted bicarbonate solution (0.01 M $NaHCO_3$ solution purged with 5% $CO_2/N_2$ mixture gas, pH 6.5).[106] HPCC is a multicomponent coating consisting of an FBE primer, a PE outer layer, and a tie layer containing a chemically modified PE adhesive, with an average thickness of 1.1 mm. Three potential regions are observed: (a) from 0 s to 1,200 s, no AC is applied and the recorded DC potential is the corrosion potential of the steel; (b) from 1,200 s to 2,400 s, the DC potential of the steel under AC (i.e., the AC-ON potential) is recorded; and (c) after 2,400 s, the AC is stopped and the AC-OFF potential is recorded. It can be seen that the corrosion potential of the steel shifts negatively upon application of AC, which is consistent with previous measurements. Moreover, the DC potential fluctuates more apparently with increasing AC current. When the AC current is stopped at the 2,400th second, the DC potentials do not return to their original values. The new steady-state DC potential is less negative than the value recorded in the period of 0 s to approximately 1,200 s.

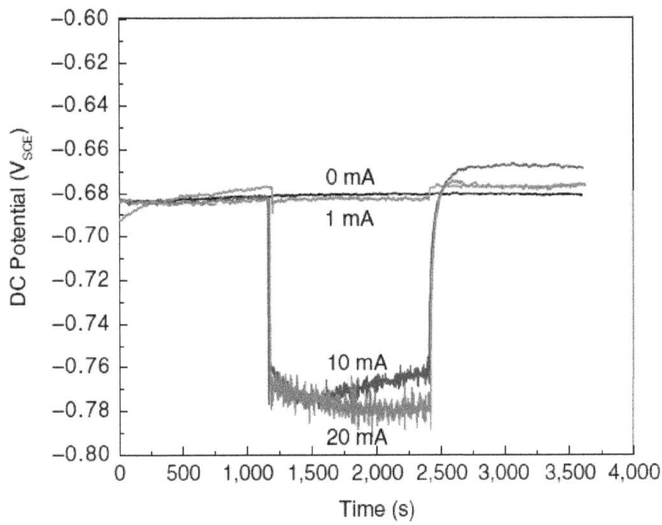


**Figure 3-18.** Time dependence of DC potential of the coated steel containing a 20 mm diameter coating defect at various AC currents in a diluted bicarbonate solution.[106]

Figure 3-19 shows the morphological view of the steel electrode at the 20 mm coating defect after 12 h of testing at various AC currents in the test solution.[106] In the absence of AC, the steel exposed at the coating defect is smooth and does not show obvious corrosion, as seen in Figure 3-19a. Upon application of AC, corrosion of the steel is apparent. Generally, the steel experiences uniform corrosion, and the increasing AC current does not obviously affect the corrosion morphology.

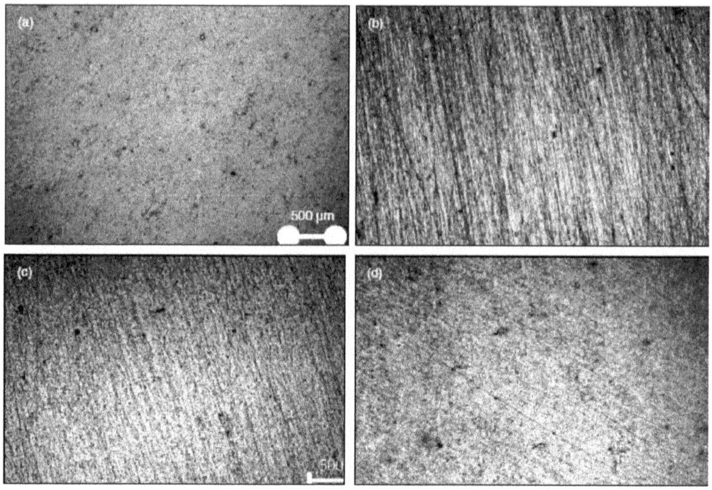


**Figure 3-19.** Optical view of the morphology of the steel electrode at the 20 mm coating defect after 12 h of testing at various AC currents in the test solution: (a) 0 mA (no AC applied), (b) 1 mA, (c) 10 mA, (d) 20 mA.[106]

Figure 3-20 shows the potentiodynamic polarization curves measured on the coated steel electrode containing a 20 mm coating defect under various AC currents in the test solution. It can be seen that the corrosion potential shifts negatively with increasing AC current, which is consistent with the measurements of DC voltage in the presence of AC. Moreover, both anodic and cathodic current densities increase with the AC current, although the increase is less apparent for the cathodic branches than for the anodic ones.

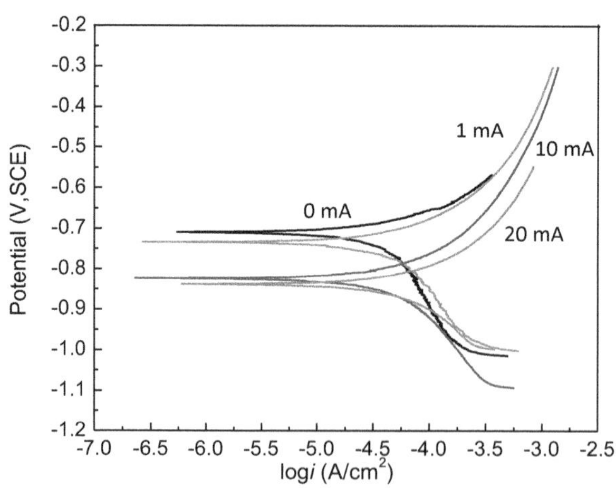


**Figure 3-20.** Potentiodynamic polarization curves measured on the coated steel electrode containing a 20 mm coating defect under various AC currents in the test solution.[106]

As a comparison, the DC potential potentiodynamic polarization curves and corrosion morphology were also measured on a HPCC-coated X65 steel electrode containing a 10 mm diameter coating defect under various AC currents in the test solution, and the results are shown in Figures 3-21 to 3-23.[106] It can be seen in Figure 3-21 that the dependence of the DC potential is identical to that observed on the coated steel electrode containing a 20 mm defect in Figure 3-18, where the DC potential shifts negatively under AC and becomes more negative with increasing AC current. When the AC is stopped, the AC-OFF potential is more negative than the original DC potential. In the absence of AC current, the steel at the 10 mm coating defect does not show apparent corrosion (Figure 3-22a). At 1 mA of AC current, uniform corrosion is observed (Figure 3-22b). When the AC current is increased to 10 mA and 20 mA, pitting corrosion occurs on the steel, as shown in Figures 3-22c and 3-22d, respectively. Moreover, the size and depth of the pits increase with the AC current. The AC dependence of the measured polarization curves of the steel electrode containing a 10 mm coating defect is also identical to that of the electrode containing a 20 mm coating defect in Figure 3-20. With increasing AC current, the corrosion potential becomes more negative and the anodic current density increases, while the cathodic current density is approximately independent of the AC current, as shown in Figure 3-23.

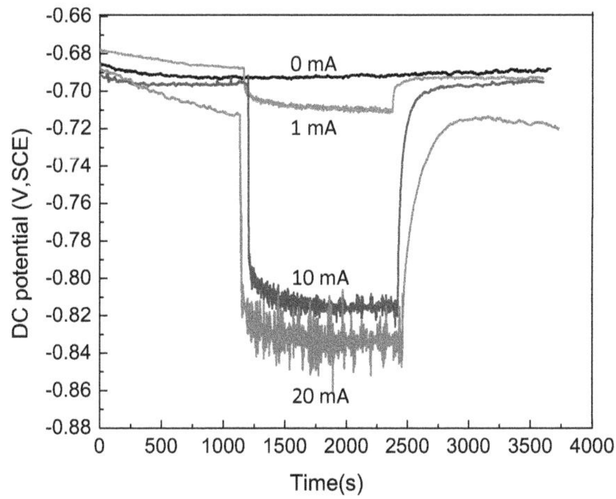


**Figure 3-21.** Time dependence of the DC potential of a coated steel electrode containing a 10 mm diameter coating defect at various AC currents in a diluted bicarbonate solution.[106]

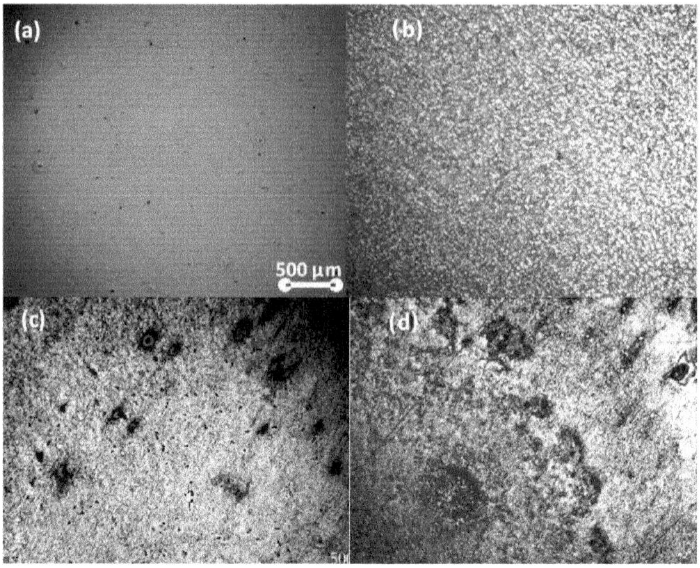


**Figure 3-22.** Optical view of the morphology of the steel electrode at a 10 mm coating defect after 12 h of testing at various AC currents in the test solution: (a) 0 mA (no AC applied), (b) 1 mA, (c) 10 mA, (d) 20 mA.[106]

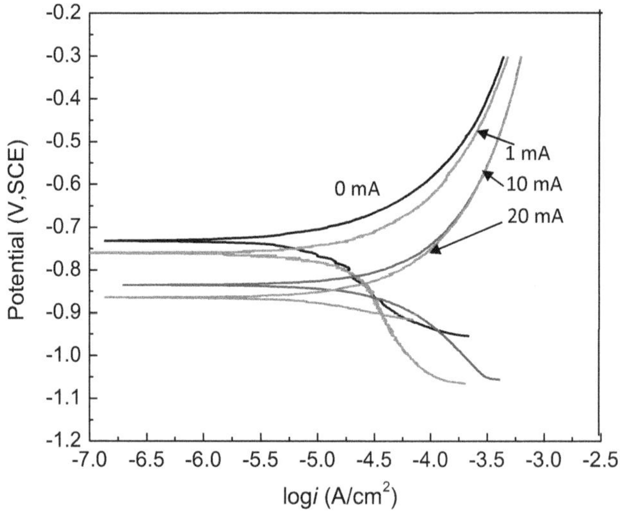


**Figure 3-23.** Potentiodynamic polarization curves measured on the coated steel electrode containing a 10 mm coating defect under various AC currents in the test solution.[106]

Figures 3-24 to 3-26 show the DC potential, the corrosion morphology, and the potentiodynamic polarization curves, respectively, of a coated steel electrode containing a 5 mm diameter coating defect in the test solution.[106] As can be seen in Figure 3-24, the DC potential shifts negatively upon AC application, and the potential value becomes more negative with increasing AC current. Moreover, the steady-state DC potential at specific AC is more negative than that measured on the steel electrode containing a 10 mm coating defect. The potential fluctuation is also more apparent. The corrosion morphology of the steel at the coating defect shows that pitting corrosion occurs under an AC current as low as 1 mA. At 10 mA or 20 mA, pitting corrosion is more pronounced. The pit depth and size increase remarkably at 20 mA of AC current, as shown in Figure 3-25d. The polarization behavior as a function of the AC current is identical to that measured on the steel electrode containing a 10 mm coating defect. The corrosion potential is shifted more negatively, and the anodic current density increases with increasing AC current. The cathodic current density does not show a consistent dependence on AC current.

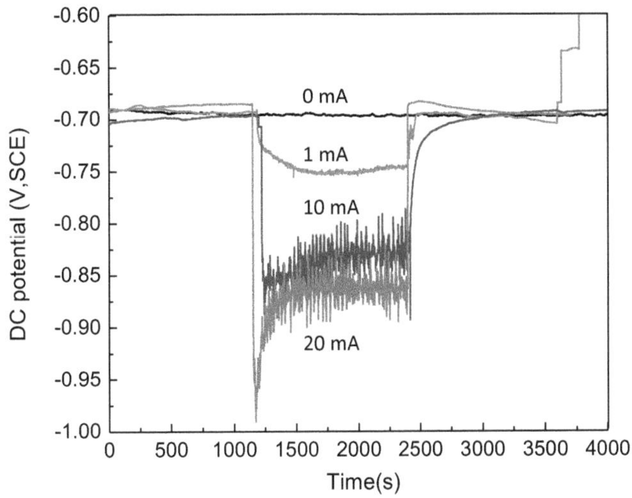


**Figure 3-24.** Time dependence of the DC potential of a coated steel electrode containing a 5 mm diameter coating defect at various AC currents in a diluted bicarbonate solution.[106]

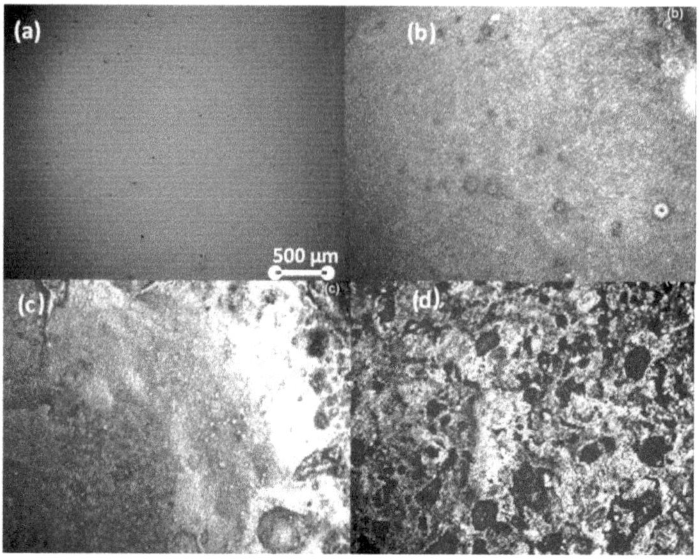


**Figure 3-25.** Optical view of the morphology of the steel electrode at a 5 mm coating defect after 12 h of testing at various AC currents in the test solution: (a) 0 mA (no AC applied), (b) 1 mA, (c) 10 mA, (d) 20 mA.[106]

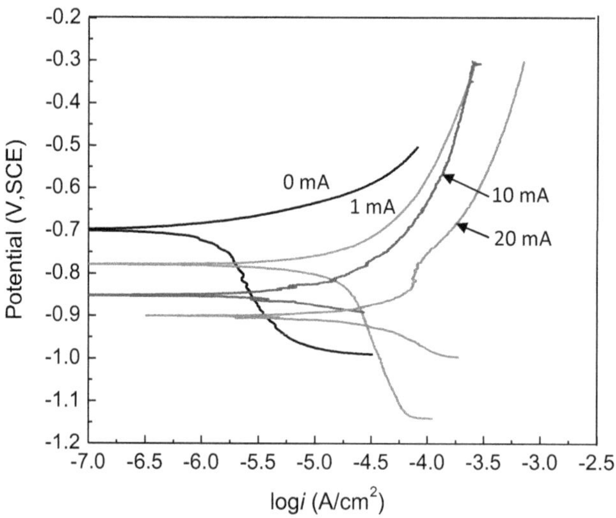


**Figure 3-26.** Potentiodynamic polarization curves measured on the coated steel electrode containing a 5 mm coating defect under various AC currents in the test solution.[106]

Generally, the presence of defects such as pinholes or holidays in pipeline coatings is inevitable. As mentioned, these defects can be introduced into the coatings at different stages of coating manufacturing and application, as well as during pipeline transportation and construction. As demonstrated, the size of the coating defects is critical to the AC corrosion of pipelines occurring at the defect base. During corrosion of pipeline steels in $CO_2$-containing, near-neutral pH soil solutions, the anodic and cathodic reactions include iron oxidation and the reduction of bicarbonate ions, respectively.[33, 89] At small defects, the corrosion products (mainly $Fe^{2+}$ ions) do not diffuse freely out of the defects owing to geometrical limitation. This would cause a local saturation of $Fe^{2+}$ ions and $CO_3^{2-}$ ions to exceed the solubility of iron carbonate, favoring the formation of $FeCO_3$ scale:

$$Fe^{2+} + CO_3^{2-} \rightarrow FeCO_3 \qquad (3\text{-}21)$$

Although the iron carbonate scale formed at ambient temperatures is not sufficiently compact and protective, the deposit inside the defects serves as a blocking barrier to separate the steel from the corrosive environment. However, at big coating defects, the generated $Fe^{2+}$ ions could diffuse out of the defects and would not contribute to the blocking effect. The resulting corrosion behavior is demonstrated by the measured polarization curves on the coated steel electrodes containing differently sized coating defects in Figures 3-27 when AC current is not applied.[106]It can be seen that with the decrease of the defect size, both anodic and cathodic current densities become smaller. This is attributed to the blocking effect of the geometry of the defects.

The application of AC current shifts the DC potential of the coated steel negatively no matter what the defect size is. Figure 3-28 shows the DC potential measured on the coated steel electrodes containing differently sized coating defects in the test solution at various AC currents.[106] It can be seen that, at specific AC currents, the DC potential is more negative with the decrease of the defect size. Moreover, with the increase of the AC current, the DC potential is more negative at a given defect size. For example, for the coating defect of 5 mm in diameter, the steady-state DC potential is about −0.75 V (SCE), −0.85 V (SCE) and −0.90 V (SCE) at AC currents of 1 mA, 10 mA, and 20 mA, respectively. The negative shift of the corrosion potential with the defect size can also be seen from the polarization curves in Figure 3-29. Both anodic and cathodic current densities decrease with the decreasing defect size, which is attributed to the blocking effect of the corrosion scale deposit in the defects.

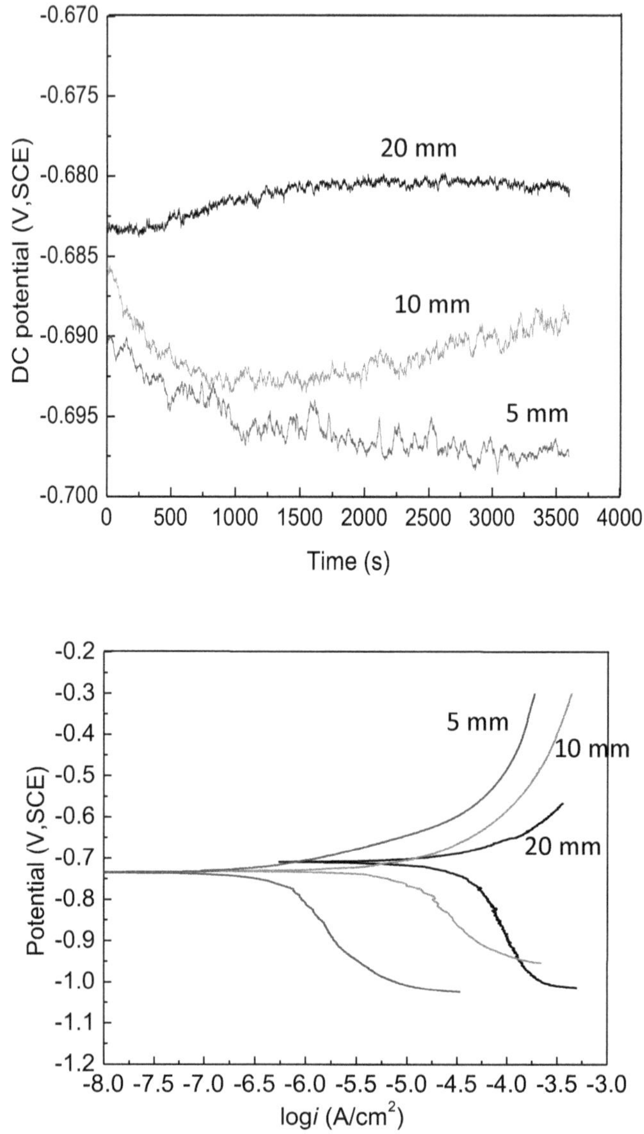


**Figure 3-27.** DC potentials and potentiodynamic polarization curves measured on coated steel electrodes containing coating defects of different sizes in the test solution when no AC current is applied.[106]

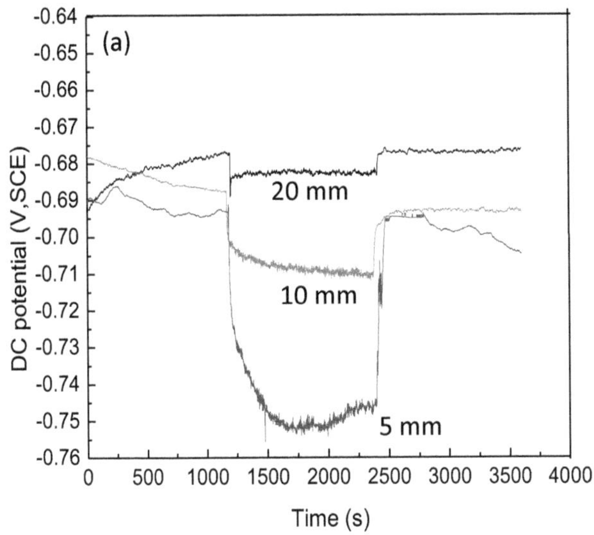


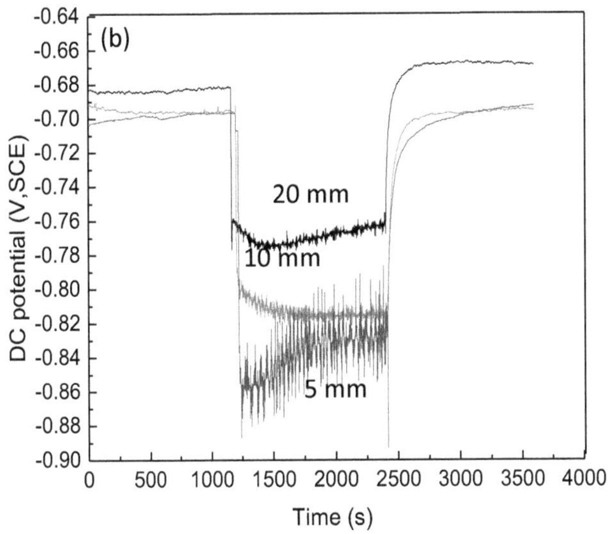


**Figure 3-28.** DC potentials measured on the coated steel electrodes containing coating defects of different sizes in the test solution at AC currents (a) 1 mA, (b) 10 mA, (c) 20 mA.[106]

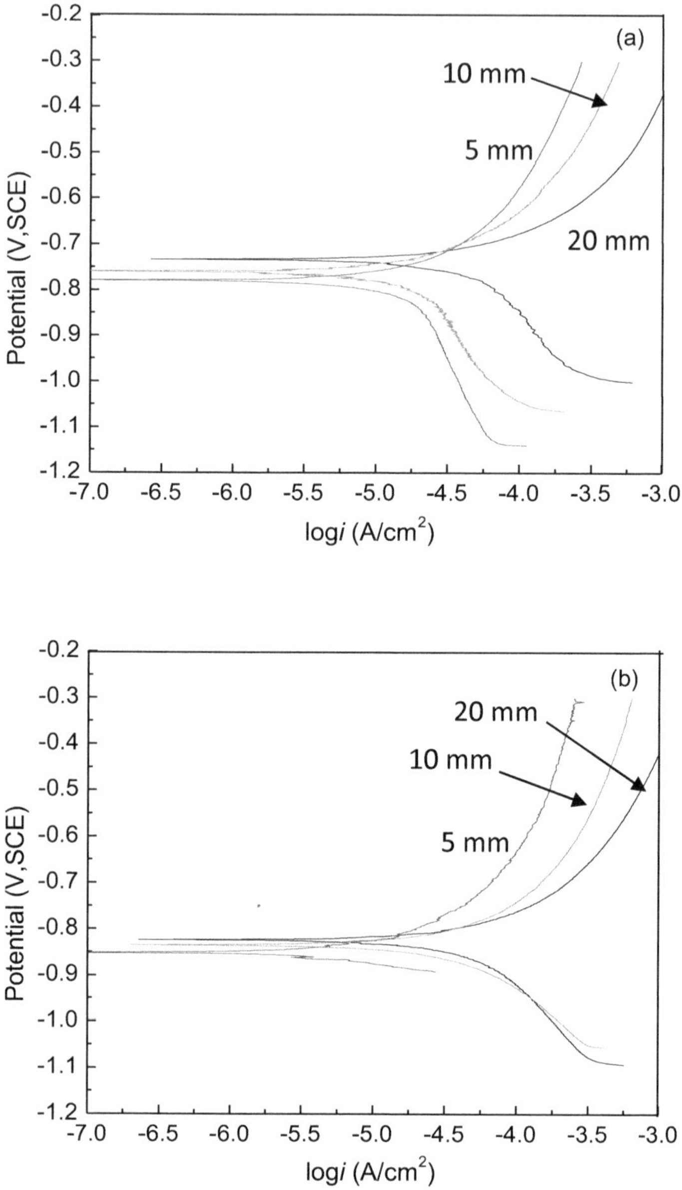


**Figure 3-29.** Potentiodynamic polarization curves measured on the coated steel electrodes containing coating defects of different sizes in the test solution at AC currents (a) 1 mA, (b) 10 mA, (c) 20 mA.[106]

Generally, upon application of AC, the AC current passes through the coating defects owing to the local high conductivity. At a given AC current, a small defect is associated with a high AC current density. It was demonstrated previously that AC current density is critical to corrosion induced by AC interference. With the decrease of the defect size, the $i_{AC}$ at the defect increases, resulting in enhanced corrosion locally. Thus, the DC potential is shifted more negatively when the defect size becomes smaller. When the AC current density at the defect reaches a certain value, pitting corrosion occurs. This is attributed to accelerated local dissolution at the defect base under a high AC current density.

### 3.6.2. Shape of coating defects

Figure 3-30 shows the DC potential of HPCC-coated X65 steel electrodes containing a circular, square, or triangular coating defect with an area of 78.5 mm² (i.e., the area of a 10 mm diameter circular defect) under 500 A/m² of AC current density in the test solution.[106] It can be seen that the DC potential of the steel electrode containing any one shape of the coating defect almost copies the others. Upon AC application, the DC potential is shifted negatively. When AC is stopped, the DC potential reaches a relatively steady-state value, which is less negative than the original potential before AC application.

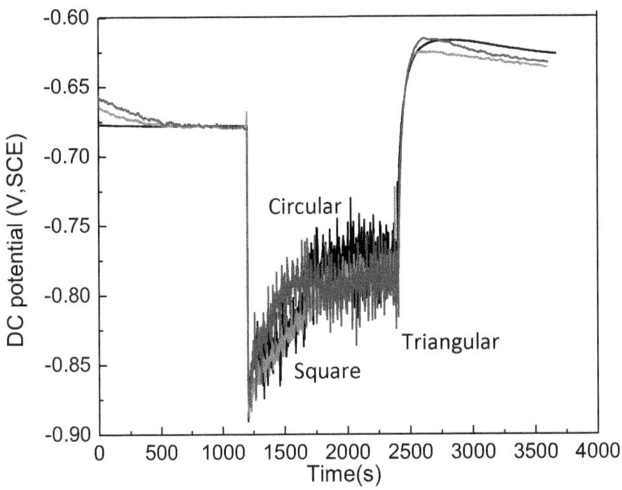


**Figure 3-30.** DC potential of HPCC-coated X65 steel electrodes containing a circular, square, or triangular coating defect with an area of 78.5 mm² under an AC current density of 500 A/m² in the test solution.[106]

Figure 3-31 shows the potentiodynamic polarization curves measured on the coated steel electrodes, each containing one of the three differently shaped coating defects, under an AC current density of 500 $A/m^2$ in the solution. There is little difference in the measured curves.

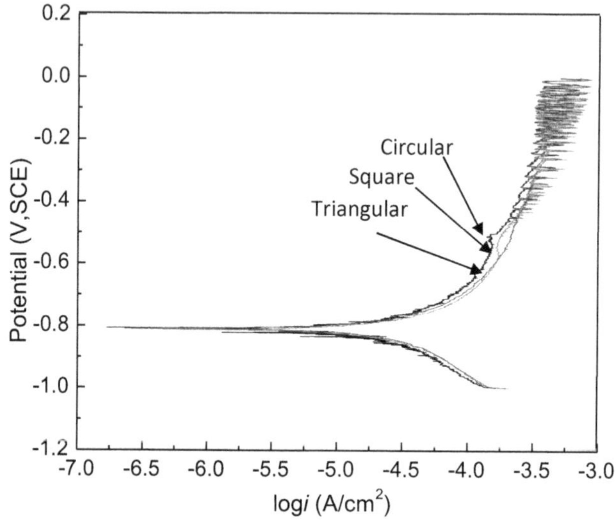


**Figure 3-31.** Potentiodynamic polarization curves of the coated steel electrodes containing a circular, square, or triangular coating defect with an area of 78.5 $mm^2$ under an AC current density of 500 $A/m^2$ in the test solution.[106]

When the defect area is reduced to 19.6 $mm^2$, which is equivalent to the area of a 5 mm diameter circular defect, the measured DC potential and potentiodynamic polarization curves under an AC current density of 500 $A/m^2$ in the test solution are shown in Figures 3-32 and 3-33, respectively. The differences in both the DC potential and the polarization curves are obvious. The circular and triangular coating defects are associated with the most and the least negative DC potentials, respectively. Moreover, the anodic current density increases by the order of triangular to square and circular defects.

It can thus be seen that the effect of the defect shape on AC corrosion depends on the size of the coating defect. The shape of the coating defect affects AC corrosion only when the defects are small, such as 19.6 $mm^2$ in area. The effect is undetectable for big defects, such as 78.5 $mm^2$ in area. At large coating defects, the local solution chemistry is identical or nearly identical to that of bulk solution. The exposed steel behaves like a macroscopic electrode in the solution. Thus, the AC-enhanced corrosion occurs independently of the defect shape. However, at small defects, the mass transfer step and the deposit of the generated corrosion products become important. Particularly, coating defects with varied geometries possess different abilities to retain corrosion products owing to geometrical effect, resulting in the dependence of AC corrosion on the defect shape.

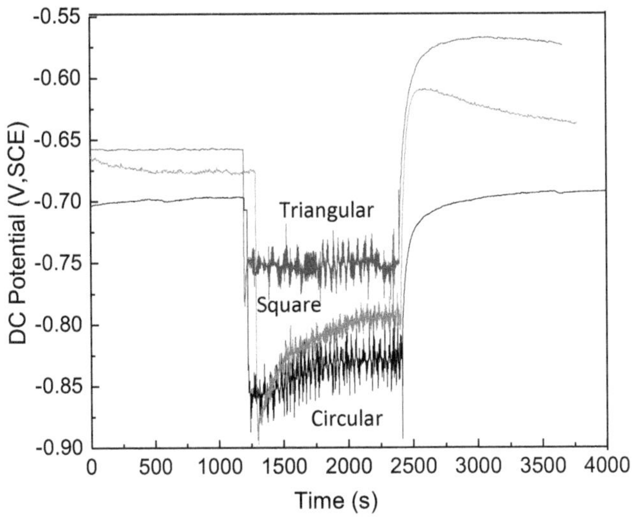


**Figure 3-32.** DC potential of coated steel electrodes containing a circular, square, or triangular coating defect with an area of 19.6 $mm^2$ under an AC current density of 500 $A/m^2$ in the test solution.[106]

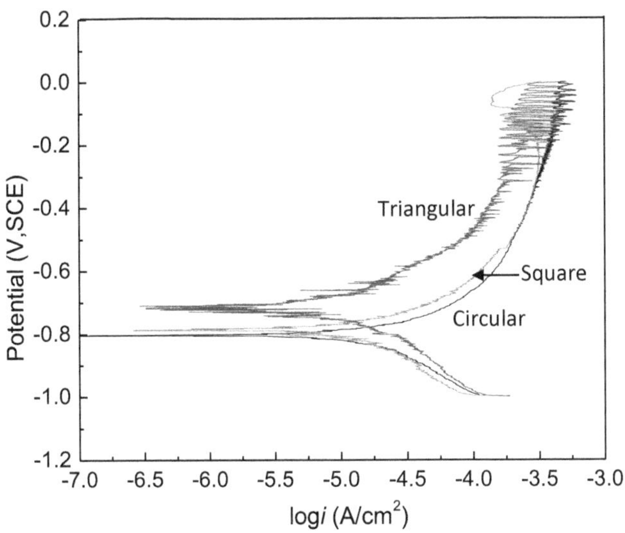


**Figure 3-33.** Potentiodynamic polarization curves of coated steel electrodes containing a circular, square, or triangular coating defect with an area of 19.6 $mm^2$ under an AC current density of 500 $A/m^2$ in the test solution.[106]

Furthermore, the circular and triangular coating defects are associated with the most and least negative DC potentials and the largest and smallest anodic current densities, respectively. As discussed, with accelerated corrosion by AC, the corrosion products deposit at the small defects owing to the limited geometry for mass transfer. At the sharp corners of the triangular defect, the geometrical space is too small for corrosion products to diffuse away. Thus, it is easy for the generated corrosion products to accumulate. In contrast, the circular defect is relatively easy for corrosion products to leave owing to its open geometry. At the same time, the AC current density at the sharp corner is expected to be larger than that at the circular edge. As a result, localized corrosion at the triangular defect generates more corrosion products to deposit locally. Figure 3-34 shows the morphology of the steel at coating defects of 19.6 mm$^2$ in area under an AC current density of 500 A/m$^2$ in the test solution. It can be seen that, owing to the deposit of corrosion products at the sharp corners of the triangular defect, there are different colors between the corner and the central part, where corrosion occurs actively. This can also be seen at the corners of the square defect, where different colors exist between the corners and the defect center. Thus, a smaller anodic current density is obtained on the steel electrode containing the triangular defect, and a larger anodic current density is measured on the electrode containing the circular defect.

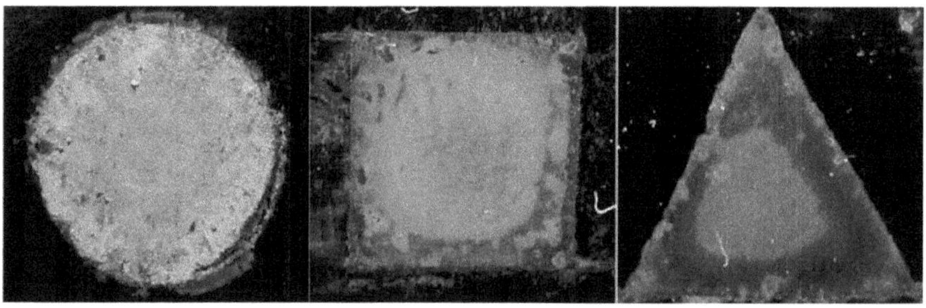

**Figure 3-34. Morphology of the steel electrodes at coating defects of 19.6 mm$^2$ in area under an AC current density of 500 A/m$^2$ in the test solution.[106]**

## References


1. McCollum, B., Ahlborn, G.H., "Influence of frequency of alternating or infrequently reversed current on electrolytic corrosion," *Transactions of the American Institute of Electrical Engineers* 35: 301–345, 1916.
2. Kajiyama, F., "Strategy for eliminating risks of corrosion and overprotection for buried modern pipelines," International Gas Union Research Conference, Seoul, 2011.
3. Galimberti, C.E., "Corrosion of lead by alternating current," *Corrosion* 20: 150*t*–157*t*, 1964.
4. Williams, J.F., "Corrosion of metals under the influence of alternating current," *Materials Protection* 5: 52–53, 1966.

5. Pookote, S.R., Chin, D.T., "Effect of alternating current on the underground corrosion of steels," *MP* 17: 9–15, 1978.
6. McIntosh, D., "Grounding where corrosion protection is required," *IEEE Transactions on Industry Applications* 18: 600–607, 1982.
7. Nielsen, L.V., Nielsen, K., Baumgarten, B., Breuning-Madsen, H., Cohn, P., Rosenberg, H., "AC induced corrosion in pipelines: Detection, characterization and mitigation," CORROSION/2004, paper no. 04211, NACE, Houston, TX, 2004.
8. Ghanbari, E., Iannuzzi, M., Lillard, R.S., "The mechanism of alternating current corrosion of API grade X65 pipeline steel," *Corrosion* 72: 1196–1210, 2016.
9. Hartmann, P., "External corrosion on a cathodically protected gas pipeline due to interference from 50 Hz alternating current," *3R International* 30: 584–589, 1991.
10. Heim, G., Peez, G., "The influence of alternating current on buried and cathodically protected high-pressure gas pipelines," *Gas-Erdgas* 133, 1992.
11. Prinz, W., "AC induced corrosion on cathodically protected pipelines," Proc. UK Corrosion '92, Vol. 1, Institute of Corrosion, Northampton, UK, 1992.
12. Gummow, R.A., Wakelin, R.G., Segall, S.M., "AC corrosion: A new threat to pipeline integrity?" International Pipeline Conference, ASME, New York, 1996.
13. Wakelin, R.G., Sheldon, C., "Investigation and mitigation of AC corrosion on a 300 MM natural gas pipeline," CORROSION/2004, paper no. 04205, NACE, Houston, TX, 2004.
14. NACE, *AC Corrosion State-of-the-Art: Corrosion Rate, Mechanism, Mitigation Requirements*, NACE Task Group 327, Houston, TX, 2010.
15. Brelsford, C., "Mitigating AC corrosion on cathodically protected pipelines," *Pipeline & Gas Journal* 242, 10: 1–3, 2015.
16. Hanson, H.R., Smart, J., "AC corrosion on a pipeline located in a HVAC utility corridor," CORROSION/2004, paper no. 04209, NACE, Houston, TX, 2004.
17. Wakelin, R., Gummow, R., Segall, S., "AC corrosion—Case histories, test procedures, and mitigation," CORROSION/98, paper no. 98565, NACE, Houston, TX, 1998.
18. Bruckner, W.H., *The Effects of 60 Cycle Alternating Current on the Corrosion of Steels and Other Metals Buried in Soils*, Engineering Experiment Station Bulletin 470, University of Illinois, Urbana. 1964.
19. Simon, P.D., Lopez-Garrity, M., Wagner, D., Garrity, K., "Identification and prioritization of potential HVAC interference locations on a 6,000-mile crude and product transmission pipeline system," CORROSION/2017, paper no. 9260, NACE, Houston, TX, 2017.
20. Kuang, D., Cheng, Y.F., "Effects of alternating current interference on cathodic protection potential and its effectiveness for corrosion protection of pipelines," *Corrosion Engineering, Science and Technology* 52: 22–28, 2017.
21. Davis, J.R., *Corrosion – Understanding the Basics*, ASM International, Materials Park, OH, 2000.
22. Linhardt, P., Ball, G., "AC corrosion: Results from laboratory investigations and from a failure analysis," CORROSION/2006, paper no. 06160, NACE, Houston, TX, 2006.

23. Nielsen, L.V., Cohn, P., "AC corrosion and electrical equivalent diagrams," CeoCor International Congress, European Committee for the Study of Corrosion and Protection of Pipes and Pipeline Systems, Brussels, 2000.
24. Ormellese, M., Goidanich, S., Bolzoni, F., Lazzari, L., Pedeferri, M.P., "Laboratory testing on the influence of alternated current on steel corrosion," CORROSION/2004, paper no. 04208, NACE, Houston, TX, 2004.
25. Ragault, I., "AC corrosion induced by V.H.V. electrical lines on polyethylene coated steel gas pipelines," CORROSION/98, paper no. 98557, NACE, Houston, TX, 1998.
26. Bertocci, U., "AC induced corrosion–The effect of an alternating voltage on electrodes under charge-transfer control," *Corrosion* 35: 211–215, 1979.
27. Jones, D.A., "Effect of alternating current on corrosion of low alloy and carbon steels," *Corrosion* 24: 428–433, 1978.
28. Chin, D.T., Fu, T.W., "Corrosion by alternating current: A study of the anodic polarization of mild steel in $Na_2SO_4$ solution," *Corrosion* 35: 514–523, 1979.
29. Hamlin, A.W., "Some effects of alternating current on pipeline operations," *MP* 19: 18–27, 1980.
30. Yunovich, M., Thompson, N.G., "AC corrosion: Mechanism and proposed model," 2004 International Pipeline Conference, Calgary, Alberta; ASME, New York, 2004.
31. Büchler, M., Schöneich, H., "Investigation of alternating current corrosion of cathodically protected pipelines: Development of a detection method, mitigation measures, and a model for the mechanism," *Corrosion* 65: 578–586, 2009.
32. Nielsen, L.V., "Role of alkalization in AC induced corrosion of pipelines and consequences hereof in relation to CP requirements," CORROSION/2005, paper no. 05188, NACE, Houston, TX, 2005.
33. Cheng, Y.F., *Stress Corrosion Cracking of Pipelines*, Wiley, Hoboken, NJ, 2013.
34. Panossian, Z., Filho, S.E., de Almeida, N.L., Pereira Filho, M.L., de L. Silva, D., Laurino, E.W., de L. Oliver, J.H., de S. Pimenta, G., de C. Albertini, J.A., "Effect of alternating current by high power lines voltage and electric transmission systems in pipelines corrosion," CORROSION/2009, paper no. 09541, NACE, Houston, TX, 2009.
35. Brenna, A., Ormellese, M., Lazzari, L., "A proposal of AC corrosion mechanism of carbon steel in cathodic protection condition," CORROSION/2013, paper no. 2457, NACE, Houston, TX, 2013.
36. Villain, S., Pischedda, M.H., Nigrelli, E., Godiart, F., Gavarri, J.R., "Degradation mechanism of electrodes subjected to alternating potentials: Modeling and protection," *Corrosion Science* 44: 675–683, 2002.
37. Lalvani, S.B., Lin, X., "A revised model for predicting corrosion of materials induced by alternating voltages," *Corrosion Science* 38, 10: 1709–1719, 1996.
38. Bosh, R.W., Bogaert, W.F., "A theoretical study of AC-induced corrosion considering diffusion phenomena," *Corrosion Science* 40: 323–336, 1998.
39. Goidanich, S., Lazzari, L., Ormellese, M., Pedeferri, M., "Influence of AC on corrosion kinetics for carbon steel, zinc and copper," Corrosion/2005, paper no. 05189, NACE, Houston, TX, 2005.
40. Xu, L.Y., Su, X., Yin, Z.X., Tang, Y.H., Cheng, Y.F. (2012), "Development of a real-time AC/DC data acquisition technique for studies of AC corrosion of pipelines," *Corrosion Science* 61: 215–223, 2012.

41. Ouadah, M., Touhami, O., Ibtiouen, R., "Diagnosis of the AC current densities effect on the cathodic protection performance of the steel X70 for a buried pipeline due to electromagnetic interference caused by HVPTL," *Progress in Electromagnetics Research* 45: 163–171, 2016.
42. Finneran, S., Krebs, B., "Advances in HVAC transmission industry and its effects on pipeline induced AC corrosion," CORROSION/2014, paper no. 4421, NACE, Houston, TX, 2014.
43. Yin, K., Tang, M., Hsiong, C., "Corrosion of buried pipeline under effect of electrical field with industrial frequency," Proc. Society of Petroleum Engineering, SPE-10578, Beijing, China, 1982.
44. Finneran, S., Krebs, B., Bensman, L., *Criteria for Pipelines Co-Existing with Electric Power Lines*, INGAA Foundation, Dublin, OH, 2015
45. Ruedisueli, R., Hager, H., Sandwith, C., "An application of a state-of-the-art corrosion measurement system to a study of the effects of alternating current on corrosion," *Corrosion* 43: 331–338, 1987, 2015.
46. Guo, Y.B., Liu, C., Wang, D.G., Liu, S.H., "Effects of alternating current interference on corrosion of X60 pipeline steel," *Petroleum Science* 12: 316–324, 2015.
47. Fernandes, S., Mehendale, S., Venkatachalam, S., "Influence of frequency of alternating current on the electrochemical dissolution of mild steel and nickel," *Journal of Applied Electrochemistry* 10: 649–654, 1980.
48. Qiu, W.W., Pagano, M., Zhang, G., Lalvani, S., "A periodic voltage modulation effect on the corrosion of Cu-Ni alloy," *Corrosion Science* 37: 97–110, 1995.
49. British Standards Institution, *Evaluation of AC Corrosion Likelihood of Buried Pipelines Applicable to Cathodically Protected Pipelines*, BS EN 15280, London, 2013.
50. NACE, *Mitigation of Alternating Current and Lightning Effects on Metallic Structures and Corrosion Control Systems*, SP0177, Houston, TX, 2007.
51. Gilroy, D.E., "AC interference-important issues for cross country pipelines," CORROSION/2003, paper no. 03699, NACE, Houston, TX, 2003.
52. Eden, D.A., "AC interference and accelerated pipeline corrosion in utility corridors: An alternative pers, 2006pective to the root causes," CORROSION/2006, paper no. 06159, NACE, Houston, TX.
53. Fu, A.Q., Cheng, Y.F., "Effect of alternating current on corrosion and the effectiveness of cathodic protection of pipelines," *Canadian Metallurgical Quarterly* 51: 81–90, 2012.
54. Helm, G., Helm, T., Heinzen, H., Schwenk, W. (1993), "Investigation of corrosion of cathodically protected steel subjected to alternating currents," *Corrosion Science* 5: 246–253, 1993.
55. German Institute for Standardization, *Corrosion of Metals, Proof of Effectiveness of Cathodic Corrosion Protection of Underground Installations*, DIN 50925, Berlin, 1992.
56. Funk, D., Prinz, W., Schöneich, H.G., "Investigations of AC corrosion in cathodically protected pipes," *3R International* 31: 336–341, 1992.
57. Baeckmann, W.V., Schwenk, W., Prinz, W., *Handbuch des Kathodischen Korrosionsschutzes*, Wiley-VCH, Weinheim, Germany, 1989.
58. Yunovich, M., Thompson, N.G., "AC corrosion: Corrosion rate and mitigation requirements," CORROSION/2004, paper no. 04206, NACE, Houston, TX, 2004.

59. Goidanich, S., Lazzari, L., Ormellese, M., "AC corrosion. Part 2: Parameters influencing corrosion rate," *Corrosion Science* 52: 916-922, 2010.
60. Xu, L.Y., Su, X., Cheng, Y.F., "Effect of alternating current on cathodic protection on pipelines," *Corrosion Science* 66: 263–268, 2013.
61. Kuang, D., Cheng, Y.F., "Understand the AC induced pitting corrosion on pipelines in both high pH and neutral pH carbonate/bicarbonate solutions," *Corrosion Science* 85: 304–310, 2014.
62. Ghanbari, E., Lillard, S., Carpenter, C., "The influence of scale formation on the AC corrosion of API grade X65 pipeline steel under cathodic protection," CORROSION/2017, paper no. 9809, NACE, Houston, TX, 2017.
63. Ghanbari, E., Lillard, R.S., Iannuzzi, M., Ortiz, M.R., "Corrosion behavior of buried pipeline in presence of AC stray current in controlled environment," CORROSION/2015, paper no. 5588, NACE, Houston, TX, 2015.
64. Cheng, Y.F., Norsworthy, R., *Pipeline Coatings*, NACE, Houston, TX, 2017.
65. Munger, C.G., Vincent, L.D., *Corrosion Prevention by Protective Coatings*, 2nd ed., NACE, Houston, TX, 1999.
66. Junker, A., Nielsen, L.V., Møller, P., "Effect of chemical environment and pH on AC corrosion of cathodically protected structures," CORROSION/2017, paper no. 9352, NACE, Houston, TX, 2017.
67. Jiang, Z., Du, Y., Lu, M., Zhang, Y., Tang, D., Dong, L., "New findings on the factors accelerating AC corrosion of buried pipeline," *Corrosion Science* 81: 1–10, 2014.
68. Collet, E., Delores, B., Gabillard, M., Ragault, I., "Corrosion due to AC influence of very high voltage power lines on polyethylene-coated steel pipelines: Evaluation of risks-preventive measures," *Anti-Corrosion Methods and Materials* 48: 221–227, 2001.
69. Norsworthy, R., "Addressing soil stress and CP shielding by using a woven geotextile fabric backed tape system with a rubber modified bituminous compound," CORROSION/2000, paper no. 00769, NACE, Houston, TX, 2000.
70. Xu, L.Y., Cheng, Y.F., "Experimental and numerical studies of effectiveness of cathodic protection at corrosion defects on pipelines," *Corrosion Science* 78: 162–171, 2014.
71. Funk, D., Prinz, W., Schöneich, H., "Investigations of AC corrosion in cathodically protected pipes", *Ochrona Przed Korozja* 36: 225–228, 1993.
72. Junker, A., Nielsen, L.V., Heinrich, C., Møller, P., "Laboratory and field investigation of the effect of the chemical environment on AC corrosion," CORROSION/2018, paper no. 10844, NACE, Houston, TX, 2018.
73. Du, Y., Xiao, Y., Xie, S., Lu, M., "Research on the effects of environmental parameters on AC corrosion behavior," CORROSION/2018, paper no. 10676, NACE, Houston, TX, 2018.
74. Schöneich, H.G., *Discussion of Criteria to Assess the Alternating Current Corrosion Risk of Cathodically Protected Pipelines*, Ruhrgas AG, Essen, Germany, 2004.
75. Yang, L., Xu, Y., Zhu, Y., Liu, L., Wang, X., Huang, Y., "Evaluation of interaction effects of sulfate and chloride ions on reinforcements in simulated marine environment using electrochemical methods," *International Journal of Electrochemical Science* 11: 6943–6958, 2016.

76. Al-Amoudi, O.S.B., Maslehuddin, M., "The effect of chloride and sulfate ions on reinforcement corrosion," *Cement and Concrete Research* 23: 139–146, 1993.
77. Ghanbari, E., Lillard, R.S., "The influence of $CaCO_3$ scale formation on AC corrosion rates of pipeline steel under cathodic protection," *Corrosion* 74: 551–565, 2018.
78. Hilleary, J., "Calculating and tracking soil resistivity change in high AC corridors," CORROSION/2017, paper no. 8984, NACE, Houston, TX, 2017.
79. Büchler, M., Schöneich, H.G., Stalder, F. (2005), "Discussion of criteria to assess the alternating current corrosion risk of cathodically protected pipelines," Joint Technical Meeting on Pipeline Research, paper no. 26, Pipeline Research Council International, Arlington, VA, 2005.
80. Nielsen, L.V., Baumgarten, B., Cohn, P., "On-site measurements of AC-induced corrosion: Effect of AC and DC parameters," CeoCor International Congress, European Committee for the Study of Corrosion and Protection of Pipes and Pipeline Systems, Brussels, 2004.
81. Büchler, M., "Alternating current corrosion of cathodically protected pipelines: Discussion of the involved processes and their consequences on the critical interference values," *Materials and Corrosion* 63: 1181–1187, 2012.
82. Zhu, M., Yuan, Y.F., Yin, S.M., Yu, G.H., Guo, S.Y., Huang, Y.Z., Du, C.W., "Corrosion behavior of pipeline steel with different microstructures under AC interference in acid soil simulation solution," *Journal of Materials Engineering and Performance* 28: 1698–1706, 2019.
83. Bolzoni, F., Goidanich, S., Lazzari, L., Ormellese, M., "Laboratory test results of AC interference on polarized steel," CORROSION/2003, paper no. 03704, NACE, Houston, TX, 2003.
84. Brenna, A., Ormellese, M., Lazzari, L., "Electromechanical breakdown mechanism of passive film in alternating current-related corrosion of carbon steel under cathodic protection condition," *Corrosion* 72: 1055–1063, 2016.
85. Zhu, M., Ou, G., Jin, H., Du, C., Li, X., Liu, Z., "Influence of AC waveforms on stress corrosion cracking behavior of pipeline steel in high pH solution," *Corrosion Engineering, Science and Technology* 51: 18–24, 2016.
86. Cheng, Y.F., Niu, L., "Mechanism for hydrogen evolution reaction on pipeline steel in near-neutral pH solution," *Electrochemistry Communication* 9: 558-562, 2007.
87. National Energy Board, *Stress Corrosion Cracking on Canadian Oil and Gas Pipelines*, MH-2-95, Calgary, Alberta, 1996.
88. Michael Baker Jr., *Stress Corrosion Cracking Study: Final Report*, TTO No. 8, Office of Pipeline Safety, U.S. Department of Transportation, Washington, DC, 2005.
89. Zhang, G.A., Cheng, Y.F., "On the fundamentals of electrochemical corrosion of X65 steel in $CO_2$-containing formation water in the presence of acetic acid in petroleum production," *Corrosion Science* 51: 87–94, 2009.
90. Lalvani, S.B., Zhang, G., "The corrosion of carbon steel in a chloride environment due to periodic voltage modulation: Part I," *Corrosion Science* 37: 1567–1582, 1995.
91. Lalvani, S.B., Zhang, G., "The corrosion of carbon steel in a chloride environment due to periodic voltage modulation: Part II," *Corrosion Science* 37: 1583–1598, 1995.

92. Bard, A.J., Faulkner, L.R., *Electrochemical Methods, Fundamentals and Applications*, 2nd ed., Wiley, Hoboken, NJ, 2001.
93. Li, M.C., Cheng, Y.F., "Mechanistic investigation of hydrogen-enhanced anodic dissolution of X-70 pipe steel and its implication on near-neutral pH SCC of pipelines," *Electrochimica Acta* 52: 8111–8117, 2007.
94. Fu, A.Q., Cheng, Y.F. (2010), "Effects of alternating current on corrosion of a coated pipeline steel in a chloride-containing carbonate/bicarbonate solution," *Corrosion Science* 52: 612–619.
95. Parkins, R.N., Zhou, S., "The stress corrosion cracking of C-Mn steel in $CO_2$-$HCO_3^-$-$CO_3^{2-}$ solutions. II: Electrochemical and other data," *Corrosion Science* 39: 175–191, 1997.
96. Xue, H.B., Cheng, Y.F., "Passivity and pitting corrosion of X80 pipeline steel in carbonate/bicarbonate solution studied by electrochemical measurements," *Journal of Materials Engineering and Performance* 19: 1311–1317, 2010.
97. Heuer, J.K., Stubbins, J.F., "An XPS characterization of $FeCO_3$ films from $CO_2$ corrosion," *Corrosion Science* 41: 1231–1243, 1999.
98. Linter, B.R., Burstein, G.T., "Reaction of pipeline steel in carbon dioxide solution," *Corrosion Science* 41: 117–139, 1999.
99. Li, Y., Cheng, Y.F., "Passive film growth on carbon steel and its nanoscale features at various passivating potentials," *Applied Surface Science* 396: 144–153, 2017.
100. Ormellese, M., Lazzari, L., Brenna, A., "Effects of AC interference on passive metal corrosion," CORROSION/2011, paper no. 11342, NACE, Houston, TX, 2011.
101. Tan, T.C., Chin, D.T., "AC corrosion of nickel in sulphate solutions," *Journal of Applied Electrochem*istry 18: 831–838, 1988, 1988.
102. Wan, H., Song, D., Du, C., Liu, Z., Li, X., "Effect of alternating current and *Bacillus cereus* on the stress corrosion behavior and mechanism of X80 steel in a Beijing soil solution," *Bioelectrochemistry* 127: 49–58, 2019.
103. Sun, D., Wu, M., Xie, F., "Effect of sulfate-reducing bacteria and cathodic potential on stress corrosion cracking of X70 steel in sea-mud simulated solution," *Materials Science and Engineering A* 721: 135–144, 2018.
104. Liu, T., Cheng, Y.F., "The influence of cathodic protection potential on the biofilm formation and corrosion behaviour of an X70 steel pipeline in sulfate reducing bacteria media," *Journal of Alloys and Compounds* 729: 180–188, 2017.
105. Qing, Y., Bai, Y., Xu, J., Wu, T., Yan, M., Sun, C., "Effect of alternating current and sulfate-reducing bacteria on corrosion of X80 pipeline steel in soil-extract solution," *Materials* 12, 1: 144, 2019.
106. Kuang, D., Cheng, Y.F., "AC corrosion at coating defect on pipelines," *Corrosion* 71: 267-276, 2015.
107. Wahl, C., Schmautzer, E. (2015), "Pipeline voltage," *Pipeline Technology Journal*, 22–31.

# 4 AC-Induced Pitting Corrosion of Pipelines

## 4.1. Conditions for initiation of pitting corrosion on pipelines

Initiation of pitting corrosion on metals depends on the electrochemical activity of the metals in the environment. In the pipeline industry, pitting corrosion is considered one of the most severe failure mechanisms due to its rapid growth and difficulty of detection. It has been estimated that over 90% of corrosion failures of oil and gas transmission pipelines in the United States between 1970 and 1984 were due to pitting corrosion.[1] The majority of pipeline stress corrosion cracking starts from corrosion pits.[2]

Pipelines are usually made of carbon steels, which show either active dissolution or passivity in near-neutral pH bicarbonate or high pH carbonate–bicarbonate solutions, respectively.[3] Thus, it is expected that pitting corrosion on pipelines would initiate and grow through different mechanisms in the specific service environment.

### 4.1.1. Near-neutral pH bicarbonate environments

In near-neutral pH aqueous environments, such as groundwater or soil solutions in most territories, pitting corrosion occurs on pipelines primarily through the galvanic coupling effect between the steel matrix and the metallurgical features, such as grain boundaries,[4, 5] nonmetallic inclusions,[6, 7] and phase interfaces.[8] Corrosion pits are initiated on pipeline steels when the potential of the steels experiences fluctuations in the soil environment. Although the pipelines are protected by CP, the CP potential would not always stay stable for reasons such as seasonal dry–wet cycling of the soil, interference of stray current, and temperature changes.[3] To simulate the conditions causing corrosion of pipeline steel in the soil under fluctuating potentials, a square wave polarization (SWP) was applied on an X70 pipeline steel electrode. Figure 4-1 shows the maps of localized electrochemical impedance spectroscopy measured on the steel under the SWP from –950 mV (SCE) to –800 mV (SCE) with varied durations in a near-neutral pH bicarbonate solution, simulating the electrolyte being trapped under a disbonded PE coating.[5] It can be seen that the impedance distribution over the scanning area on the steel electrode depends on the SWP time. After 48 h of SWP (Figure 4-1a), the impedance distribution is uniform on the steel, with some sites–such as points A, B, and C–having lower impedance values than the adjacent area. After 96 h of SWP (Figure 4-1b), the impedance fluctuates apparently. The marked points A, B, C, and D show drops in the local impedance. The localized electrochemical impedance map measured after 144 h of SWP (Figure 4-1c) indicates a further decrease of the impedance at these points. These sites present the active locations on the steel under SWP, which simulates the CP potential fluctuations in the field.

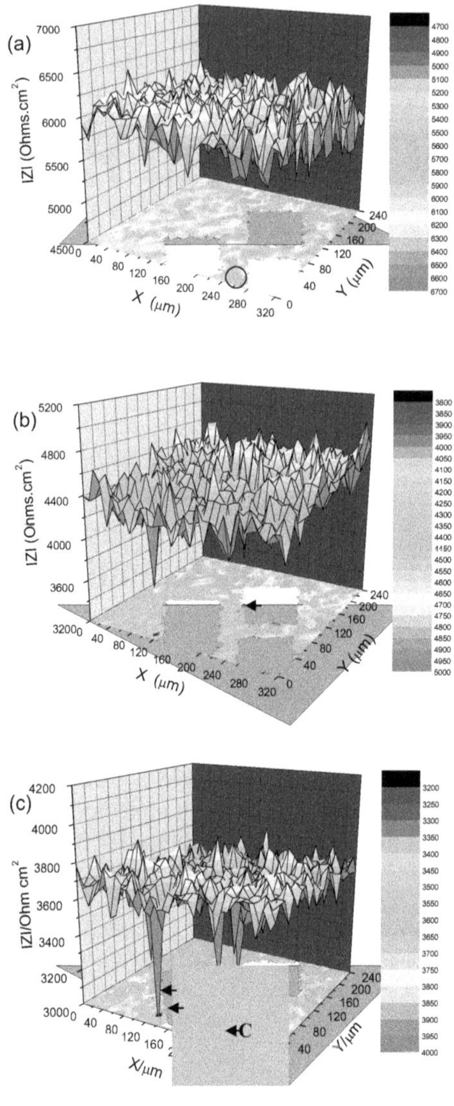


**Figure 4-1.** Localized electrochemical impedance maps measured on X70 steel electrode in a near-neutral pH bicarbonate solution under SWP from –950 mV (SCE) to –800 mV (SCE) for (a) 48 h, (b) 96 h, (c) 144 h.[5]

The surface morphology of the steel electrode after 144 h of SWP between -950 mV (SCE) and -800 mV (SCE) shows that corrosion pits initiate on the steel surface, as seen in Figure 4-2. Most pits are observed along the grounding lines introduced during the electrode preparation. As a comparison, the surface morphologies of the steel electrode after 144 h of potentiostatic polarization at -950 mV (SCE) and -800 mV (SCE), respectively, are shown in Figure 4-3. There is no general corrosion or pitting corrosion observed on the steel that is under constant CP potentials. Therefore, the CP potential fluctuations could initiate corrosion pits on pipeline steel.

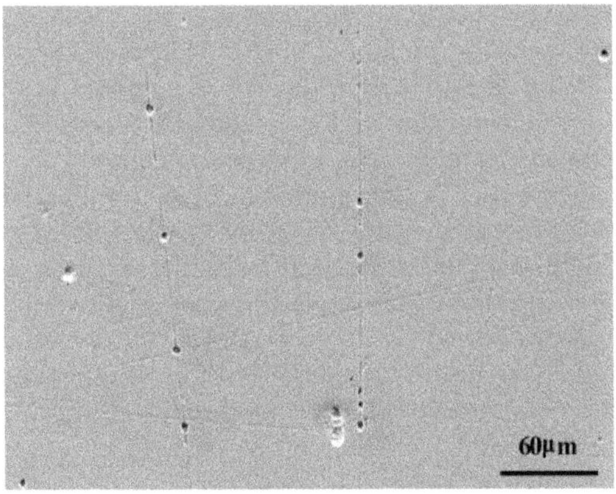


**Figure 4-2.** Corrosion pits initiating on X70 steel electrode after 144 h of SWP between -950 mV (SCE) and -800 mV (SCE) in the near-neutral pH bicarbonate solution.[5]

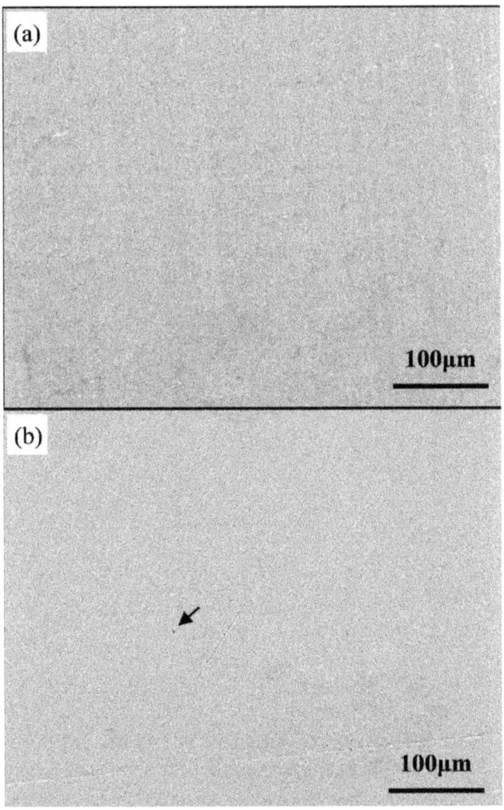


**Figure 4-3.** Surface morphologies of X70 steel electrode after 144 h of potentiostatic polarization at (a) -950 mV (SCE) and (b) -800 mV (SCE) in the solution.[5]

An electrochemical state conversion model was developed to explain the occurrence of pitting corrosion on pipelines that are under CP potential fluctuations.[9] According to the model, various metallurgical microfeatures existing on the steel surface would introduce an additional cathodic potential, enhancing the cathodic reactions locally.[10] If the cathodic reaction is controlled by a mass transfer step, the cathodic reactants, such as $H^+$, $HCO_3^-$, and $H_2CO_3$, would be consumed more rapidly at the features than at the intact area. Figure 4-4a shows schematically the double-charge layer at the steel–solution interface under a constant cathodic potential, where both the features and the intact areas are associated with a uniform cathodic electric field. All cathodic reactions are in equilibrium with the mass transfer process of the reactants. The steel is under a full CP and does not corrode. When potential fluctuations occur, the double-charge layer is disturbed, and the reactants consume more rapidly at the microfeatures than at any other area, as shown in Figure 4-4b. As a result, the features are under a temporary anodic potential field, resulting in local anodic dissolution, while the intact area is still protected by the applied cathodic potential. If the potential fluctuations are sufficiently large that the cathodic reactants are consumed rapidly over the whole surface of the steel (Figure 4-4c), the double-charge layer can become anodized on the steel, which thus experiences an anodic dissolution with a higher corrosion rate at the features. If the potential fluctuations occur repeatedly for a sustainable time period, the pits grow self-catalytically.

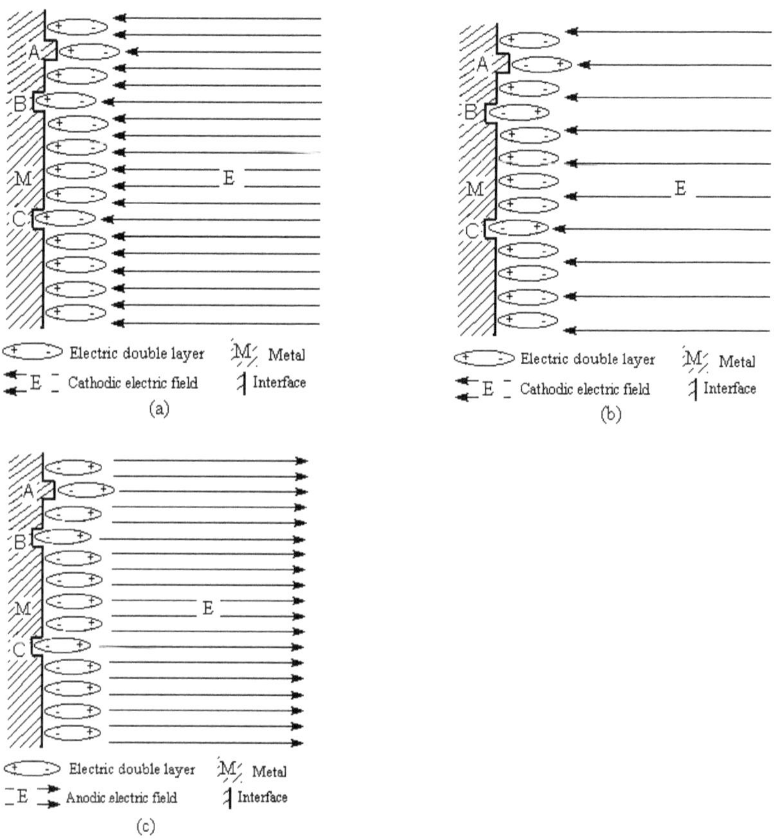


**Figure 4-4.** Schematic view of the double-charge layer under (a) a constant cathodic potential, (b) a potential fluctuation resulting in a temporary anodic state at surface features, and (c) a potential fluctuation causing anodic reaction over the whole steel surface, where points A, B, and C refer to the metallurgical microfeatures on the steel surface.[9]

### 4.1.2. High pH carbonate-bicarbonate environments

In high pH aqueous environments, pitting corrosion would initiate on pipelines through a mechanism that is different from that in the near-neutral pH solutions. Upon application of CP, the CP current would permeate through the permeable coatings such as FBE and reach the steel surface, generating $OH^-$ ions by electrochemical reduction of dissolved oxygen or water. This results in an elevated pH of the trapped electrolyte under a disbonded coating. Simultaneously, $CO_2$ in the soil generated by the decay of organic matter permeates and dissolves in the trapped electrolyte. At a pH of about 9–11, a concentrated carbonate–bicarbonate electrolyte is formed owing to the hydrolysis of $H_2CO_3$ and $HCO_3^-$ in the alkaline environment.[11] A small amount of $Cl^-$ ions from soils are also contained in the trapped solution.[12]

Figure 4-5 shows the potentiodynamic polarization curves of an X70 pipeline steel in the high pH carbonate–bicarbonate solutions with varied solution layer thicknesses, where the low, intermediate, and high concentrations refer to 0.05 M $Na_2CO_3$ + 0.1 M $NaHCO_3$ + 0.1 M NaCl (pH 9.6, conductivity 21.8 mS/cm); 0.25 M $Na_2CO_3$ + 0.5 M $NaHCO_3$ + 0.1 M NaCl (pH 9.4, conductivity 49.6 mS/cm); and 0.5 M $Na_2CO_3$ + 1 M $NaHCO_3$ + 0.1 M NaCl (pH 9.4, conductivity 61.5 mS/cm), respectively. It can be seen that the solution layer thickness has a significant influence on the electrochemical polarization behavior of the steel. In the low-concentration solution (Figure 4-5a), the steel can be passivated, with a high active–passive transition current density. With a decrease of the solution layer thickness, the transition current density decreases. When the solution layer is reduced to 30 μm in thickness, the active–passive transition phenomenon is not observed, and a stable passivity develops on the steel. Although a well-defined dependence of the passive current density, $i_p$, and pitting potential, $E_{pit}$, on the solution layer thickness is not established, it can be observed that the smallest $i_p$ and the most negative $E_{pit}$ are measured in the solution with a thickness of 30 μm. In the intermediate concentration carbonate–bicarbonate solution (Figure 4-5b), the passive current density is approximately independent of the solution layer thickness. The pitting potential is shifted negatively when the solution layer thickness decreases from the bulk solution to 30 μm. With a further increase of the solution concentration, the stable passivity is developed on the steel in all solutions, as can be seen in Figure 4-5c.

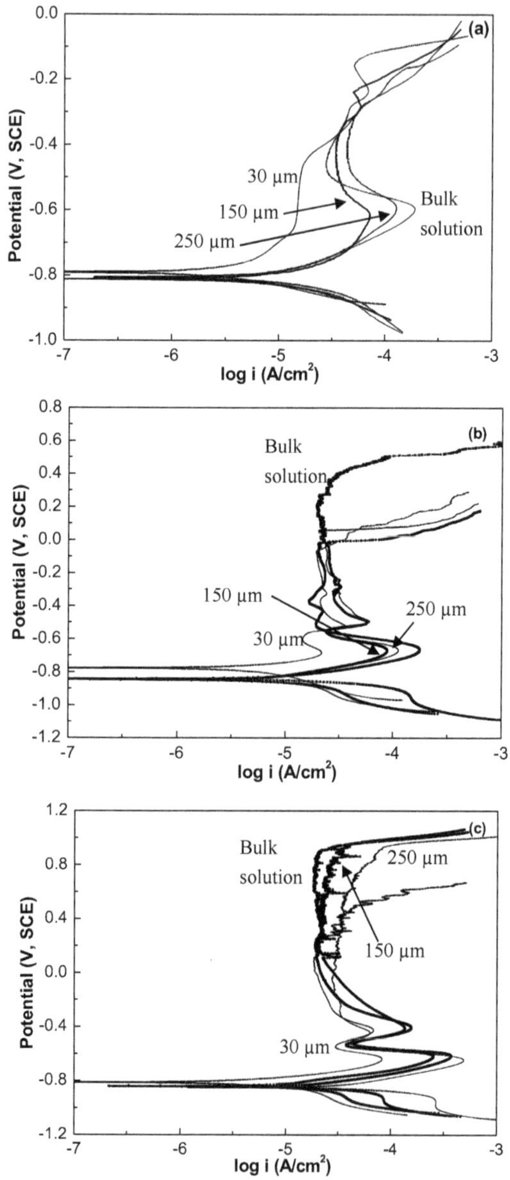


**Figure 4-5.** Potentiodynamic polarization curves of X70 pipeline steel in (a) low, (b) intermediate, and (c) high concentrations of high pH carbonate-bicarbonate solutions with various thicknesses.[13]

When the external coating is initially disbonded from the pipe steel, the crevice under the disbonded coating is narrow. The trapped solution experiences an elevation of the carbonate–bicarbonate concentration owing to the $CO_2$ permeation. In both low and high concentrations of the solution, there is a lower $E_{pit}$ in the thinner solution layer than in the thicker solution layer.[13] In the presence of $Cl^-$, corrosion pits initiate at the early stage of the coating disbondment, where the trapped solution is usually thin, and the solution concentration is relatively low. The role of chloride ions in pitting corrosion is versatile, including prevention of stable passivation and increase of anodic activity of the steel.[14] With the increase of the solution layer thickness and the solution concentration, the $E_{pit}$ shifts positively, and the resistance of the steel to pitting corrosion increases. Furthermore, a pre-application of CP on the coated steel increases the anodic current density.[13] Thus, a CP application history on pipelines would degrade the formed passive film, making it more active. As a result, the passive film would be more easily attacked by chloride ions, resulting in initiation of corrosion pits.

## 4.2. AC-induced pitting corrosion

As demonstrated, pipelines experience accelerated corrosion in both near-neutral pH and high pH environments under AC interference. Moreover, AC can result in localized corrosion on pipelines in the form of extensive corrosion pits.[15] It has also been noted that AC corrosion is localized, and mass loss measurement is not ordinarily recommended to provide information about the corrosion penetration depth.[16] As shown in Chapter 3, in high pH carbonate–bicarbonate solutions where the pipeline steel is in passivity, frequent oscillations of the passive current density occur in the presence of AC interference (Figure 3-17). The amplitude of the current oscillation increases with increasing AC current density. When the AC current density is up to 100 A/m$^2$, corrosion pits are formed on the steel surface (Figure 3-15). In near-neutral pH bicarbonate solutions, extensive corrosion pits are observed on the steel when the AC current density is increased to 800 A/m$^2$, as seen in Figure 3-8. The mechanisms that result in pitting corrosion of pipelines in the presence of AC interference and the threshold AC current density at which pits are initiated depend on the environments in which the pipeline steels corrode.

### 4.2.1. Near-neutral pH aqueous environments

Figure 4-6 shows the time dependence of the DC potential of an X65 pipeline steel electrode at various AC current densities in a near-neutral pH bicarbonate solution (0.01 M $NaHCO_3$) purging with 5% $CO_2/N_2$ gas (solution pH of 7.2).[17] It can be seen that, when AC is applied at the 1,200th second, the DC potential shifts negatively, with the most negative value observed at 0.02 A/cm$^2$ (i.e., 200 A/m$^2$). As the AC is stopped at the 2,400th second, the DC potential shifts positively toward the original corrosion potential. However, all AC-OFF potentials are less negative than the original corrosion potential.

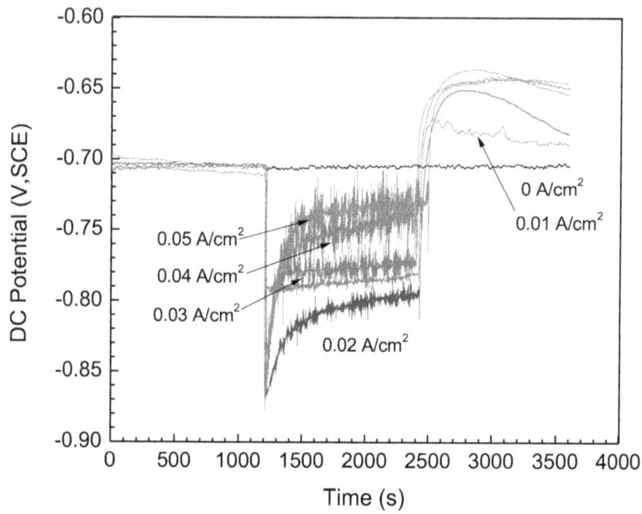


**Figure 4-6.** Time dependence of DC potential of an X65 pipeline steel electrode at various AC current densities in a near-neutral pH, 0.01 M $NaHCO_3$ solution purging with 5% $CO_2/N_2$ gas.[17]

Figure 4-7 shows the potentiodynamic polarization curves of an X65 steel electrode in 0.01 M $NaHCO_3$ solution purging with 5% $CO_2/N_2$ gas under various AC current densities. It can be seen that the steel is in active dissolution before and after the AC application. The cathodic curves also show an active polarization state. Moreover, the cathodic current density increases with the AC current density, but the anodic current densities almost copy one another at varied AC current densities.

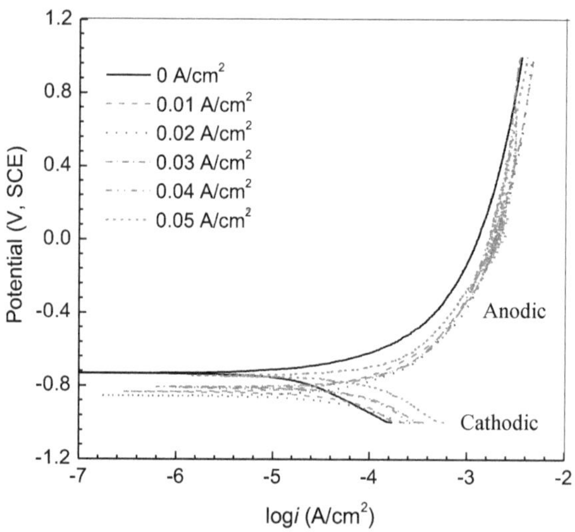


**Figure 4-7.** Potentiodynamic polarization curves of an X65 steel electrode measured at various AC current densities in 0.01 M $NaHCO_3$ solution purging with 5% $CO_2/N_2$ gas.[17]

Figure 4-8 shows the optical view of the steel electrode after different testing times in 0.01 M $NaHCO_3$ solution purging with 5% $CO_2/N_2$ gas.[17] It can be seen that no pit is generated after 12 h of testing in the absence of AC. However, corrosion pits are formed under 0.01 A/cm$^2$ of AC current density after 12 h of testing. When the AC current density is increased to 0.02 A/cm$^2$, pits are formed after 5 min only. With the AC current density increasing to 0.03–0.05 A/cm$^2$, the number and size of the pits clearly increase. Moreover, with the increase in AC current density, the pits become bigger and deeper, but the number of the pits decreases when the time increases from 5 min to 12 h.

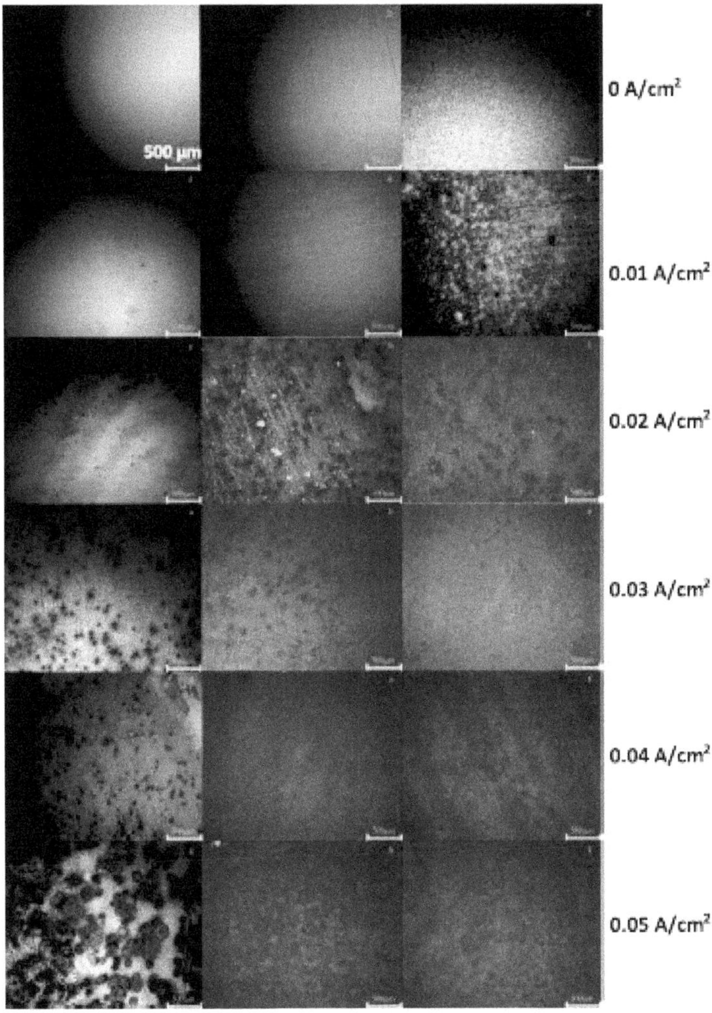


**Figure 4-8.** Optical views of an X65 steel electrode after different testing times in 0.01 M $NaHCO_3$ solution purging with 5% $CO_2/N_2$ gas under various AC current densities. From left to right: 5 min, 2 h, 12 h.[17]

## 4.2.2. High pH aqueous environments

Figure 4-9 shows the time dependence of DC potential of an X65 steel electrode at various AC current densities in a high pH, 0.05 M $Na_2CO_3$ + 0.1 M $NaHCO_3$ solution (pH 9.6).[17] Three potential regions are observed: (a) from 0 s to 1,200 s, no AC is applied on the steel electrode, and the recorded DC potential is the corrosion potential of the steel; (b) from 1,200 s to 2,400 s, the DC potential of the steel electrode under AC (i.e., the AC-ON potential) is recorded; and (c) after 2,400 s, the AC application is stopped, and the AC-OFF potential is recorded. It can be seen that the corrosion potential of the steel reaches the value of −0.2 V (SCE) after 1,200 s of immersion in the solution. Upon AC application, the DC potential shifts positively at AC current densities of 0.01 A/cm² and 0.02 A/cm². When the AC current density is up to 0.03 A/cm², the DC potential shifts negatively. Moreover, with the increase of AC current density, the AC-ON potential becomes more negative. When AC is stopped at the 2,400th second, the DC potentials cannot be recovered back to their original value of −0.2 V (SCE). At 0.01 A/cm² and 0.02 A/cm², the AC-OFF potentials are less negative than the original corrosion potential, while at AC current densities from 0.03 A/cm² to 0.05 A/cm², the AC-OFF potentials are much more negative than the corrosion potential. Clearly, the DC potential of the steel under AC interference depends on the AC current density.

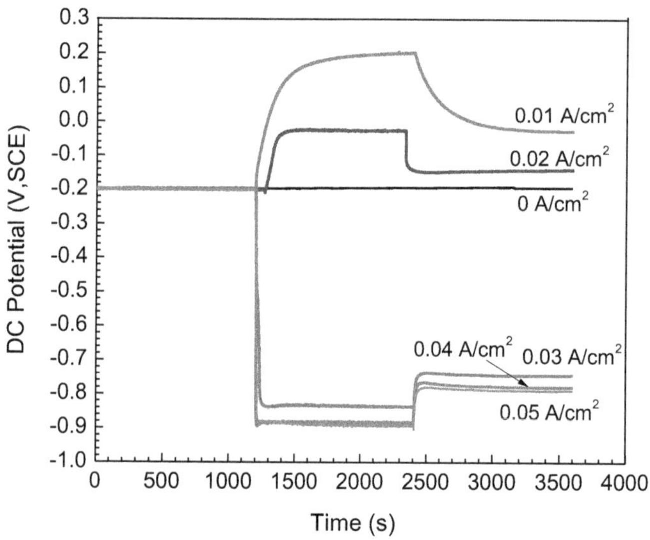


**Figure 4-9.** Time dependence of DC potential of an X65 steel electrode at various AC current densities in a high pH, 0.05 M $Na_2CO_3$ + 0.1 M $NaHCO_3$ solution (pH 9.6).[17]

Figure 4-10 shows the potentiodynamic polarization curve of the steel in the absence and presence of AC in the high pH carbonate-bicarbonate solution. It can be seen that the steel can be passivated in the solution when AC is not applied. Two oxidative current peaks are observed: at about −0.6 V (SCE) and −0.3 V (SCE). At −0.6 V (SCE), the current peak is attributed to the formation of $FeCO_3$ and $Fe(OH)_2$ on the steel surface:[11, 18, 19]

$$Fe + CO_3^{2-} \rightarrow FeCO_3 + 2e \quad (4\text{-}1)$$

$$Fe + HCO_3^{-} \rightarrow FeCO_3 + H^{+} + 2e \quad (4\text{-}2)$$

$$Fe + 2H_2O \rightarrow Fe(OH)_2 + 2H^{+} + 2e \quad (4\text{-}3)$$

The second current peak at about −0.3 V (SCE) is due to oxidation of ferrous species to $Fe_2O_3$ and $Fe_3O_4$.[13, 20; 21]

$$4Fe(OH)_2 + O_2 \rightarrow 2Fe_2O_3 + 4H_2O \quad (4\text{-}4)$$

$$4FeCO_3 + O_2 + 4H_2O \rightarrow 2Fe_2O_3 + 4HCO_3^{-} + 4H^{+} \quad (4\text{-}5)$$

$$6FeCO_3 + O_2 + 6H_2O \rightarrow 2Fe_3O_4 + 6HCO_3^{-} + 6H^{+} \quad (4\text{-}6)$$

For the transitional zone in between, it is believed that two steps are involved: (a) the dissolution of $FeCO_3$ to complex $Fe(CO_3)_2^{2-}$ and (b) the oxidation of ferrous ions to ferric substances.[3, 22]

When the AC interference is present, the passive potential region becomes small compared to the curve in Figure 4-10a. Moreover, the anodic current density increases with the increasing AC current density, while the cathodic curves show an active polarization feature and almost copy one another.

Figure 4-11 shows the optical view of the steel electrode after different testing times in the high pH carbonate–bicarbonate solution under various AC current densities. No pit is observed on the steel surface when the AC current density is smaller than 0.02 A/cm$^2$ for a testing time up to 12 h. However, when the AC current density is increased to 0.03 A/cm$^2$, corrosion pits are observed after only 5 min. The size of the pits increases with time. This trend is also observed at the AC current densities of 0.04 A/cm$^2$ and 0.05 A/cm$^2$.

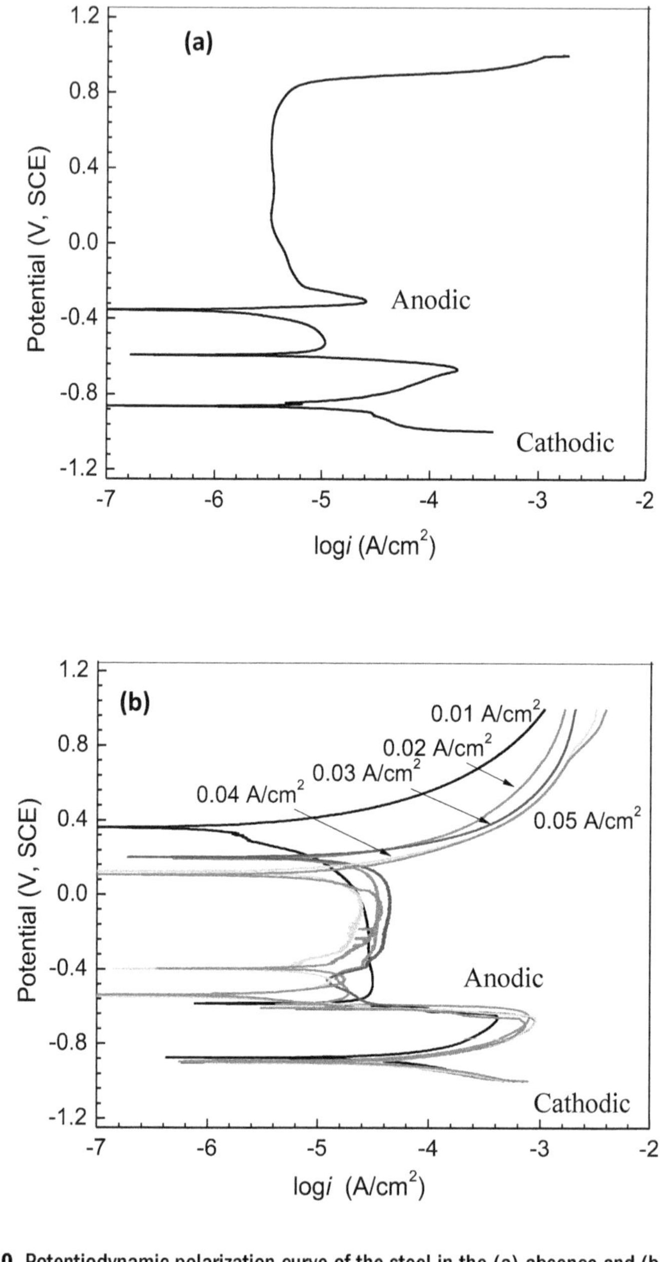


**Figure 4-10.** Potentiodynamic polarization curve of the steel in the (a) absence and (b) presence of AC at various AC current densities in the high pH carbonate–bicarbonate solution.[17]

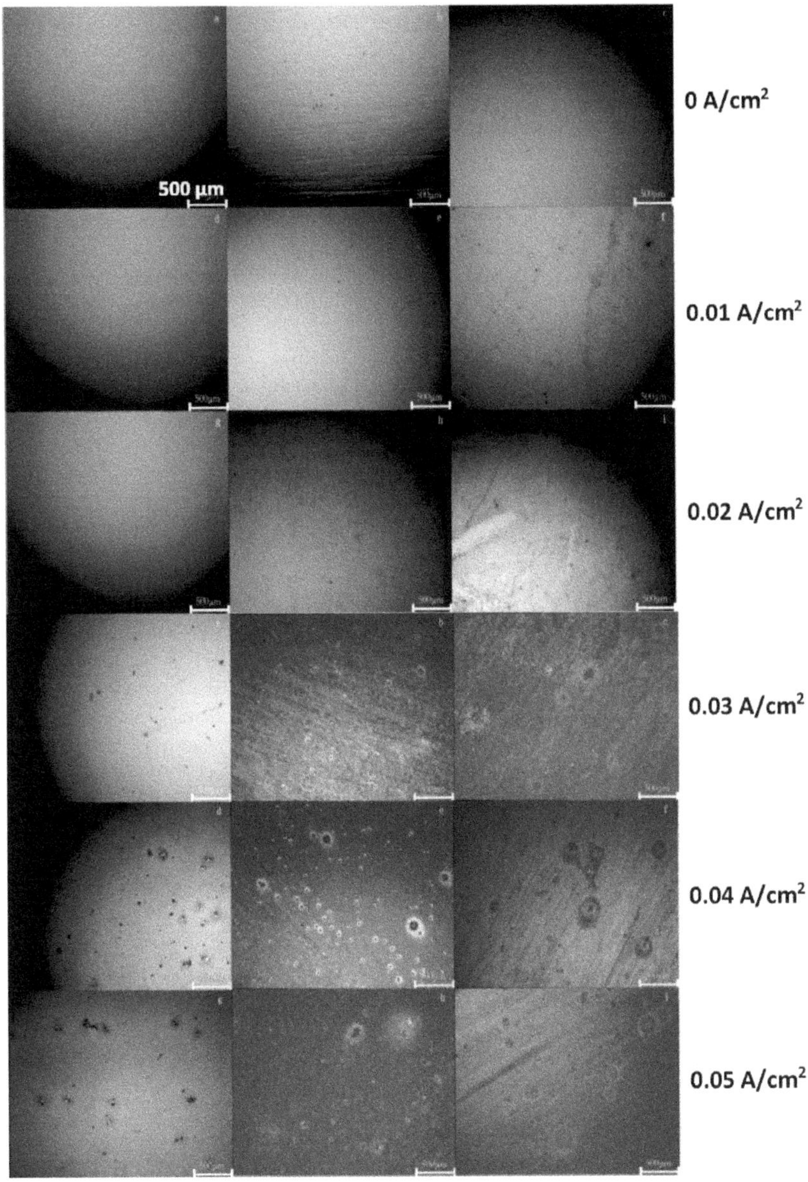


**Figure 4-11.** Optical view of the steel electrode after different testing times in the high pH carbonate-bicarbonate solution under various AC current densities.[17]

### 4.2.3. Threshold conditions for initiation of pitting corrosion under AC interference

The measured polarization curves in the absence and presence of AC interference show that AC greatly affects the anodic polarization behavior of steel in both high and near-neutral pH solutions. In the near-neutral pH bicarbonate solution, the pipeline steel shows an active dissolution behavior, as shown in the measured polarization curves in Figure 4-7. The corrosion products, such as $Fe(OH)_2$ and $FeCO_3$, deposit on the steel surface but would not have much protective ability owing to the porous structure.[23] Upon AC application, the DC potential shifts negatively and reaches the most negative value at 0.02 $A/cm^2$ (i.e., 200 $A/m^2$). After AC is stopped, the DC potential reaches a steady state that is less negative than the original corrosion potential, as seen in Figure 4-6. At the same time, the morphological observations in Figure 4-8 show that the critical AC current density at which corrosion pits are initiated on the steel is 0.02 $A/cm^2$, below which general corrosion occurs. Above the critical AC current density, pits can initiate after 5 min only. The pits become bigger and deeper with time. At individual times, the size and depth of the pits increase with the AC current density.

In the high pH carbonate–bicarbonate solution, the passivity of the steel and the formation of ferric oxides are retarded by AC, and the steel becomes de-passivated. Under low AC current densities, such as those smaller than 0.02 $A/cm^2$ (i.e., 200 $A/m^2$), the AC-induced polarization is not enough to cause localized dissolution of the steel. With the increase of AC current density to 0.03 $A/cm^2$, the dissolution current density is sufficiently large and small pits are formed locally only after 5 min. Thus, the AC current density of 0.03 $A/cm^2$ (i.e., 300 $A/m^2$) is considered the threshold value to initiate corrosion pits on the steel in the high pH solution. When the AC current density is up to 0.04 $A/cm^2$ and 0.05 $A/cm^2$, the anodic dissolution current density further increases and corrosion pits form extensively on the steel surface. Moreover, the number and size of the pits increase with time, showing the AC-induced growth of corrosion pits. When AC is removed, the DC potential cannot return to the original corrosion potential, indicating that the AC-induced activation effect on the steel remains even after AC is stopped. Thus, the AC effect on the steel is irreversible, as mentioned previously.

## 4.3. Mechanisms for AC-induced pitting corrosion on pipelines

The mechanistic model illustrating the effect of AC interference on corrosion and pitting corrosion of pipeline steels in high pH solutions is shown in Figure 4-12. Generally, during corrosion, charge-transfer reactions occur primarily in the double-charge layer at the steel–solution interface owing to an extremely strong electric field being exerted locally, which, however, can become disturbed in the presence of AC. The steel is in passivity in the high pH carbonate–bicarbonate solution, generating a passive film with varied compositions of $FeCO_3$, $Fe(OH)_2$,

$Fe_2O_3$, and $Fe_3O_4$, depending on the environment. In the absence of AC, the formed passive film protects the substrate steel from further corrosion, as illustrated in Figure 4-12a.

At small AC current densities such as 0.01 A/cm², the passive film is still effective in protecting the steel, preventing cations (mainly $Fe^{2+}$ ions generated at the steel–film interface) from flowing into the solution, but contributing to the film growth (Figure 4-12b). The double-charge layer serves as a current reservoir for the faradic process. The extra positive charges accumulating in the double-charge layer shift the DC potential of the steel positively, as measured in Figure 4-9. With the increase of the AC current density to 0.02 A/cm², the anodic polarization during AC cycling generates more ferrous ions due to the iron oxidation, and the cathodic polarization results in reduction of the passive film. As a result, some cations leak out of the film toward the solution by concentration gradient, as shown in Figure 4-12c. The DC potential thus becomes more negative than that measured at 0.01 A/cm².

When the AC current density further increases to 0.03 A/cm², the enhanced cathodic polarization can break the passive film and decrease the electric field strength of the double-charge layer, accelerating the flow of positive charges out of the film. This results in fewer positive charges being reserved and a decreased DC potential. As shown in Figure 4-12e, the higher the AC current density, the more breakage of the passive film and thus the more extensive the pitting corrosion of the steel. A further increase of AC current density to 0.05 A/cm² would not result in a DC potential that is much different from that measured at 0.04 A/cm², which means that the positive charges remaining in the double-charge layer achieve a balance with those releasing into the solution.

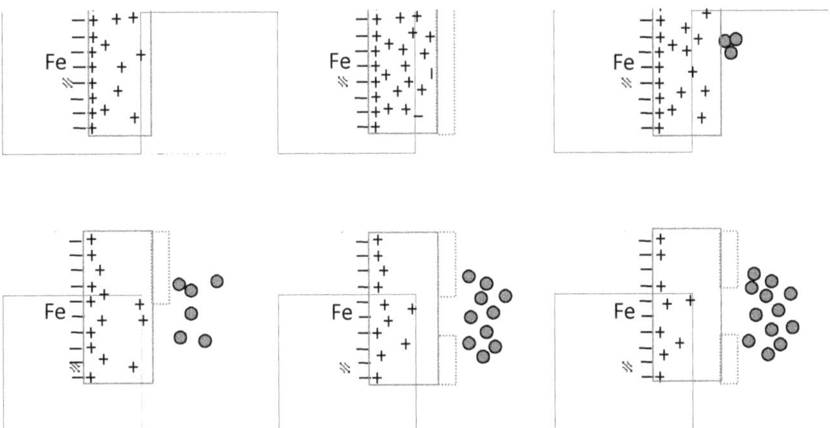


**Figure 4-12.** Schematic diagram of the mechanistic model for AC corrosion and pitting corrosion of pipeline steels in high pH solutions, where the red box refers to the passive film formed on the steel surface.[17]

As the steel cannot be passivated in near-neutral pH, diluted bicarbonate solutions, AC-induced pitting corrosion of pipeline steels occurs through a different mechanism from that described above. Figure 4-13 shows the mechanism for corrosion and pitting corrosion of pipeline steels in near-neutral pH solutions. In the absence of AC, a layer of corrosion products forms on the steel surface. Different from the passive film formed in high pH solutions, the corrosion product layer is usually porous and loose, providing some protection to the steel. When an AC current density of 0.01 A/cm$^2$ is applied, the AC-enhanced electric field accelerates the diffusion of cations that are generated during corrosion of the steel through the porous corrosion product layer, as shown in Figure 4-13b. Thus, the number of positive charges in the double-charge layer decreases, resulting in a negative shift of the DC potential, as shown in Figure 4-6. When the AC current density increases to 0.02 A/cm$^2$, more ferrous ions generate and diffuse through the corrosion product layer toward the solution, causing a further negative shift of the corrosion potential. At the same time, the thickness of the corrosion product layer increases owing to the enhanced corrosion of the steel under AC interference. The length of the diffusive pathway for positive charges increases, reducing the leakage of cations. Thus, with the further increase of AC current density, the DC potential shifts positively since more cations stay in the double-charge layer.

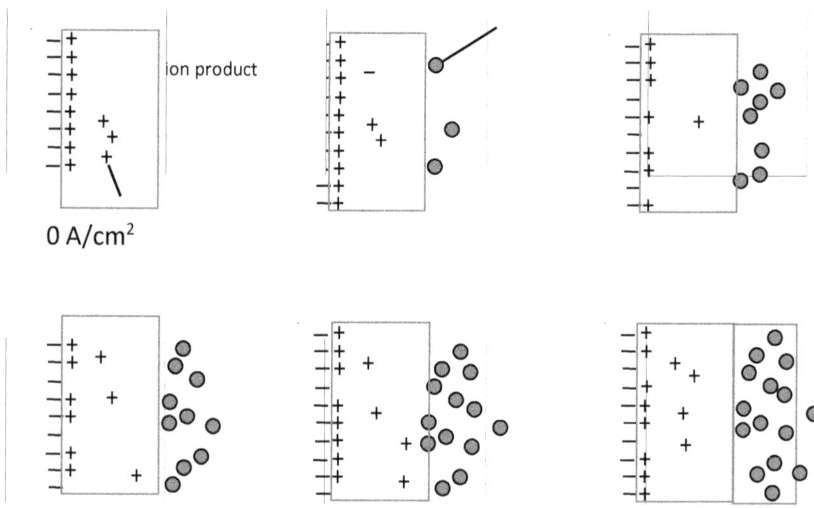


**Figure 4-13.** Schematic diagram of the mechanistic model for AC corrosion and pitting corrosion of pipeline steels in near-neutral pH solutions, where the red box refers to the corrosion product layer.[17]

Generally, theories and models have been developed to illustrate the pitting corrosion of steels that become passivated in the environment. These include depassivation–repassivation theory,[22] local acidification theory,[24] point defect models,[25] and chemical–mechanical models.[2, 27] In these classic theories, a common requirement for initiation of corrosion pits is the adsorption of aggressive ions, such as chloride ions, on the metal surface. This is different from the AC-induced pitting corrosion, where chloride ions are not required or at a quite low content in the environment. Moreover, carbon steels are not passivated in the near-neutral pH, diluted bicarbonate solution. Thus, the initiation of pitting corrosion on pipeline steels cannot be explained and described by the classic theories. Instead, the critical role of AC in the interfacial double-charge layer and the properties of the surface film formed on the steel in the environment is the key to understanding the pitting corrosion phenomenon on pipelines under AC interference.

## References


1. Papavinasam, S., *Corrosion Control in the Oil and Gas Industry*, Elsevier, London, 2013.
2. Michael Baker Jr., *Stress Corrosion Cracking Study: Final Report*, TTO No. 8, Office of Pipeline Safety, U.S. Department of Transportation, Washington, DC, 2005.
3. Cheng, Y.F., *Stress Corrosion Cracking of Pipelines*, Wiley, Hoboken, NJ, 2013.
4. Chu, R., Chen, W., Wang, S.H., King, F., Jack, T.R., Fessler R.R., "Microstructure dependence of stress corrosion cracking initiation in X-65 pipeline steel exposed to a near-neutral pH soil environment," *Corrosion* 60: 275–283, 2004.
5. Liu, Z.Y., Li, X.G., Cheng, Y.F., "In-situ characterization of the electrochemistry of grain and grain of an X70 steel in a near-neutral pH solution," *Electrochemistry Communications* 12: 936–938, 2012.
6. Elboujdaini, M., Wang, Y.Z., Revie, R.W., Parkins, R.N., Shehata, M.T., "Stress corrosion crack initiation processes: pitting and microcrack coalescence," CORROSION/2000, paper no. 00379, NACE, Houston, TX, 2000.
7. Jin, T.Y., Cheng, Y.F., "In-situ characterization by localized electrochemical impedance spectroscopy of the electrochemical activity of microscopic inclusions in an X100 steel," *Corrosion Science* 53: 850–853, 2011.
8. Kushida, T., Nose, K., Asahi, H., Kimura, M., Yamane, Y., Endo, S., Kawano, H. (2001), "Effects of metallurgical factors and test conditions on near-neutral pH SCC of pipeline steels," CORROSION/2001, paper no. 01213, NACE, Houston, TX.
9. Liu, Z.Y., Li, X.G., Cheng, Y.F., "Electrochemical state conversion model for occurrence of pitting corrosion on a cathodically polarized carbon steel in a near-neutral pH solution," *Electrochimica Acta* 56: 4167–4175, 2011.
10. Liu, Z.Y., Li, X.G., Du, C.W., Cheng, Y.F., "Local additional potential model for effect of strain rate on SCC of pipeline steel in an acidic soil solution," *Corrosion Science* 51: 2863–2871, 2009.
11. King, F., Jack, T., Chen, W., Wilmott, M., Fessler, R.R., Krist, K., "Mechanistic studies of initiation and early stage crack growth for near-neutral pH SCC on pipelines," CORROSION/2000, paper no. 361, NACE, Houston, TX, 2000.
12. Song, F.M., "Overall mechanisms of high pH and near-neutral pH SCC, models for forecasting SCC susceptible locations, and simple algorithms for pre-

dicting high pH SCC crack growth rates," CORROSION/2008, paper no. 8129, NACE, Houston, TX, 2008.

13. Fu, A.Q., Cheng, Y.F., "Electrochemical polarization behavior of X70 steel in thin carbonate/bicarbonate solution layers trapped under a disbonded coating and its implication on pipeline SCC," *Corrosion Science* 52: 2511–2518, 2010.
14. Jelinek, J., Neufeld, P., "Temperature effect on pitting corrosion of mild steel in de-aerated sodium bicarbonate-chloride solutions," *Corrosion Science* 20: 489–496, 1980.
15. Lalvani, S.B., Lin, X.A., "A theoretical approach for predicting AC-induced corrosion," *Corrosion Science* 36: 1039–1046, 1994.
16. Brenna, A., *A Proposal of AC Corrosion Mechanism of Carbon Steel in Cathodic Protection Condition*, PhD thesis, Politecnico di Milano, Italy, 2012.
17. Kuang, D., Cheng, Y.F., "Understand the AC induced pitting corrosion on pipelines in both high pH and neutral pH carbonate/bicarbonate solutions," *Corrosion Science* 85: 304–310, 2014.
18. Armstrong, R.D., Coates, A.C., "The passivation of iron in carbonate/bicarbonate solutions," *Journal of Electroanalytical Chemistry* 50: 303–313, 1974.
19. Zhang, G.A., Cheng, Y.F., "On the fundamentals of electrochemical corrosion of X65 steel in $CO_2$-containing formation water in the presence of acetic acid in petroleum production," *Corrosion Science* 51: 87–94, 2009.
20. Heuer, J.K., Stubbins, J.F., "An XPS characterization of $FeCO_3$ films from $CO_2$ corrosion," *Corrosion Science* 41: 1231–1243, 1999.
21. Linter, B.R., Burstein, G.T., "Reactions of pipeline steels in carbon dioxide solutions," *Corrosion Science* 41: 117–139, 1999.
22. Davis, D.H., Burstein, G.T., "The effect of bicarbonate on the corrosion and passivation of iron," *Corrosion* 36: 416–422, 1980.
23. Meng, G.Z., Zhang, C., Cheng, Y.F., "Effects of corrosion product deposit on the subsequent cathodic and anodic reactions of X-70 steel in near-neutral pH solution," *Corrosion Science* 50: 3116–6122, 2008.
24. Galvele, J.R., "Transport process and the mechanism of pitting of metals," *Journal of the Electrochemical Society* 123: 464–474, 1976.
25. Chao, C.Y., Lin, L.F., Macdonald, D.D. (1981), "Point defect model for anodic passive films," *Journal of the Electrochemical Society* 128: 1187–1194.
26. Hoar, T.P., "The production and breakdown of the passivity of metals," *Corrosion Science* 7: 341–355. 1967.
27. Sato, N., "A theory for breakdown of anodic oxide films on metals," *Electrochimica Acta* 16, 1683–1692, 1971.

# 5 Coating Degradation and Failure under AC Interference

## 5.1. Essential coating properties for pipeline integrity maintenance

Coatings in conjunction with CP provide the primary means of protecting buried pipelines from corrosion attack, mechanical damage, and geotechnical threats, while maintaining the integrity of the pipeline system in service. The principle of the strategy is straightforward: the coatings, when they are intact and adhere well to the pipe steel, form the first line of defense by effectively separating the pipelines from the soil environment and, at the same time, reducing the CP current demand. At areas where the coatings have failed, the CP acts as a backup to further protect the pipelines from corrosion.

A wide variety of coatings have been used for pipeline corrosion protection and integrity maintenance over the past decades.[1, 2] These include coal tar, asphalt, and PE-based coatings; FBE or FBE-containing multilayered coatings; multilayered polyolefin (PE or polypropylene) coatings; and composite coatings. Although most of them have successfully served the pipeline industry to maintain pipeline integrity, challenges have been encountered in many exceptional applications, as well as in oil and gas production activities conducted in remote, geographically difficult areas. These areas may feature extremely cold weather; unstable geotechnical conditions such as slopes, earthquakes, permafrost or semi-permafrost; microbial activity; and permeation of water and gases over the long term. Therefore, the industry has always pursued novel and effective pipeline coating technologies to meet specific challenges.

Coating technology and products have experienced unprecedented development, accompanying the progress in petroleum and chemical industries. From the numerous coating products, a proper selection of coatings is critical for the sustainability of pipeline operation and safety in practice. It has been demonstrated that inadequate coating performance is a major contributing factor to increased corrosion and stress corrosion cracking of buried pipelines.[3, 4, 5]

The performance of pipeline coatings is dependent on their properties and characteristics. Coatings possess a wide variety of physical, chemical–electrochemical, and mechanical properties, affecting the function, performance, and service life of the coatings in the field. It is essential to have a complete understanding of the coating properties before a proper product is chosen and used. Primarily, the desired characteristics of a pipeline coating include:

- Effective electrical insulation
- Effective moisture barrier
- Good adhesion to the pipe steel
- Applicable by a method that will not adversely affect the properties of the pipe
- Applicable with a minimum of defects
- Ability to resist generation and development of holidays with time
- Ability to resist damage and deterioration during handling, storage, and installation
- Ability to maintain substantially constant resistivity with time
- Resistance to disbonding and cathodic disbonding
- Resistance to chemical degradation
- Ease of repair
- Retention of physical characteristics
- Nontoxicity to environments

## 5.2. Effect of AC on the resistance of coatings to cathodic disbonding

### 5.2.1. Concept and principle of cathodic disbonding of pipeline coatings

Adhesion is a measure of how firmly a coating is bonded to the substrate surface. Irrespective of most of other properties, a coating with an extremely strong adhesion to the metal substrate usually retains its integrity much longer than one with less adhesion but possessing other strong characteristics.[2] Indeed, the property of coating adhesion is essential in preventing water permeation from affecting the performance of the coating and its service life, and in preventing the problems caused by osmosis, electroendosmosis, and temperature gradient across the coating. Moreover, a high adhesion helps prevent mechanical damage during handling and service, as well as disbondment in service by factors such as soil stress and CP current. Loss of adhesion is usually the first step that allows corrosion and stress corrosion cracking to occur underneath the coating.

Many factors affect the adhesion of the coating to the substrate material, including the chemistry and physics of the substrate surface, coating materials and properties, stresses in coating and/or the substrate, the application procedure, and the service environment. In particular, cathodic disbonding is a type of coating failure related to electroendomosis, where the loss of adhesion between the coating and the metal substrate is due to the products of cathodic reductive reactions that take place on the metal surface. For buried pipelines, CP is always applied to prevent the corrosion of pipe steels. The CP current penetrates the coating or coating defects and reaches the steel surface under the coating. Electrochemical cathodic reactions, which could be the reduction of dissolved oxygen or water, depending on the environmental aeration, are driven by the applied cathodic polarization to generate hydroxyl ions, elevating the pH of the local solution. Since polymeric coatings are vulnerable to alkaline attack, disbonding occurs at the steel–coating interface.

For pipeline coatings, the CP potential recommended by NACE is –850 mV (CSE). Generally, the coating withstands CP of approximately –1,000 mV (CSE) without occurrence of cathodic disbonding. A potential of –1,100 mV (CSE), or a more negative potential, can create conditions causing cathodic disbonding of the coatings, depending on the coating properties and the thickness, dielectric strength, and water resistance.[6] The coatings must have a low moisture vapor transfer rate and a very high adhesion to resist cathodic disbonding.

Cathodic disbonding usually starts at coating defects (e.g., pinholes, holidays), where the CP current passes through the defects. The amount of the current flow depends on the size of the defects. With an increase in the defect size, the CP current drives the coating disbondment to extend away from the defects, causing an increase of the disbonding distance. The depth of the coating defect is also a critical parameter affecting the cathodic disbonding. Generally, a deep, narrow defect would shield the CP current from reaching the defect base, mitigating or even avoiding occurrence of cathodic disbonding. However, corrosion may occur at the bottom of the defect locally as the CP current would be shielded from reaching the steel at the defect base for corrosion protection.

### 5.2.2. Testing for resistance of coatings to cathodic disbonding under AC interference

As stated previously, testing of pipeline corrosion and coating disbondment in the presence of AC interference is subject to a reasonable experimental setup, where at least two electrical circuits (i.e., an electrochemical corrosion measurement circuit and an AC circuit) are included. The mutual interference of the two circuits should be minimized. At the same time, any impact and even damage to the corrosion measurement equipment by the applied AC must be avoided. An experimental setup similar to the one used in AC corrosion measurements in Figure 3-6 was used to test the cathodic disbonding of a coating applied on a pipeline steel specimen, as shown in Figure 5-1. The coated steel specimen serving as a working electrode was connected to a DC power supply, where the steel was connected to the negative pole and a platinum sheet was connected to the positive pole. A DC voltage (e.g., –3.5 V versus SCE) was applied on the working electrode for a certain time. A carbon rod was connected to the AC signal source through a slide rheostat, and the AC current flowing between the working electrode and the carbon rod was varied by adjusting the AC voltage applied on the rheostat. A capacitor was used to block DC signal from flowing into the AC circuit. A holiday with a given diameter was drilled at the center of the coated steel specimen, through the coating, to the steel substrate. The coated specimen was tested in a 3 wt% NaCl solution. The test cell was placed in a water bath maintained at 65°C. Through this setup, the cathodic disbonding tests were conducted on the coated steel specimen containing a coating holiday under simultaneous application and control of DC voltage and AC current density.

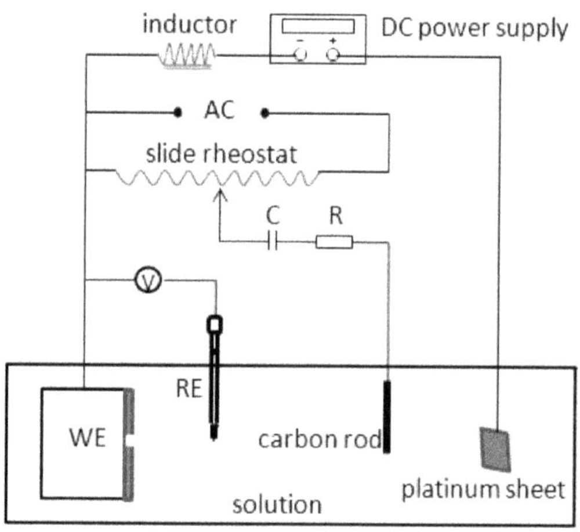


**Figure 5-1.** Schematic diagram of the experimental setup for cathodic disbonding testing conducted on a coated steel specimen.[7]

After testing, the coated steel specimen was taken out of the cell and cooled in air to ambient temperature. A utility knife was used to make eight radial cuts, starting at the holiday through the coating to the steel substrate, at 45° between the cuts. The coating was then peeled off manually until a resistance was met against this force. The peel-off distances of the eight directions were measured, and the average distance was defined as the cathodic disbonding radius.

### 5.2.3. Cathodic disbonding of pipeline coatings under AC interference

To investigate the effect of AC interference on cathodic disbonding resistance, two commonly used pipeline coatings (FBE and liquid epoxy) were applied on an X65 pipeline steel specimen. While FBE represents the primary coating on mainlines, liquid epoxy is the important repair coating applied in the field. Figure 5-2 shows the surface morphology of the FBE-coated steel specimen after cathodic disbonding tests under various AC current densities in 3 wt% NaCl solution. It can be seen that, generally, the disbonding area on the specimen increases with increasing AC current density.

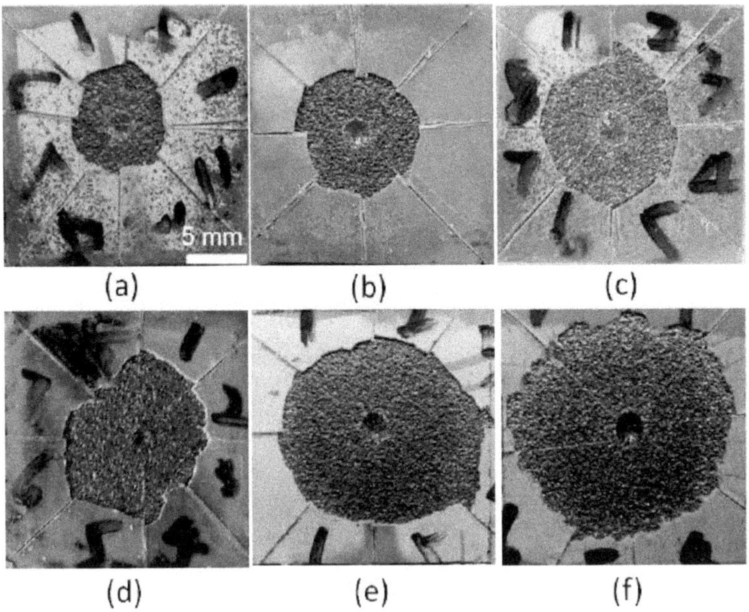


**Figure 5-2.** Surface morphology of FBE-coated steel specimen after cathodic disbonding tests under various AC current densities in 3 wt% NaCl solution: (a) 0 A/m$^2$, (b) 100 A/m$^2$, (c) 200 A/m$^2$, (d) 300 A/m$^2$, (e) 400 A/m$^2$, (f) 500 A/m$^2$.[7]

To quantitatively determine the effect of AC current density on coating disbondment, the disbonding radius from the center of the holiday as a function of AC current density was measured, and the results are shown in Figure 5-3. It can be seen that the disbonding radius of the FBE coating is 3.37 mm in the absence of AC, which is similar to results obtained previously (8). Upon AC application, the cathodic disbonding radius increases. At an AC current density of 500 A/m$^2$, the radius is up to 7.98 mm, more than twice that measured without AC. Thus, AC promotes the coating disbondment.

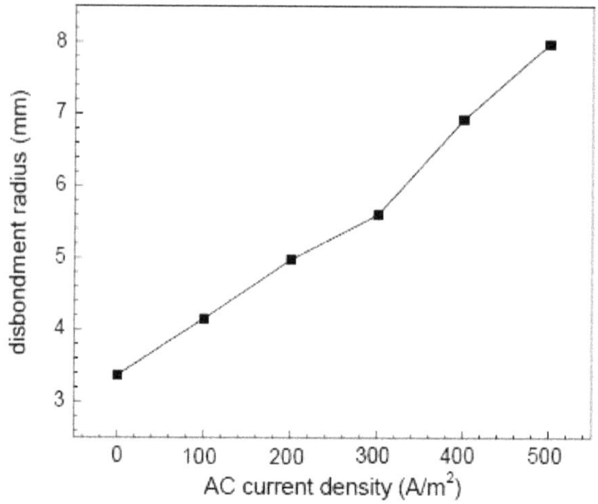


**Figure 5-3. Coating disbondment radius as a function of AC current density measured on an FBE-coated steel specimen.**[7]

As a comparison, the cathodic disbonding tests were conducted on a liquid epoxy–coated X65 steel specimen, and the disbonding radius was measured after testing. Figures 5-4 and 5-5 show the surface morphology of the liquid epoxy–coated steel specimen after cathodic disbonding tests under various AC current densities in 3 wt% NaCl solution and the measured disbonding radius as a function of AC current density, respectively.[7] It can be seen that the disbonding area of the coated specimen increases with the AC current density. The disbonding radius is 4.07 mm in the absence of AC. After AC is applied, the disbonding radius increases. At the AC current density of 500 $A/m^2$, the radius is up to 10.09 mm. Obviously, the AC facilitates the cathodic disbondment of liquid epoxy coating as well. As the cathodic disbonding tests were conducted under identical conditions on both coatings, it can be seen that the liquid epoxy has a larger disbonding radius than FBE at specific AC current densities. Although AC increases the coating disbondment, the FBE shows a higher resistance to cathodic disbonding than does liquid epoxy under AC interference.

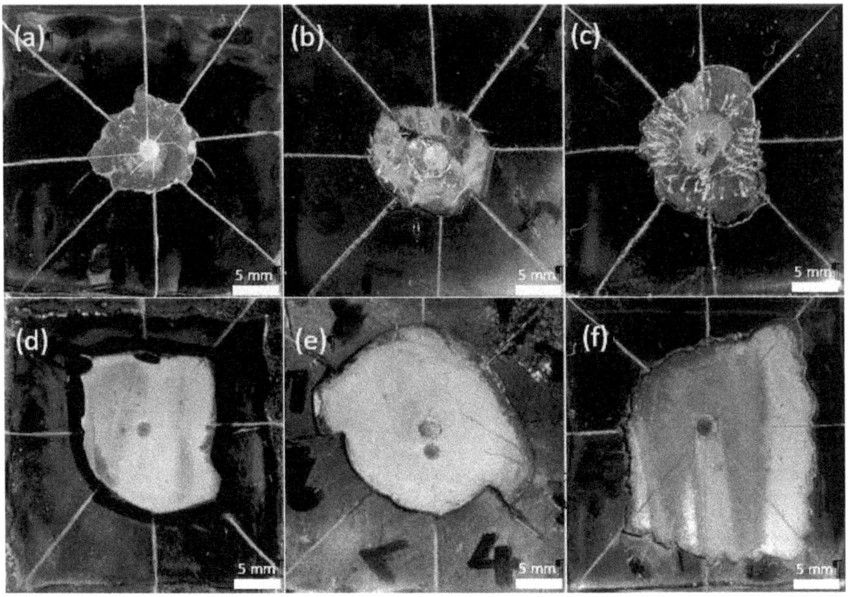


**Figure 5-4.** Surface morphology of liquid epoxy-coated steel specimen after cathodic disbonding tests under various AC current densities in 3 wt% NaCl solution: (a) 0 A/m², (b) 100 A/m², (c) 200 A/m², (d) 300 A/m², (e) 400 A/m², (f) 500 A/m².[7]

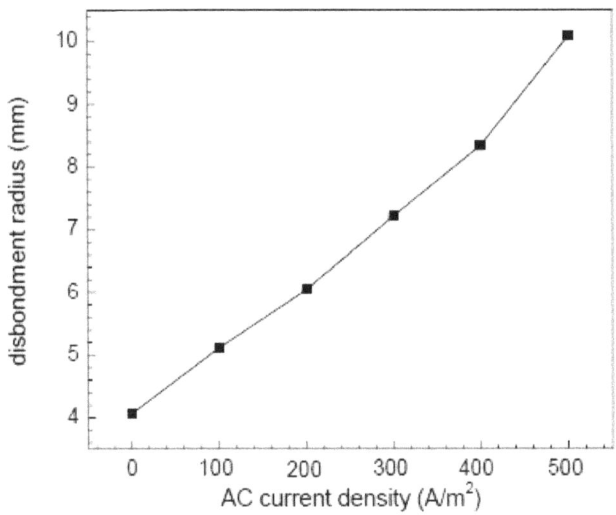


**Figure 5-5.** Coating disbonding radius as a function of AC current density measured on liquid epoxy-coated steel specimen.[7]

# 5.3. Effect of AC on water uptake by pipeline coatings

## 5.3.1. Coating properties associated with water resistance

Water resistance is probably the most important characteristic for a coating, since most other coating properties are related to this characteristic.[6] Generally, organic materials can be penetrated by small molecules such as water molecules. Water passes through the intermolecular spaces of the organic materials, such as polymeric coatings, either remaining there in an absorbed state or passing through the materials.[6] Eventually, moisture comes to an equilibrium, with the content of water entering the coating identical to that of the water evaporating out of the coating surface. Thus, the water content inside the coating tends to achieve a constant value, depending on the temperature and the moisture vapor pressure. As most coatings, including pipeline coatings, are organic materials, they should possess the highest possible resistance to water and moisture permeation to maintain their properties and be effective over a long period of time.

Frequently, coated structures are used in environments containing various types of chemical ions, molecules, gases, and so on. Water permeation also affects the permeability of other species if their sizes are sufficiently small. For example, small molecules such as carbon dioxide or ammonia can be aided by water vapor during their penetration into organic coatings.[6] The permeation of these species would degrade the coating properties remarkably, especially the adhesion and resistance of the coatings. Important properties that are relevant to the water resistance of coatings are discussed in the following paragraphs.

**Water absorption.** Water absorption refers to the amount of water that is absorbed and retained within the intermolecular spaces of a coating. Under certain temperature–pressure conditions, the water content in the coating is in equilibrium with the atmosphere, desorbing water under dry conditions and absorbing water in environments with high humidity or in immersion. All coatings have their own levels of water absorption. When a coating is strongly adhered to substrate metal and there is no disbondment between the coating and the substrate, the moisture absorbed into the coating would not cause corrosion of the metal. At given moisture vapor pressures, the number of absorbed water molecules in the coating is in equilibrium with the number of water molecules desorbing out of the coating. Thus, the content of water molecules in the coating remains constant. Generally, the best corrosion-resistant coating has the lowest water absorption.

**Moisture vapor transfer rate.** The moisture vapor transfer rate is the rate at which moisture vapor transfers through a coating when there is a difference in moisture vapor pressures on one side of the coating compared with the other side. Each coating has its own characteristic moisture vapor transfer rate. In general, the lower the moisture vapor transfer rate, the more protective the coating. The transfer of water or moisture through a coating film refers to the transfer through the intermolecular space of the resin material, rather than through physical imperfections such as pinholes or holidays. When coatings contain these imperfections, water or moisture enter and directly access the substrate metal, causing corrosion or degrading the coating adhesion or both.

**Blistering.** The moisture transfer through a coating depends on the difference of pressures between two sides of the coating. If the coating is tightly adhered to the substrate metal, there is no difference in pressure between the two sides, and the coating is in equilibrium with the moisture in the air. However, if the coating is disbonded from the substrate, a gap is generated under the coating. Moisture vapor tends to transfer into the gap and condense locally. At elevated temperatures, the moisture vapor within the gap results in enough pressure to create a blister. The blistering area could expand with the penetration of moisture vapor into the disbonded gap.

**Osmosis.** Osmosis refers to the passage of water through a coating film from a solution of less concentration to one of greater concentration. Since all organic materials are somewhat permeable to water or moisture, this mechanism applies for all organic coatings, especially when they are subject to immersion in aqueous solutions, condensation, or high humidity. The surface condition of the substrate metal affects the osmosis phenomenon. Generally, the moisture can be pulled through the coating toward the area of contamination. Thus, a clean substrate surface can reduce the osmosis when a coated metal is in immersion or subject to high humidity.

**Electroendosmosis.** Electroendosmosis is defined as the passage of water through a coating film accelerated by an electric potential in the direction with the same charge as the coating. Coatings are generally negatively charged. At coating breakages or defects, excessive negatively charged electrons accumulate, making the metal a negatively charged surface. Water tends to permeate through the coating toward the metal (i.e., the cathode). This mechanism particularly applies to metal structures that are under CP, such as buried pipelines. The cathodically polarized pipeline steel is negatively charged for corrosion protection. However, this could facilitate water existing in the soil environment to penetrate through the coating toward the steel surface.

**Dielectric strength.** Wherever a coating is used for corrosion protection, an important characteristic the coating possesses is good dielectric strength. A coating with excellent dielectric strength can break the electric circuit setup that would otherwise allow corrosion to occur. In general, the lower the moisture absorption, the more favorable the dielectric strength of the coating. A high dielectric strength of the coating can also break the electric circuit that would otherwise allow CP current to flow, shielding the CP current from reaching the substrate metal. Thus, the effect of the dielectric property of a coating is dual.[2]

## 5.3.2. Testing setup to measure water uptake by coatings under AC interference

The water uptake by a coating can be measured using the gravimetric cup method,[9] where a test cell facility is used to determine the water absorption in coatings films such as FBE and liquid epoxy, as shown in Figure 5-6. To conduct the test, a coating film is mounted as a lid on a glass container, which is pre-filled with groundwater or soil solution, simulating the aqueous environment the coating is directly exposed to. Various AC voltages from 0 V to 50 V are applied between a carbon rod installed in the glass container and the coating film, where a conductive adhesive is bonded with the coating. The container is kept in an oven at 65°C to achieve a relative humidity of nearly zero. The container with the coating sealing is weighed before the test and then weighed at 24 h intervals until the end of the test. The obtained weight change is used to calculate the water uptake as a function of time.

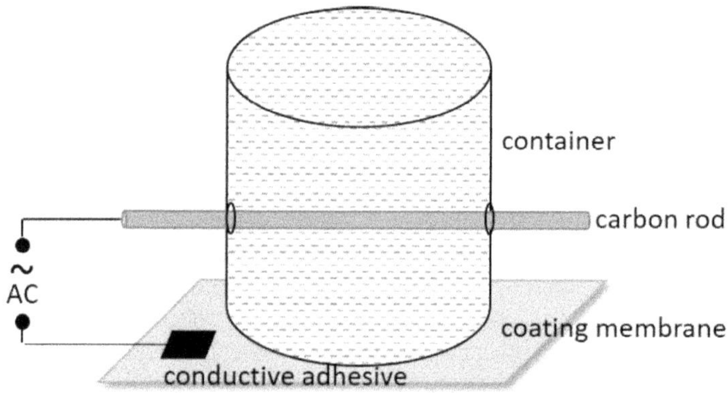


**Figure 5-6.** Experimental setup for measurements of water uptake by a coating membrane.[9]

## 5.3.3. Dependence of the water resistance of coatings on AC interference

Figure 5-7 shows the weight change of the testing containers sealed by an FBE or a liquid epoxy membrane as a function of time under various AC voltages. It can be seen that the total weight of the container decreases with time, indicating the evaporation of water from the solution. Moreover, the weight loss increases with increasing AC voltage. Thus, the application of AC enhances the water uptake by both coatings. Particularly, there is a larger weight loss for the liquid epoxy coating than for the FBE at individual AC voltages.

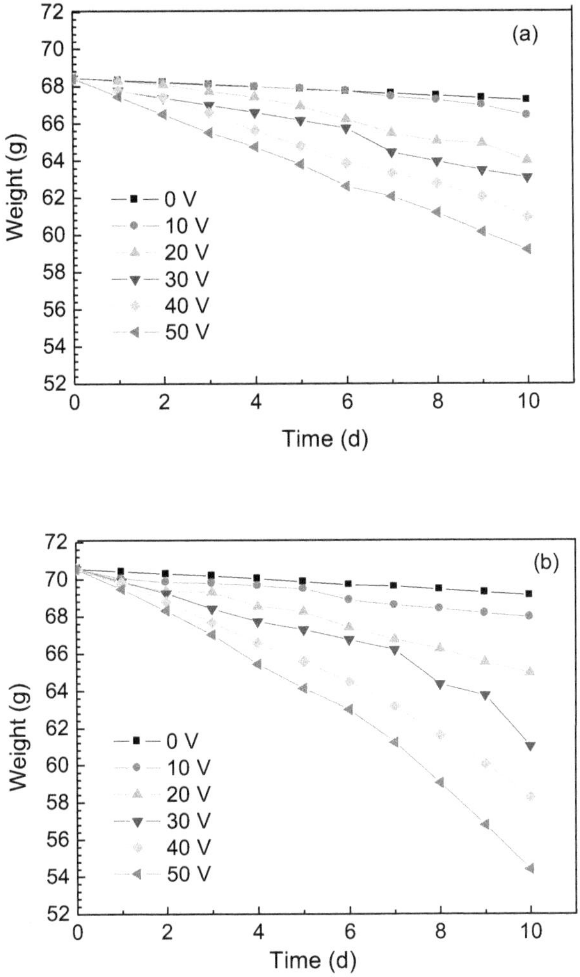


**Figure 5-7.** Time dependence of the weight change of the testing container sealed by a coating membrane under various AC voltages: (a) FBE, (b) liquid epoxy.[7]

The water transmission rate (WTR) has been used to quantitatively characterize the water permeability of polymeric coatings.[10] The WTR is equal to the slope of weight loss versus time, divided by the exposed area of the coating:

$$WTR = \frac{\Delta G}{tA} \tag{5-1}$$

where $\Delta G$ is the weight loss (g), $t$ is time (h), and $A$ is the exposed area of the coating ($m^2$). The WTRs of the two coatings under various AC voltages were calculated, and the results are shown in Figure 5-8. It can be seen that the WTR increases with the AC voltage, indicating an increased water uptake of the coatings in the presence of AC interference. At specific AC voltages, the liquid epoxy coating has a larger water permeation than the FBE.

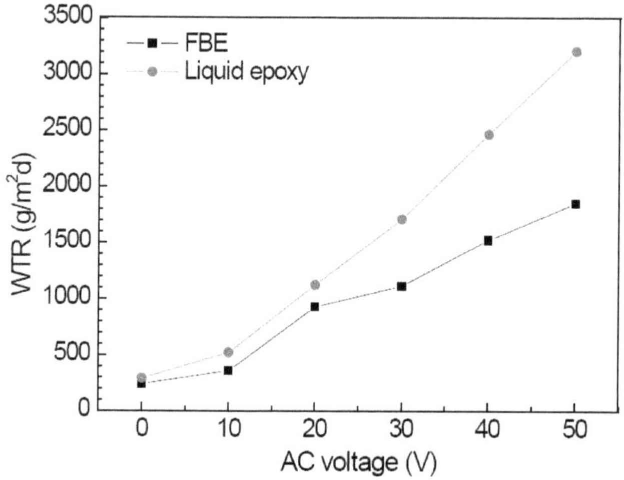


**Figure 5-8.** WTR of FBE and liquid epoxy coatings under various AC voltages.[7]

## 5.4. Effect of AC on electrochemical impedance of the coatings

Figures 5-9 and 5-10 show the Nyquist diagrams measured on FBE-coated and liquid epoxy–coated X65 steel specimens after, respectively, five days and 30 days of immersion in an extracted soil solution (mg/L: $Ca^{2+}$ 20, $Mg^{2+}$ 15, $Na^{+}$ 17, $K^{+}$ 3, $SO_4^{2-}$ 45, $Cl^{-}$ 26, $HCO_3^{-}$ 104, pH 6.7) under various AC voltages.[7] In the absence of AC interference, the Nyquist diagram measured on the FBE-coated specimen features a big, incomplete semicircle. This indicates typical capacitive behavior associated with a high coating resistance of more than $10^9$ Ω·cm². Upon application of various AC voltages, depressed semicircles are recorded. With an increase of the AC voltage, the size of the semicircle decreases, suggesting decreasing coating resistance under AC interference. Compared with FBE, the liquid epoxy coating shows lesser resistance, as indicated by the smaller size of the semicircles under specific AC voltages. After 30 days of immersion, the size of the semicircle further decreases, a result of the barrier property of the coating decreasing with the extended immersion time. The coating resistances decrease to $10^7$ Ω·cm² and $10^6$ Ω·cm² for FBE and liquid epoxy coatings, respectively, in the absence of AC. Moreover, with increases in the AC voltage, the semicircle becomes smaller. Similarly, the semicircle for liquid epoxy is slightly smaller than that for FBE at each AC voltage.

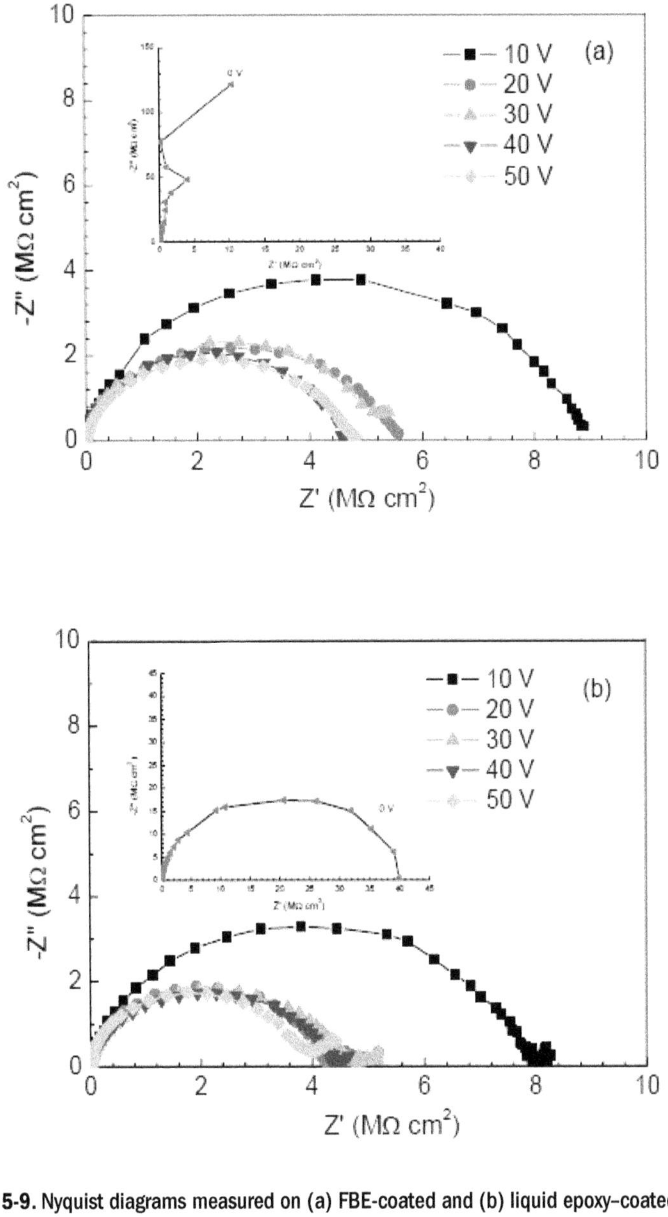


**Figures 5-9.** Nyquist diagrams measured on (a) FBE-coated and (b) liquid epoxy-coated X65 steel specimens after five days of immersion in an extracted soil solution under various AC voltages.[7]

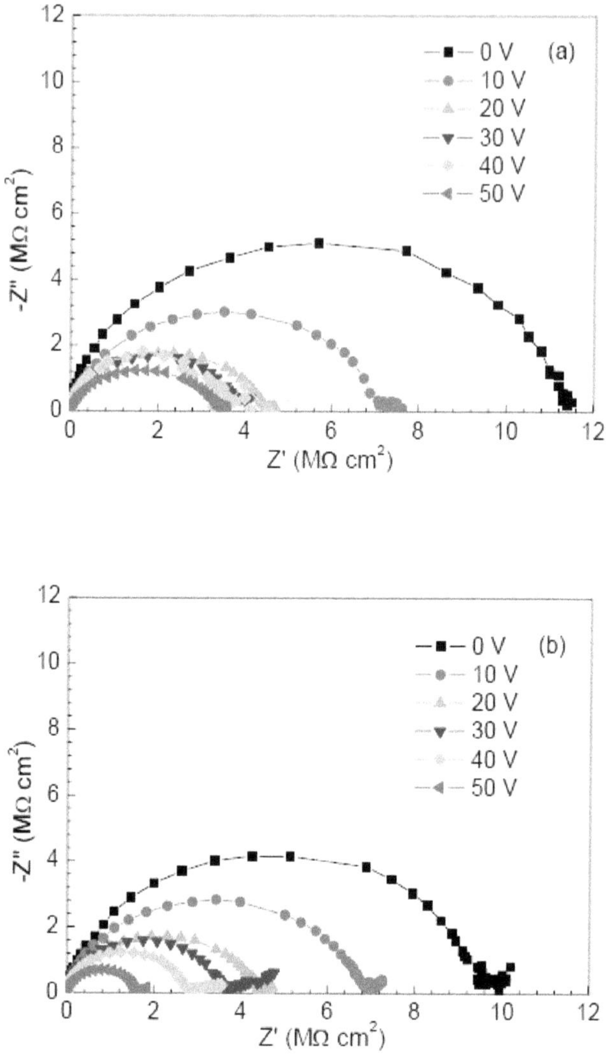


**Figures 5-10. Nyquist diagrams measured on (a) FBE-coated and (b) liquid epoxy-coated X65 steel specimens after 30 days of immersion in an extracted soil solution under various AC voltages.[7]**

Another important phenomenon is that one more semicircle is observed in the low-frequency range after five days of immersion for the liquid epoxy–coated specimen (Figure 5-9b) and after 30 days of immersion for both coatings (Figure 5-10). Generally, the high-frequency semicircle in Nyquist diagrams is associated with the coating properties, and the low-frequency impedance is attributed to the corrosion reaction at the steel–coating interface.[2] Thus, with an increase of immersion time, corrosion occurs on the steel under the coating owing to the permeation of water and corrosive species. Moreover, the steel coated with liquid epoxy starts to have two semicircles after only five days of immersion, indicating a poorer corrosion resistance than FBE.

## 5.5. Coating structure and performance degradation under AC interference

As measured, AC can increase the coating disbondment from the steel substrate, as seen in Figure 5-2. The applied AC facilitates the solution alkalization by electrochemical reductive reactions, elevating the solution pH and inducing loss of the coating adhesion to the substrate steel. The hydrogen evolution and the resulting bubbling due to the enhanced cathodic reaction under AC also contribute to the coating disbondment.

The performance of pipeline coatings in the field is highly dependent on their resistance to water permeation. However, the presence of AC interference can increase the water permeation through both FBE and liquid epoxy coatings. As shown in Figure 5-8, the WTR of FBE under 50 V of AC is about four times that in the absence of AC interference, while the WTR of liquid epoxy coating under the same AC voltage is nearly seven times more than without AC. Moreover, the coating resistance decreases under AC application. Thus, the AC interference can degrade the properties and performance of pipeline coatings. Compared with FBE, the liquid epoxy coating degrades more rapidly under AC interference, as demonstrated by the poorer resistance to cathodic disbonding, more rapid water permeation, and a lower impedance under identical conditions.

To investigate the influence of AC on coating structure, the morphological features of the liquid epoxy coating after 30 days of immersion in a soil-extracted solution in the absence and presence of applied AC voltage of 50 V are characterized by SEM in Figure 5-11. It can be seen that the coating generally has a loose structure, with microcracks present in the coating. Under the effect of AC, the coating becomes looser, and some tortuous, deep microvoids can be observed.

**Figure 5-11.** SEM images of the cross-section of liquid epoxy coating after 30 days of immersion in a soil-extracted solution: (a) no AC, (b) AC voltage of 50 V.[7]

Figure 5-12 shows the cross-sectional view of the FBE coating after 30 days of immersion in the solution in the absence and presence of an applied AC voltage of 50 V. Generally, the FBE has a compact structure, with some micropores. Upon application of the AC voltage, more micropores of a larger size are generated in the coating. Thus, the AC can adversely affect the morphological structure of both coatings, reducing their structural compactness.

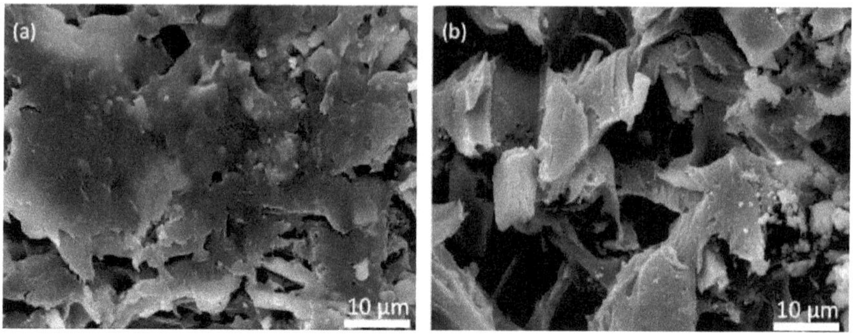


**Figure 5-12.** SEM images of the cross-section of the FBE coating after 30 days of immersion in a soil-extracted solution: (a) no AC, (b) AC voltage of 50 V.[7]

Figures 5-13 and 5-14 show the Fourier transform infrared (FTIR) spectra of the FBE and liquid epoxy coatings, respectively, under various AC voltages after 30 days of immersion in the soil solution. The characteristic functional groups identified are listed in Table 5-1. It can be seen that three characteristic absorption peaks are observed in the spectrum measured on the FBE-coated specimen. The peak at 831 $cm^{-1}$ is attributed to the carbon–oxygen bond in the oxirane group, the peak at 2,873 $cm^{-1}$ is the carbon–hydrogen tension of the methylene group, and the peak at 4,623 $cm^{-1}$ corresponds to a combination bond of the second overtone of epoxy ring stretching with the fundamental carbon–hydrogen at 2,873 $cm^{-1}$.[11] As for the liquid epoxy coating, no peak is observed at 4,623 $cm^{-1}$ due to the introduction of inclusions, which makes the structure of the liquid epoxy coating different from typical epoxy resin coatings. Moreover, two types of water bonds are observed in the spectrum of the FBE coating: highly mobile free water molecules at about 3,600 $cm^{-1}$ and the combination of asymmetric stretching and bending of hydroxyl vibrations located at 5,215 $cm^{-1}$.[12, 13] This combined bond is not observed in the spectrum of liquid epoxy coating.

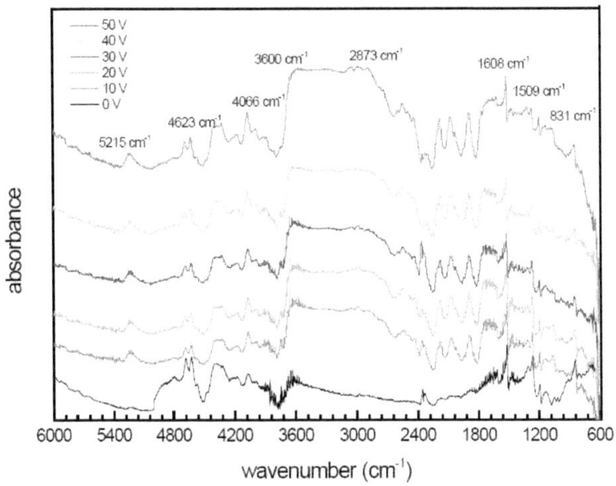


**Figure 5-13.** FTIR spectrum of FBE coating after 30 days of immersion in the soil-extracted solution under various AC voltages.[7]

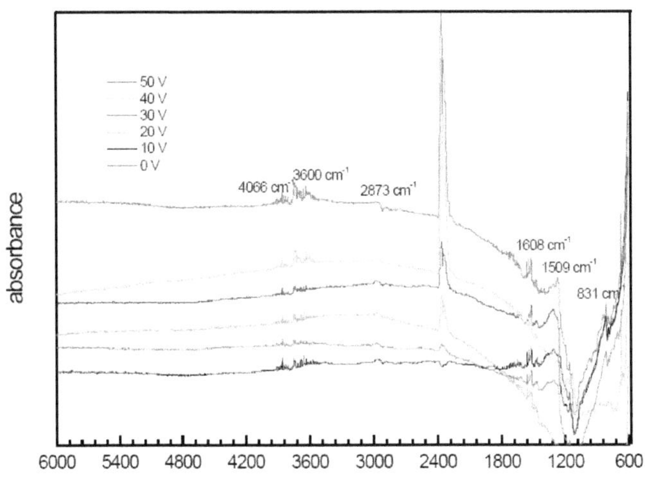


**Figure 5-14.** FTIR spectrum of liquid epoxy coating after 30 days of immersion in the soil-extracted solution under various AC voltages.[7]

| Band ($cm^{-1}$) | Assignment |
|---|---|
| 5215 | Combination asymmetric stretching and bending of O-H [Blanco et al., 2006] |
| 4623 | Overtone of C-H stretching of the aromatic ring [Mijović and Andjelić, 1995] |
| 4066 | Stretching C-H of aromatic ring [George et al., 1991] |
| 3600 | O-H stretching [González et al., 2012] |
| 2873 | C-H stretching [Chike et al., 1993] |
| 1608 | Stretching C=C of aromatic rings [Mertzel and Koenig, 1986] |
| 1509 | Stretching C-C of aromatic rings [Meure et al., 2010] |
| 831 | Stretching C-O-C of oxirane group [Meure et al., 2010] |

**Table 5-1.** Characteristic bonds of epoxy coatings in the FTIR spectrum.[7]

It can be seen from Figure 5-13 that the bonds at 3,600 $cm^{-1}$ and 5,215 $cm^{-1}$, which represent the absorbed water in the FBE coating, are enhanced in their absorbance intensities with increasing AC voltage. This indicates that the water uptake by FBE is increased under the AC application. The result provides direct evidence that the water permeation into FBE coatings increases with increasing AC voltage. Moreover, the absorption peaks located between 1,800 $cm^{-1}$ and 2,400 $cm^{-1}$ are not the characteristic functional groups of epoxy resin but are observed under AC application. It is thus assumed that the structure of FBE coatings is changed in the presence of AC interference. These peaks are related to C≡C stretching, indicating degradation of the coating under AC. For liquid epoxy coating, the bond at 3,600 $cm^{1}$ represents the absorbed water and is increased in its intensity with increasing AC voltage. Thus, AC facilitates water uptake in the coating. Similar to the observations for FBE, a new absorption peak is observed at 2,400 $cm^{-1}$, which is associated with the C≡N functional group. The intensity of the peak increases with the AC voltage. Thus, AC application changes the structure of liquid epoxy coating, affecting its performance.

## 5.6. A conceptual model for coating degradation under AC interference

The mechanistic aspect of the AC-induced coating degradation can be illustrated by a conceptual model, as shown in Figure 5-15. Generally, epoxy resin is a highly polar material, containing hydroxyl groups.

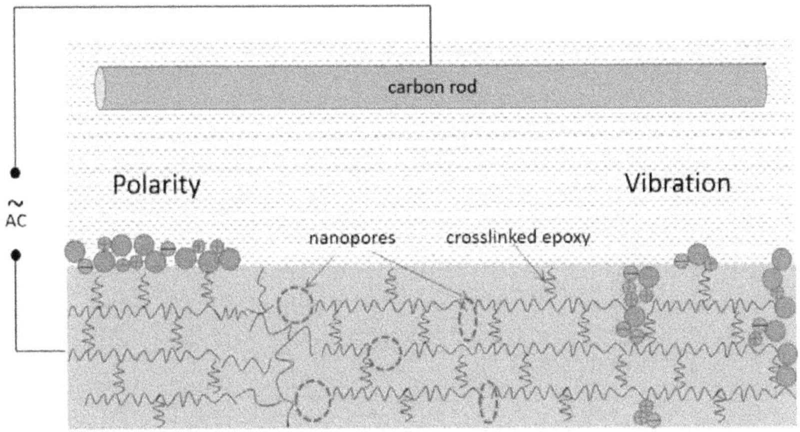


**Figure 5-15. Schematic diagram of the conceptual model illustrating coating degradation in the presence of AC interference.**[7]

Water is also a polar molecule and can form hydrogen bonding with the hydroxyl groups. Thus, the epoxy coatings are permeable to water. In the presence of AC, an electric field is generated on the coating, which is able to polarize the functional groups, increase the coating polarity and enhance the water permeation. Thus, a layer of water molecules and ions from the aqueous solution is formed by bonding with the polar groups on the coating surface, which is called the "polar effect" of the AC interference. It has been proposed that the AC application could generate a "vibrating effect" due to the low-frequency alternating electric field exerting on the reactants in the solution.[14] The mobility of solvated ions in the solution is usually slow and is thus sensitive to the low-frequency electric field force. Upon application of AC, the ions are accelerated by the electric field. As a result, the kinetic energy of polarized ions and water is increased, enhancing their permeation into the coating through nanopores and microvoids. The accelerated ions, once they enter the coating, could impact the functional groups and change the epoxy structure, as shown in the FTIR spectra, where new absorption peaks are observed in the presence of AC. With the increase in AC voltage, the enhanced AC electric field polarizes more functional groups in the coating to absorb water and ions, resulting in coating degradation.

## References


1. Niu, L., Cheng, Y.F., "Development of innovative coating technology for pipeline operation crossing the permafrost terrain," *Construction and Building Materials* 22: 417–422, 2008.
2. Cheng, Y.F., Norsworthy, R., *Pipeline Coatings*, NACE, Houston, TX, 2017.
3. National Energy Board, *Stress Corrosion Cracking on Canadian Oil and Gas Pipelines*, Report of the Inquiry, MH-2-95, Calgary, Alberta, 1996.
4. National Energy Board, *Focus on Safety and Environment: A Comparative Analysis of Pipeline Performance 2000–2009*, Calgary, Alberta, 2011.
5. Cheng, Y.F., *Stress Corrosion Cracking of Pipelines*, Wiley, Hoboken, NJ, 2013.
6. Munger, C.G., Vincent, L.D., *Corrosion Prevention by Protective Coatings*, 2nd ed., NACE, Houston, TX, 1999.
7. Kuang, D., Cheng, Y.F., "Coating degradation in the presence of alternating current interference," CORROSION/2016, paper no. 07153, NACE, Houston, TX, 2016.
8. Howell, G., Cheng, Y.F., "Characterization of high performance composite coating for the northern pipeline application," *Progress in Organic Coatings* 60: 148–152, 2007.
9. Fu, A.Q., Cheng, Y.F., "Characterization of the permeability of a high performance composite coating to cathodic protection and its implications on pipeline integrity," *Progress in Organic Coatings* 72: 423–428, 2011.
10. Barbucci, A., Delucchi, M., Cerisola, G., "Organic coatings for concrete protection: liquid water and water vapour permeabilities," *Progress in Organic Coatings* 30: 293–297, 1997.
11. Chike, K., Myrick, M., Lyon, R., Angel, S., "Raman and near-infrared studies of an epoxy resin," *Applied Spectroscopy* 47: 1631–1635, 1993.
12. Musto, P., Mascia, L., Ragosta, G., Scarinzi, G., Villano, P., "The transport of water in a tetrafunctional epoxy resin by near-infrared Fourier transform spectroscopy," *Polymer* 41: 565–574, 2000.
13. Blanco, I., Cicala, G., Costa, M., Recca, A., "Development of an epoxy system characterized by low water absorption and high thermomechanical performances," *Journal of Applied Polymer Science* 100: 4880–4887, 2006.
14. Xu, L.Y., Su, X., Yin, Z.X., Tang, Y.H., Cheng, Y.F., "Development of a real-time AC/DC data acquisition technique for studies of AC corrosion of pipelines," *Corrosion Science* 61: 215–223, 2012.

**Other References**

George, G.A., Cole-Clarke, P., St. John, N., Friend, G. (1991), "Real-time monitoring of the cure reaction of a TGDDM/DDS epoxy resin using fiber optic FT-IR," *Journal of Applied Polymer Science* 42: 643–644.

González, M.G., Baselga, J., Cabanelas, J.C. (2012), *Applications of FTIR on Epoxy Resins-Identification, Monitoring the Curing Process, Phase Separation and Water Uptake*, IntechOpen, New York.

Mertzel, E., Koenig, J.L. (1986), "Application of FT-IR and NMR to epoxy resins," *Advances in Polymer Science* 75: 73–81.

Meure, S., Wu, D.Y., Furman, S.A. (2010), "FTIR study of bonding between a thermoplastic healing agent and a mendable epoxy resin," *Vibrational Spectroscopy* 52: 10–21.

Mijović, J., Andjelić, S. (1995), "A study of reaction kinetics by near-infrared spectroscopy: 1. Comprehensive analysis of a model epoxy/amine system," *Macromolecules* 28: 2787–2796.

# 6 Cathodic Protection Performance in the Presence of AC Interference

Buried pipelines are protected from corrosion attack by coatings and CP. The performance of the coatings and CP are highly related. Generally, the properties and performance condition of a coating should be considered when the required CP current is too high to protect a structure. A high-performance coating can reduce the CP current demand for corrosion protection. The amount of the required CP current is dependent on the coating properties and the percentage of the coating damage. When the coating has failed (e.g., the coating contains defects or is disbonded from the substrate steel), the CP current should reach the steel surface for corrosion protection.

With the increasing collocation of buried pipelines with HVAC power lines or AC-powered rail transit systems, the interference of AC on CP performance has been paid particular attention. In 1986, AC corrosion caused the failure of a high-pressure gas pipeline in Germany, although the CP current density was maintained at 1.5 $A/m^2$ to 2 $A/m^2$ and the ON potential ($E_{ON}$) was between −1.8 V and approximately −2.0 V (CSE).[1] In 2002, an AC corrosion failure occurred on a buried pipeline in the U.S., where the pipeline was protected with the OFF potentials ($E_{OFF}$) between −1.230 V (CSE) and approximately −1.056 V (CSE) on the steel coupon.[17] However, the coupon AC current density was as high as 144.13 $A/m^2$. In 2003, AC corrosion was reported on a buried pipeline with a CP potential of −1.2 V (CSE) and an AC current density of 130 $A/m^2$.[2] In addition to the listed events, numerous evidence has demonstrated that AC corrosion can occur on cathodically protected pipelines where the CP protection criterion was followed. The mutual effect of AC and CP on pipeline corrosion has also been demonstrated in field investigations conducted over 10 months.[3] It was shown that AC corrosion of the pipeline could be prevented if the average $E_{ON}$ over time was more positive than −1.2 V (CSE) and the average AC voltage was below 15 V, as well as the $i_{AC}$ being smaller than 30 $A/m^2$ or the protection current density being below 1 A/

m$^2$. It was concluded that a high DC current density ($i_{DC}$) in combination with $i_{AC}$ > 30 A/m$^2$, and a high $i_{AC}$ in combination with $i_{DC}$ > 1 A/m$^2$ led to AC corrosion. Accordingly, a negative $E_{ON}$ and a high VAC led to AC corrosion. It was possible to prevent AC corrosion at high or low $i_{DC}$, while significant AC corrosion occurred at intermediate $i_{DC}$ above 1 A/m$^2$. Therefore, the AC would affect the CP performance, compromising the capability of the applied CP under the industry-accepted criteria for corrosion protection of pipelines. Development of new CP criteria for corrosion protection in the presence of AC interference is critical to maintaining the integrity of the pipelines.

## 6.1. Principle of impressed current CP on pipelines

CP is a technique to control the corrosion of a metal structure by making it the cathode of an electrochemical corrosion cell.[4] The electrochemical cell for a corrosion reaction includes an anode (i.e., the segment of the metal structure where oxidation [or corrosion] of the metal occurs); a cathode (i.e., the segment of the structure where the electrochemical reductive reaction occurs); and an electrically conductive pathway, such as an electrolyte. The CP is an electrochemical technique enabling change of the metal (which corrodes in service environments in the absence of CP or other protective measures) from anode to cathode, thus preventing corrosion. Mechanistically, CP is a process to cathodically polarize the protected metallic structure.

CP is probably one of the most important measures used for corrosion control for engineering structures. Principally, a metal surface that is under CP can be kept in a corrosive environment without deterioration for an indefinite time if the CP is appropriately maintained.[5] For example, CP application can reduce the corrosion rate of pipeline steels due to soil corrosiveness below 10 μm/y, which is the maximum accepted damage regardless of the pipe wall thickness, according to international standard BS EN 12954.[6] There are two types of CP systems: impressed current cathodic protection (ICCP) and sacrificial anode cathodic protection (SACP).

### 6.1.1. Impressed current cathodic protection

The ICCP system uses an external source of electrical power, which is often a transformer–rectifier connected to AC power, where its negative terminal is connected to the structure to be protected. Its positive terminal is connected to an auxiliary electrode (i.e., the anode), which is usually made of iron, graphite, or metal oxides, located some distance away from the protected structure. The current flows from the auxiliary electrode through the electrolyte (e.g., seawater, soils) to the protected structure. In ICCP, the applied voltage is not critical; it just needs to supply an adequate DC current density via the rectifier to all parts of the protected structure. The main advantage of the ICCP system is its automatic control feature, which continuously monitors and varies the current required for corrosion protection of the structure. The protection capacity is usually sufficient for large, complicated engineering structures. However, ICCP systems generate high inspection and maintenance costs and have a high risk of causing both stray current interference and overprotection.

### 6.1.2. Sacrificial anode cathodic protection

The method to apply SACP is to connect the metal structure to be protected with another metal that is more easily corroded in the environment and thus acts as the anode. When the two metals are electrically connected to each other in the environment, electrons flow from the more active (i.e., the more easily corroded) metal to the other (i.e., the metal to be protected) owing to the difference in their electric potentials. When the more active metal (anode) supplies current, it dissolves (corrodes) into ions, which go to the electrolyte and, at the same time, produces electrons, which are received by the less active metal (i.e., the protected metal structure) through the metallic connection. As a result, the protected metal is negatively polarized and, hence, protected against corrosion. For SACP to work, the anode must possess a more negative electrode potential than that of the cathode (i.e., the structure to be protected). Galvanic series of metals, where the electromotive series of various metals is listed, can be used to select the anode metal for SACP application.[7] The advantages of the SACP system include that an external power source is not needed, it is easy to install the system, the overprotection usually does not happen, and the inspection and monitoring is simple for trained personnel. There are a few disadvantages of SACP, including a limited current capacity based on the mass of the anode, ineffectiveness of corrosion protection in high-resistivity environments, and the need for periodic reinstallation of new anodes to replace the consumed (corroded) anodes.

### 6.1.3. Use of ICCP in pipeline corrosion protection

The first applications of an ICCP system for the protection of underground structures took place in the U.K. and the U.S. in 1910–1912.[8] The criterion for CP applied on steel pipelines (i.e., –850 mV [CSE]), was proposed by Kuhn.[9] Since then, it has been accepted and used worldwide on steel pipelines and structures in various soils and water bodies. Kuhn's criterion refers to the potential criterion for DC corrosion protection.[8] Up to the mid-1980s, corrosion failure on pipelines had not been attributed to AC corrosion. It was believed that AC corrosion, once happening on pipelines, could be controlled by the application of CP in accordance with the protection potential criterion. In 1986, corrosion failure on a PE-coated pipeline caused by induced AC interference resulting from a nearby AC-powered rail transit system was reported in Europe, although the applied CP on the pipeline satisfied the protection potential criterion.[8] It is now widely accepted that pipelines under the –850 mV (CSE) protection potential criterion would suffer, at least partially, from corrosion attack under AC interference.

In practice, ICCP is more effective at mitigating the AC corrosion of buried pipelines owing to its flexible voltage and current outputs,[10, 11] where the cathodic current is impressed on pipelines through a DC power source by two mechanisms. In the first mechanism, the voltage or current of the output terminals of the DC power source keeps constant, while the external resistance of the CP circuit varies to affect the CP potential. For example, the resistivity of soils changes during dry–wet cycles. As a result, the DC power source needs to adjust to maintain the protection potential value. In the second mechanism, a constant potential or current from the output terminals is controlled by maintaining the pre-set pipe-to-soil potential versus a reference electrode. If this potential changes with environmental conditions, the output current either increases or decreases. The first mechanism is termed an "internal feedback" CP system, where the DC power source unit maintains a constant output of voltage or current by an internal stabilization circuit. The desired CP potential on pipelines can be achieved by manual or semiautomatic adjustment only, as shown in Figure 6-1. In comparison, the second type of CP system can monitor CP potential in real time and maintain the value automatically by an "external feedback" circuit through an on-site reference electrode. However, the "external feedback" CP system is more complex in practical application for long-distance transmission pipelines, while the "internal feedback" CP system is economic and convenient for installation in the field.

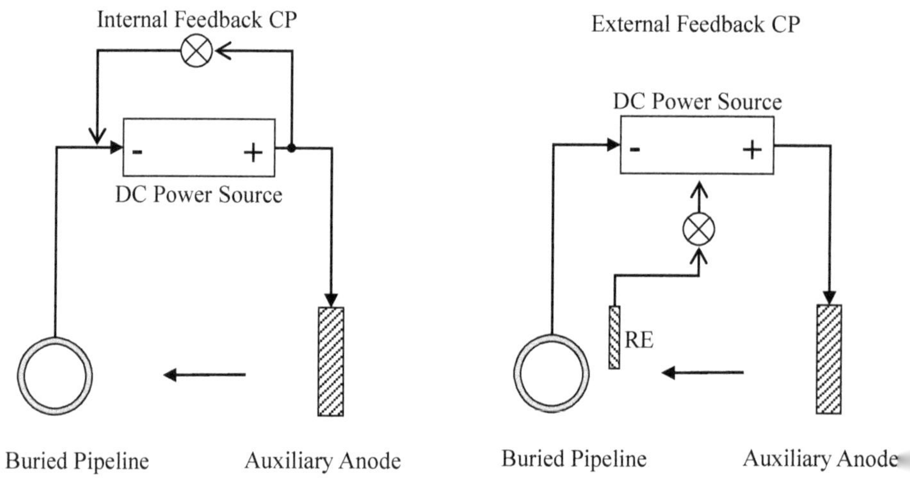


**Figure 6-1.** Schematic illustration of the "internal feedback" and "external feedback" CP systems.[12]

## 6.2. CP shielding under disbonded coating in the presence of AC interference

No matter what type of coating is applied on pipelines, it is important that the entire pipeline is checked for coating imperfections, which include isolated or linked defects (i.e., pinholes or holidays), disbondment of a defect-free coating, disbondment starting at a defect, and missing coating over a certain area.[13] This will ensure a minimum number of coating defects or disbondment and much better performance from the designed CP system. Particularly, the compatibility of the coating with CP becomes critical for pipeline corrosion protection, especially in the presence of AC interference.

### 6.2.1. What is CP shielding?

When a coating is disbonded at small defects, such as pinholes or holidays, the CP current can become shielded, either fully or partially, from reaching the disbonding crevice, especially at the crevice bottom. As a result, the CP is not able to protect the steel under the disbonded area where a corrosive environment is generated. This is called "CP shielding." Another scenario causing CP shielding on pipelines in the field is the disbondment of a defect-free coating membrane due to either an inadequate application process or the lost adhesion of the coating to the steel substrate during service. For example, spirally wrapped tape coatings were found frequently to disbond over pipeline welds. In this situation, the CP shielding is attributed to the coating structure and property,[14] where the coating itself blocks the CP current from reaching the underneath substrate metal. Generally, nonshielding coatings are those that do not prevent permeation of the CP current to the steel substrate through the disbonded coating.[15]

Norsworthy reported that up to 85% of all external corrosion of pipelines was under disbonded coatings with CP shielding. Thus, the coatings used for pipeline corrosion protection should be compatible with CP, as advised by a number of national and international codes and standards.[16] In other words, pipeline coatings should not create cathodic shielding. The relevant regulations from the Code of Federal Regulations (CFR) of the U.S. Department of Transportation are 49 CFR 192[18] and 49 CFR 195,[19] which apply to gas and liquid pipelines, respectively. It is stated in 49 CFR 192 that, under certain conditions, the pipe must be protected against external corrosion by a nonshielding coating. The regulation further indicates that the type of nonshielding coating that should be considered is FBE and, in some cases, liquid epoxy. However, all coatings, including pipeline coatings, are somewhat electrically resistant, especially the so-called good coatings.

Munger acknowledged that water resistance is one of the most important characteristics possessed by corrosion-resistant coatings.[20] High water resistance in a coating is associated with a number of other properties or performance measures, including a high resistance to water absorption, a low moisture vapor transfer rate, a high resistance to osmosis and electroendosmosis, a strong dielectric strength, and resistance to ionic passage. A good coating, based on its property requirement, will cause CP shielding.

Generally, a coating ranked as "good" depends on whether the coating possesses a combination of good properties. In addition to water resistance, the coating should be excellent in adhesion, chemical resistance, flexibility, weather resistance, and resistance to dirt pickup and bacterial growth, among other factors. A strong adhesion to the substrate is always the key for a coating to retain its integrity during the service period. For pipeline coatings, CP compatibility must be considered in conjunction with their adhesion to the steel substrate. If the coating's adhesive property is ignored, it does not make sense to discuss its CP shielding or nonshielding behavior. It has been determined that corrosion and stress corrosion cracking do not occur on pipelines when the coating maintains proper adhesion to the substrate, regardless of the CP current permeability. Therefore, whether a coating is compatible with CP matters only when the coating is disbanded.[21]

### 6.2.2. Effect of AC on CP permeation into the coating disbondment: Testing results

To investigate the CP current permeation into disbonded coating in the presence of AC interference, the experimental setup shown in Figure 6-2 was used,[22] where an artificial disbondment was created under an FBE membrane (200 μm in thickness) using a double-sided self-adhesive tape. The gap between the FBE and the steel was defined as the disbonding thickness. Tape with a known thickness was layered to establish the desired gap, which was verified with a coating thickness gauge. The FBE membrane was applied to the tape, which was then applied on the steel surface. The tape was removed to form an artificial FBE disbondment over the steel. The boundaries of the steel–coating assembly were sealed with an epoxy resin. A 10-mm diameter hole was opened on the coating to simulate the holiday. Six potential and pH microprobes were installed at distances of 30, 60, 90, 120, 150, and 180 mm from the holiday. The distance of the probing position to the open holiday was defined as the disbonding depth. A near-neutral pH, 0.01 M bicarbonate solution was used to simulate the electrolyte trapped under the disbonded coating. The solution was purged with 5% carbon dioxide–nitrogen continuously to achieve a pH of 6.5. CP and AC were applied at the holiday with the same circuit connection shown in Figure 5-1.

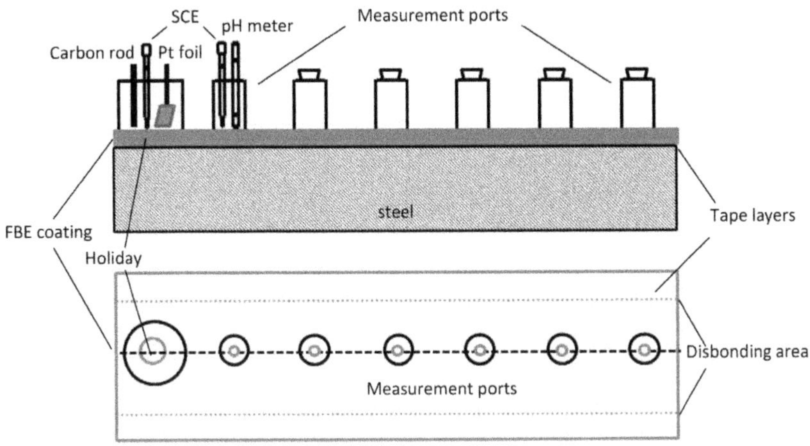


**Figure 6-2. Schematic diagram of the experimental setup simulating a disbonding crevice under FBE coating and the potential and solution pH measurements ports.**[22]

Figure 6-3 shows the potential distribution under the disbonded coating with a disbonding thickness of 120 μm at a CP potential of –875 mV (SCE) (i.e., -850 mV [CSE]) in the absence and presence of various AC current densities in the solution.[23] It can be seen that, in the absence of AC, CP can be shielded from reaching the disbonding crevice along the depth direction. At the open holiday (the distance to the holiday is 0 mm in the figure), the measured potential is the applied CP potential of –875 mV (SCE) up to 48 h of testing. However, at 30 mm from the holiday, the potential is shifted less negatively to about –800 mV (SCE). With increasing distance from the holiday, the potential becomes less negative. When the distance is up to 60 mm, the potential does not change with further increase in the disbonding depth. It can thus be seen that the coating disbondment shields the CP permeation, which is mainly due to the geometrical limitation of the crevice, decreasing the CP performance at the bottom of the crevice. The critical depth determined under the present condition in the absence of AC is 60 mm, above which CP would not be able to reach, and the steel is at the corrosion potential of –755 mV (SCE).

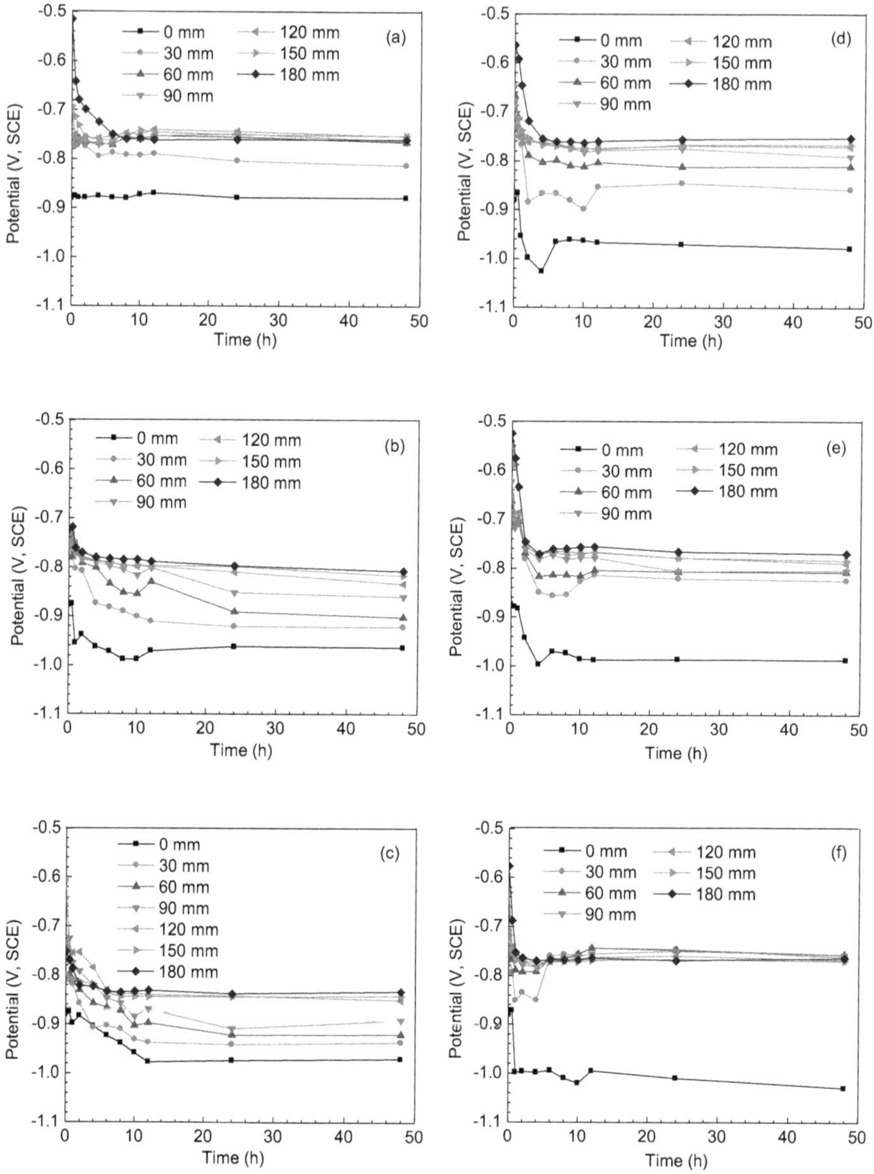


**Figure 6-3.** Local potential profile under the disbonded coating with a disbonding thickness of 120 μm at the CP potential of –875 mV (SCE) (i.e., –850 mV [CSE]) in the absence and presence of various AC current densities in 0.01 M sodium bicarbonate solution: (a) 0 A/m$^2$, (b) 100 A/m$^2$, (c) 200 A/m$^2$, (d) 300 A/m$^2$, (e) 400 A/m$^2$, (f) 500 A/m$^2$.[23]

Upon AC application, the DC potential of the steel shifts negatively. As the AC current density increases, the potential shifts more negatively, as can be seen from the potential measurements at the holiday (i.e., the distance of 0 mm) in Figures 6-3b to 6-3f. Moreover, in the direction of the disbonding depth, the potential is less negative. Obviously, the CP shielding due to the geometry of the disbonding crevice also applies when AC interference is present. It is interesting to find that at small AC current densities, such as 100 $A/m^2$, the critical crevice depth where the CP cannot reach is about 150 mm from the open holiday. With increasing AC current density, the critical crevice depth decreases. When the AC current density is up to 200 $A/m^2$, 300 $A/m^2$, and 400 $A/m^2$, the critical depths are 120 mm, 90 mm, and 30 mm, respectively. When the AC current density is 500 $A/m^2$, no CP can permeate into the potential monitoring regions. Thus, at small AC current densities, the AC results in more negative potentials at specific disbonding depths than the potentials measured in the absence of AC. However, this effect is diminished with increasing AC current density. When comparing the potential profiles measured in the absence of AC (Figure 6-3a) and at the AC current density of 400 $A/m^2$, there is a high similarity under the disbonding crevice, except that the latter has a more negative potential at the open holiday. At a large AC current density, such as 500 $A/m^2$, the CP is completely shielded from reaching the disbonding crevice.

Figure 6-4 shows the solution pH profile under the disbonded coating with a disbonding thickness of 120 μm at the CP potential of –875 mV (SCE) in the absence and presence of various AC current densities in a 0.01 M sodium bicarbonate solution.[23] It can be seen that, in the absence of AC, the CP application enhances the solution pH at the open holiday to about 8.4, compared with the original solution pH of 7.5. Under the disbonded crevice, there is a lower solution pH than at the holiday. With the increase of the disbonding depth, the pH is decreased. Upon AC application, the solution pH at the open holiday is elevated obviously, but the solution pH inside the disbonding crevice is lower. As the distance to the holiday increases, the solution pH further drops. Moreover, with the increase in AC density, the solution pH under the disbonding crevice decreases.

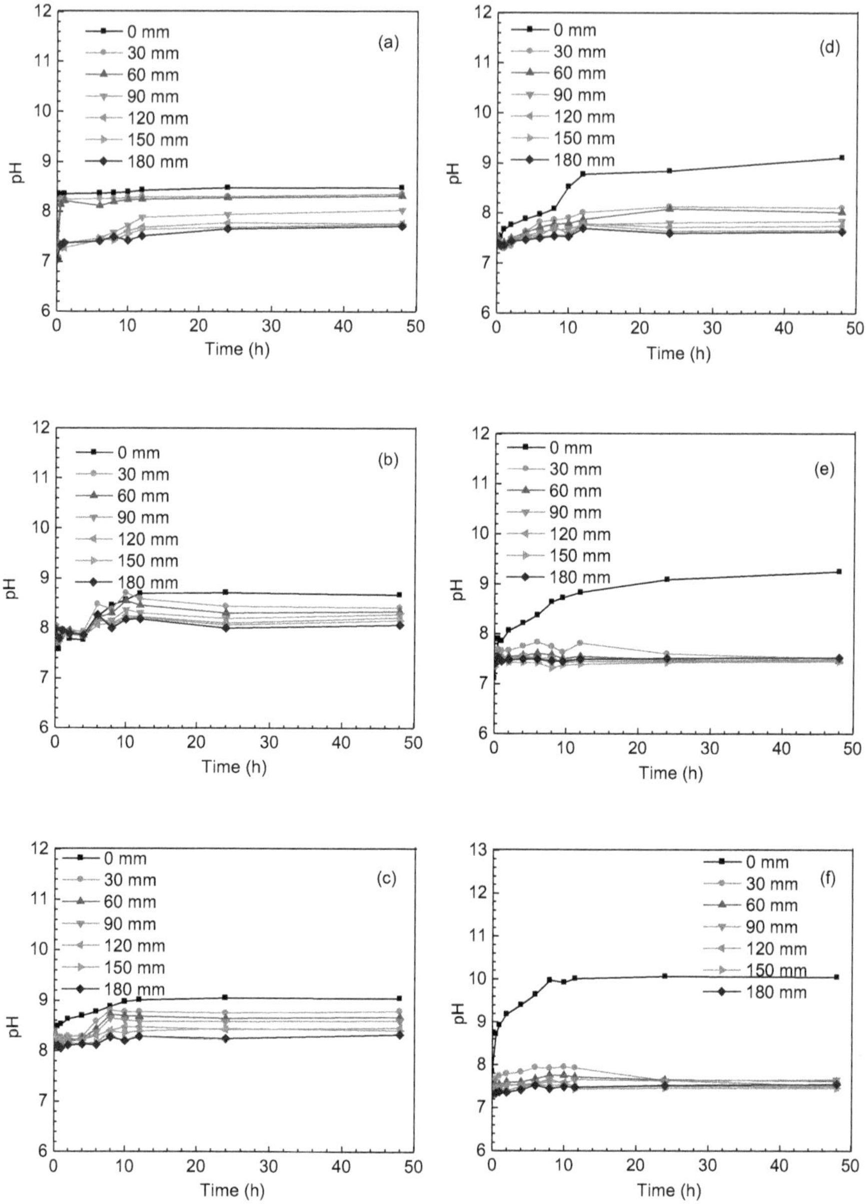


**Figure 6-4.** Solution pH profile under the disbonded coating with a disbonding thickness of 120 μm at the CP potential of –875 mV (SCE) in the absence and presence of various AC current densities in a 0.01 M sodium bicarbonate solution: (a) 0 A/m$^2$, (b) 100 A/m$^2$, (c) 200 A/m$^2$, (d) 300 A/m$^2$, (e) 400 A/m$^2$, (f) 500 A/m$^2$.[23]

## 6.2.3. Effect of AC on CP permeation into the coating disbondment: Mechanistic analysis

To look at the effect of AC on CP permeation into the disbonded coating in more detail, the local solution pH inside the disbonding crevice at 180 mm from the open holiday as a function of the AC current density is shown in Figure 6-5.[23] It can be seen that the solution pH at the crevice is elevated by the increasing AC current density. Upon application of AC, anodic oxidation and cathodic reduction reactions occur during the positive and negative half cycles, respectively. During corrosion of pipeline steel in near-neutral pH bicarbonate solution trapped under the disbonded coating, the anodic and cathodic reactions are the iron oxidation and reduction of water (in the absence of oxygen in the environment), respectively.[21] The cathodic reaction generates $OH^-$ ions to elevate the solution pH. Moreover, as the AC current density increases, the water reduction is enhanced at more negative potentials in AC negative cycles. Thus, the solution becomes more alkaline (Figure 6-4).

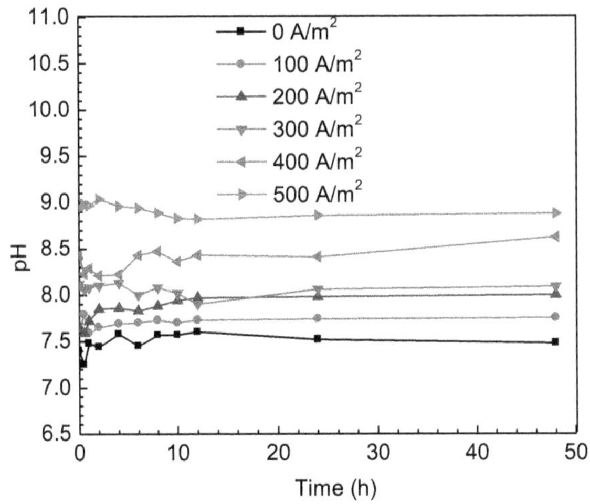


**Figure 6-5. Effect of AC current density on local solution pH inside the disbonding crevice at 180 mm from the open holiday (disbondment thickness: 120 μm) in 0.01 M $NaHCO_3$ solution.[23]**

The potential of the steel at the open holiday shifts negatively upon AC application, as seen in Figure 6-3. This is attributed to the accelerated diffusion of cations generated during the corrosion of the steel under the AC-enhanced electric field, as discussed previously in Chapters 3 and 4. As a result, the number of positive charges in the double-charge layer decreases, resulting in a negative shift of the potential. The AC-induced negative shift of the potential is due to enhanced corrosion of the steel, rather than cathodic polarization of the steel.

In both the absence and presence of AC interference, CP current can be shielded by the geometry of the coating disbondment, where the local potential inside the disbonding crevice is less negative than the potential at the open holiday. To show clearly the effect of AC current density on CP permeation into the disbonding crevice, the local potentials at the disbonding sites of 30 mm and 180 mm from the holiday are plotted as a function of AC current density, as shown in Figure 6-6.[23]

It can be seen that at small AC current densities such as 100 $A/m^2$ and 200 $A/m^2$, the local potential at both sites is shifted negatively by AC. As discussed, this shift is associated with the enhanced corrosion of the steel under AC interference. When the AC current densities increase from 300 $A/m^2$ to 500 $A/m^2$, the potentials are shifted in the positive direction. At high AC current densities, the iron dissolution and water reduction are further enhanced in the positive and negative cycles of AC. There are more ferrous and hydroxyl ions generated during anodic and cathodic reactions, respectively. Owing to the limited space under the disbonding crevice, the corrosion products are not able to diffuse outward freely. The positive shift of the steel potential is associated with the deposit of corrosion products, such as $Fe(OH)_2$, on the steel.

Furthermore, at small AC current densities, the AC-enhanced steel corrosion generates ferrous and hydroxyl ions, improving the conductivity of the solution trapped under the disbonded coating. This results in enhancement of the CP current permeation into the crevice. At AC current densities of 100 $A/m^2$, 200 $A/m^2$, and 300 $A/m^2$, the critical disbonding depths are 120 mm, 90 mm, and 60 mm, respectively, while the critical disbonding depth in the absence of AC is only 30 mm. The reason that the critical disbonding depth decreases with increasing AC current density is associated with the blocking effect of the generated corrosion products in the disbonding crevice. The corrosion products generate and deposit at high AC current densities, blocking the ionic diffusion and decreasing the CP current permeation. When the AC current density is increased to 500 $A/m^2$, the CP current is blocked completely to reach the disbonding crevice. Thus, under high AC current densities, the applied CP can be shielded completely from reaching the coating disbondment.

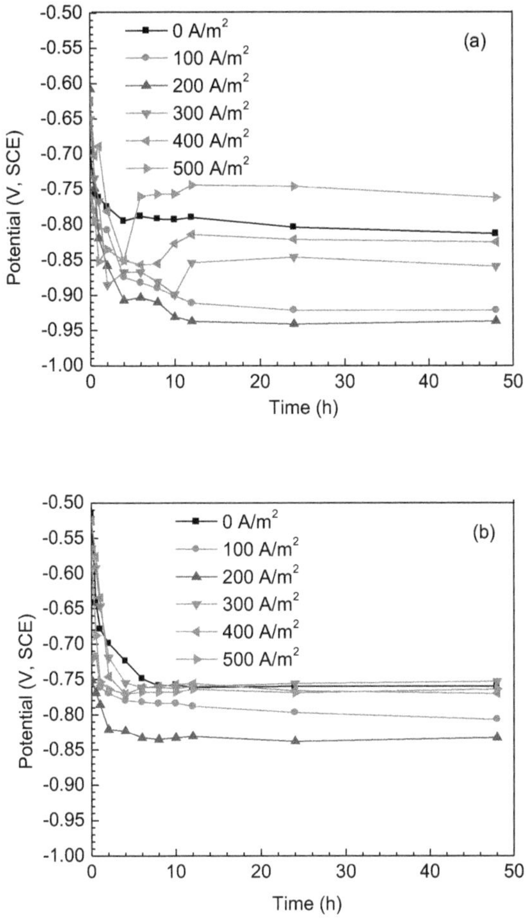


**Figure 6-6.** Local potentials at the disbonding sites of (a) 30 mm and (b) 180 mm from the holiday as a function of AC current density (the applied CP potential is –875 mV [SCE] and the disbonding thickness is 120 μm).[23]

# 6.3. Shift of CP potentials under AC interference

AC interference can deviate the potential of the protected pipelines from the applied CP value.[24] For example, Camitz et al. studied AC corrosion on steel coupons under standard CP and various AC voltages in a long-term field investigation.[25] Two series of tests were performed: one at 10 V AC for almost two years and the other at 30 V AC for 1.5 years. It was found that the potential varied according to the applied AC voltage. During the positive half cycles, the OFF potential was shifted in the anodic direction to values less negative than the applied CP value, indicating that CP was periodically lost owing to AC interference.

In several cases, the potential was shifted to values even less negative than the corrosion potential. Furthermore, it was reported that CP monitoring in the presence of AC interference could lead to erroneous measurements, since the ohmic drop contribution due to AC could heavily affect the potential readings.[24] It has been acknowledged that one of the most serious consequences of AC interference is that corrosion can occur even if CP levels satisfy the standards, despite the induced AC having been reduced to less than 15 V as required by other standards.[26, 27] For effective corrosion protection by CP for buried pipelines that are under AC interference, it is mandatory to determine the "true" CP potential exerted on the pipelines by measuring the shift of potential caused by AC from the applied CP value. AC interference can be induced from either HVAC power lines or AC-powered railway systems. It has been reported that the CP potential fluctuated on a cathodically protected pipeline subject to AC interference from the parallel rail line.[28]

In addition to the CP potential shift induced by AC interference, the presence of alternating voltage and current on buried pipelines can also cause malfunctioning or even damage to the transformer–rectifier units of the ICCP system, generating false potential information. This is particularly true for the transformer–rectifier units that are not equipped with suitable AC filter and surge protection.[29]

### 6.3.1. At NACE-recommended CP potential

The CP potential recommended for control of external corrosion on underground or submerged metallic piping systems is –850 mV (CSE) or a negative (cathodic) polarization voltage shift of 100 mV.[30] For a 16Mn carbon steel pipeline in a simulated soil solution containing 8.933 g/L KCl, 1.170 g/L $MgSO_4 \cdot 7H_2O$, and 5.510 g/L $NaHCO_3$ (pH 8.95), the corrosion potential is –750 mV (SCE), or –830 mV (CSE).[12] Thus, the standard of a minimum 100 mV cathodic polarization (i.e., –850 mV [SCE] or –930 mV [CSE]), is followed to select the potential as the recommended standard CP potential. This potential is enough to protect the steel from corrosion in the solution in the absence of AC interference.

Figure 6-7 shows the real-time DC potential recorded on a 16Mn steel electrode at various AC current densities under the CP potential of –850 mV (SCE) applied galvanostatically in the simulated soil solution.[12] It can be seen that, generally, four potential regions are observed: (a) CP region: from 0 s to 500 s, the CP is applied on the steel; (b) CP + AC region: from 500 s to 1,000 s, various AC current densities are applied while the CP is maintained; (c) AC region: from 1,000 s to 1,500 s, the AC is maintained, but the CP is stopped; and (d) open-circuit potential region (OCP in the figure): after 1,500 s, both the CP and the AC are stopped. With the CP application, the DC potential of the steel reaches the steady value of –850 mV (SCE) in the first region. When AC is applied at the 500th second, the DC potential shifts negatively. Moreover, the higher the $i_{AC}$, the more negative the DC potential. In the AC region (where CP is stopped), the DC potential shifts positively to reach a steady-state value, which becomes more negative with increasing $i_{AC}$. In the open-circuit potential region, when the AC is stopped, the DC potentials are shift positively and then decrease gradually, reaching the value of -750 mV (SCE), which is the corrosion potential of the steel in the solution.

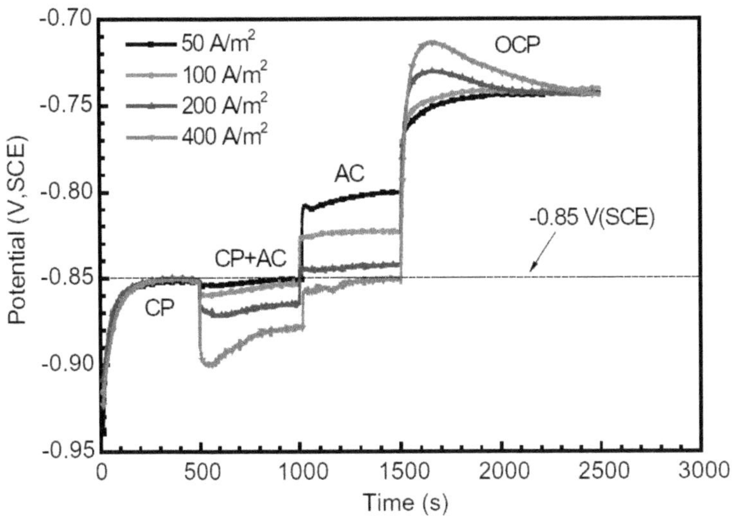


**Figure 6-7. DC potentials recorded on 16Mn steel electrode at various AC current densities under the CP potential of –850 mV (SCE) applied galvanostatically in the simulated soil solution.[12]**

To confirm the measured results, similar testing was conducted on an X65 pipeline steel in a near-neutral pH bicarbonate solution containing 0.483 g/L $NaHCO_3$, 0.122 g/L KCl, 0.181 g/L $CaCl_2 \cdot 2H_2O$, and 0.131 g/L $MgSO_4 \cdot 7H_2O$, which was purged with 5% $CO_2/N_2$ to maintain a stable pH of 6.50. Figure 6-8 shows the DC potential of X65 steel electrode at various AC current densities under the CP potential of –850 mV (CSE) in the near-neutral pH bicarbonate solution.[31] Identical to Figure 6-7, four potential regions are included in all plots: (a) CP potential region: from 0 s to 600 s, the CP is applied on the steel; (b) CP + AC region: from 600 s to 1,200 s, various AC current densities are applied on the steel while the CP is maintained; (c) AC region: from 1,200 s to 1,800 s, the CP is stopped, and the AC is applied; and (d) open-circuit potential region (OCP in the figure): after 1,800 s, both the CP and the AC are stopped. In the first region up to 600 s, the DC potential of the steel is at the applied CP potential (i.e., –850 mV [CSE]). The DC potential shifts negatively upon application of the AC in the CP + AC region. Moreover, the DC potential becomes more negative with increasing AC current density. The most negative value is observed at an AC current density of 200 A/m$^2$.

In the AC region (where the CP is stopped), the DC potential shifts positively. The steady-state value of the DC potential is more negative with increasing AC current density. In the open-circuit potential region, when both the CP and the AC are stopped, the DC potentials tend to reach a steady value of –785 mV (CSE) (i.e.

the corrosion potential of X65 steel in the solution). It can thus be seen that for both 16Mn and X65 pipeline steels in the two near-neutral pH solutions, the DC potential profiles recorded in the presence of AC interference are identical. Du et al. followed the testing approach while controlling the applied cathodic current density on pipeline steel and obtained similar results.[32] It was found that for the relatively low CP current density of 0.023 A/m$^2$ to achieve the CP potential of –0.85 V (CSE), the CP potential shifted instantly in the negative direction once the AC interference was applied.

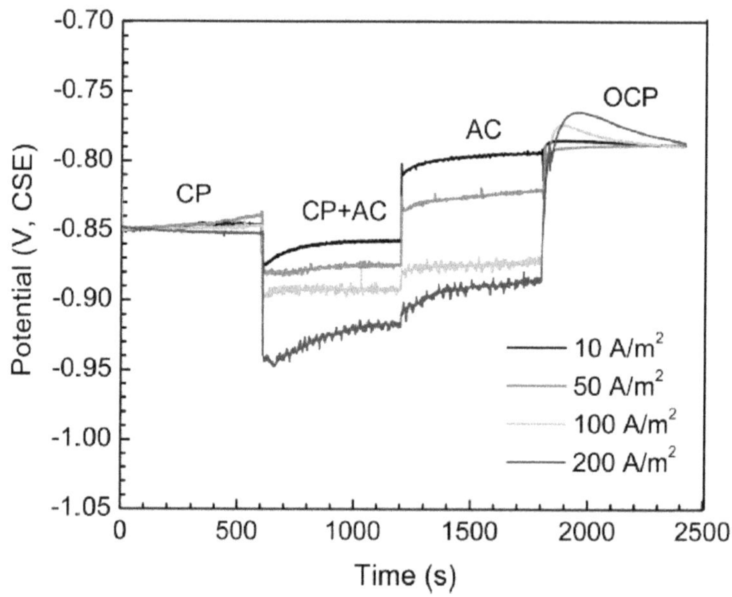


**Figure 6-8. DC potentials recorded on X65 steel electrode at various AC current densities under the CP potential of –850 mV (CSE) applied galvanostatically in the near-neutral pH bicarbonate solution.[31]**

When the CP potential is applied potentiostatically, the current densities of 16Mn and X65 steel electrodes at various AC current densities under a CP potential of –850 mV (CSE) in the near-neutral pH bicarbonate solution are shown in Figures 6-9 and 6-10.[12, 31] Generally, three regions are observed in both figures: (a) CP-I region, where the CP is applied on the steel and the current density is recorded; (b) CP + AC region, where various AC current densities are applied while the CP is maintained and the CP current density is affected by the applied $i_{AC}$; and (c) CP-II region, where $i_{AC}$ is stopped and the CP is applied, and the CP current density is recorded. In the CP-I region, the current density is close to zero, indicating that the CP is functional and able to protect the steel from corrosion. In the CP + AC region, when the AC is applied, the recorded anodic current density shows that corrosion occurs even when the CP is maintained. With the increase in $i_{AC}$, the current densities increase anodically. The higher the $i_{AC}$ applied, the more anodic the current density is. When AC is stopped in the CP-II region, the recorded current densities recover back to values identical to those in the CP-I region. Apparently, the presence of AC makes anodic reaction occur on the steels when they are under a CP potential of –850 mV (CSE).

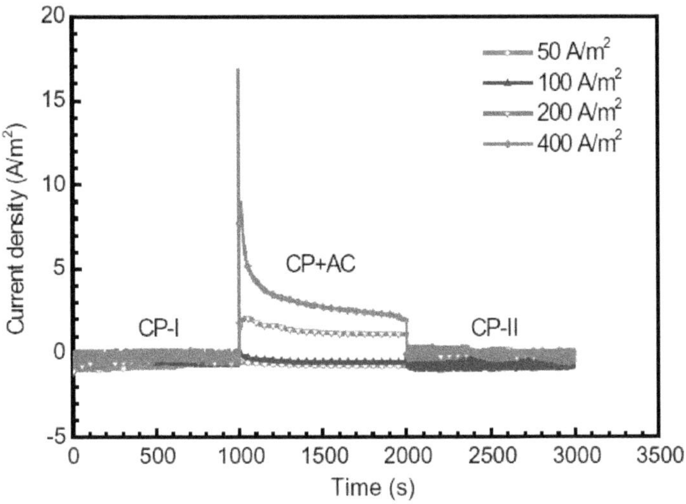


**Figure 6-9.** Current densities recorded on 16Mn steel electrode at various AC current densities under the CP potential of –850 mV (SCE) applied potentiostatically in the near-neutral pH bicarbonate solution.[12]

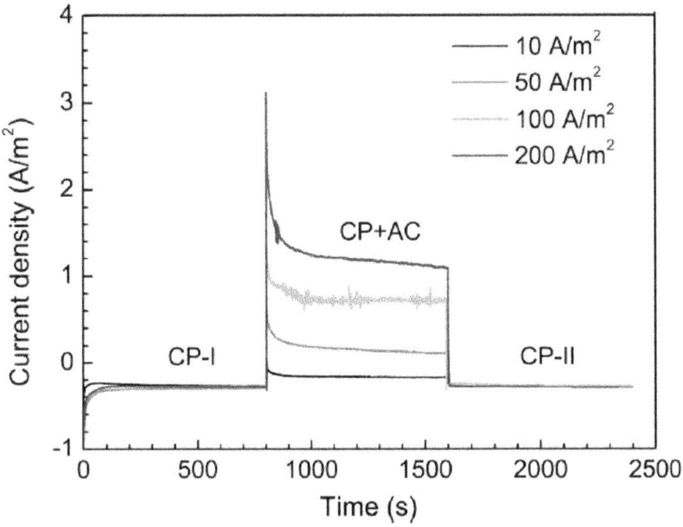


**Figure 6-10.** Current densities recorded on X65 steel electrode at various AC current densities under the CP potential of –850 mV (SCE) applied potentiostatically in the near-neutral pH bicarbonate solution.[31]

## 6.3.2. At potentials more negative than the NACE-recommended value

As the CP potential applied on pipelines may experience fluctuations in the field, similar measurements were also conducted at a potential more negative than the NACE-recommended value to investigate the effect of AC on the CP potential shift. Figure 6-11 shows the DC potential profiles recorded on 16Mn steel electrode at various AC current densities under the CP potential of –1,000 mV (SCE) (i.e., –1,080 mV [CSE]), applied galvanostatically in the simulated soil solution.[12] Four potential regions mark the application and removal of the AC or the CP on the steel electrode. However, the changes of the potential are different. In the CP region, the DC potential is maintained at the steady value of –1,000 mV (SCE). When AC is applied at the 500th second, the DC potentials shift positively, which is opposite to those shown in Figure 6-7.

In the AC region, the DC potential reaches a steady-state value and decreases with increasing $i_{AC}$. In the open-circuit potential region (OCP in the figure), the DC potential tends to be back to the open-circuit potential value of –750 mV (SCE). Du et al.[32] also found that when the applied CP current density was higher than 0.3 A/m$^2$, to achieve a cathodic polarized potential more negative than –1.0 V (CSE), the CP potential shifted positively once the AC interference was applied. Similar results were reported by researchers who found that when the pipeline was under a CP potential of –0.90 V (SCE) (i.e., –0.98 V [CSE]), the increase in the AC current density caused a positive shift of the protection potential of X70 steel pipeline and an increase in the protection current density.[33]

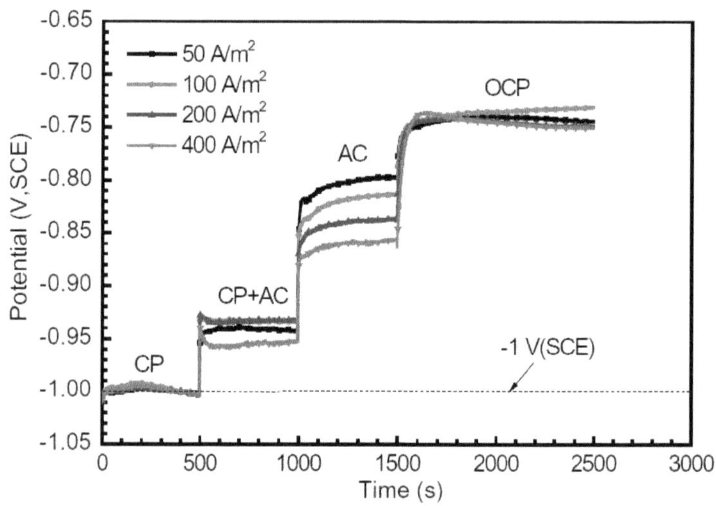


**Figure 6-11.** DC potentials recorded on 16Mn steel electrode at various AC current densities under the CP potential of –1,000 mV (SCE) applied galvanostatically in the simulated soil solution.[12]

The comparison of the DC potential shift measured at the applied CP potentials of –850 mV (SCE) and –1,000 mV (SCE) under $i_{AC}$ of 200 A/m$^2$ is shown in Figure 6-12.[12] It can be seen that although the DC potential shifts negatively under the CP potential of –850 mV (SCE) upon AC application, it shifts positively under –1,000 mV (SCE) of CP. The same phenomenon was also observed by Brenna, where an applied potential of –1.173 V (CSE) under the galvanostatic condition shifted 0.034 V positively in the presence of 10 A/m$^2$ AC and 0.179 V in the presence of 200 A/m$^2$ AC.[34] It can also be seen in Figure 6-12 that the DC potential shifts positively when the CP is stopped. Furthermore, when the AC is removed in the open-circuit potential region, the DC potential of the steel increases positively and gradually reaches the corrosion potential of about –750 mV (SCE). The difference occurs primarily in the CP+AC region.

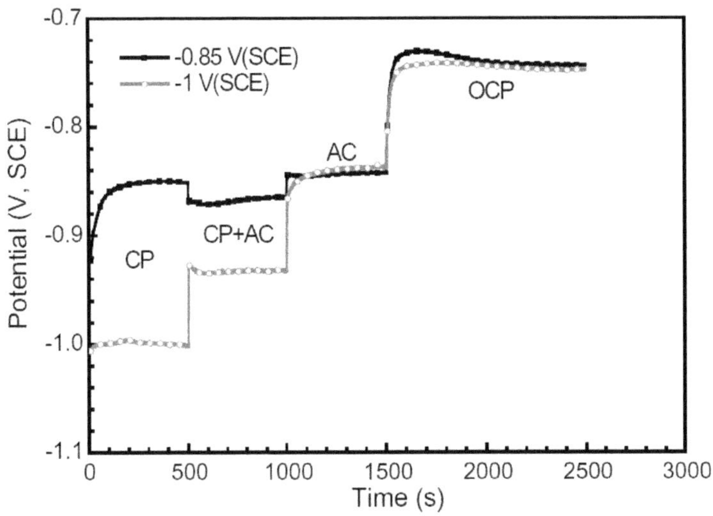


**Figure 6-12. Comparison of the DC potentials as a function of time under $i_{AC}$ of 200 A/m$^2$ and the CP potentials of –850 mV (SCE) and –1,000 mV (SCE), applied galvanostatically in the simulated soil solution.[12]**

Testing under the more negative CP potential was also conducted on X65 pipeline steel in the near-neutral pH bicarbonate solution. Figure 6-13 shows the DC potential profiles recorded at the applied CP potential of –1,000 mV (CSE).[31] It can be seen that the DC potential also shifts in the positive direction when AC is applied. As the AC current density increases, the DC potential becomes more negative. In the AC region, the DC potential is further shifted positively and reaches a steady-state value, which is more negative with the increase of AC current density. When both CP and AC are stopped, the DC potential changes toward the open-circuit potential value. With the measurement results on both 16Mn and X65 steels, it is confirmed that the CP potential shift upon AC application depends on the applied CP potential values.

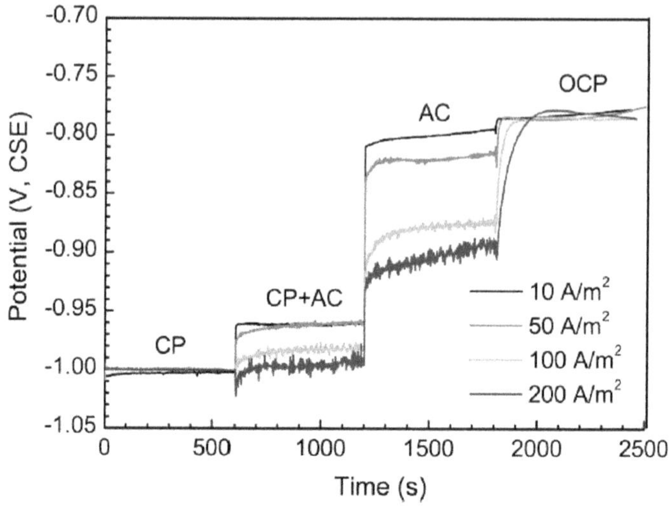


**Figure 6-13.** DC potentials recorded on X65 steel electrode at various AC current densities under the CP potential of −1,000 mV (CSE) applied galvanostatically in the near-neutral pH bicarbonate solution.[31]

The current densities of 16Mn and X65 steel electrodes at various AC current densities under the CP potential of −1,000 mV (CSE) in the near-neutral pH bicarbonate solution are shown in Figures 6-14 and 6-15. In these measurements, the CP effect on the steels is obtained by applying a constant potential on the electrode (i.e., a potentiostatic polarization method is used).. As in Figures 6-9 and 6-10, three regions are observed. In the CP-I and CP-II regions, the recorded current densities are cathodic, showing that the steels are fully protected. In the CP + AC region, the current densities drop instantly and then reach a steady value that is more negative than that recorded before the AC application. Moreover, the higher the $i_{AC}$, the more negative the current density. Obviously, the effect of AC on CP performance on the steels depends on the applied CP potential level. A detailed discussion about the mechanistic aspect of the interaction between AC and CP is included in the following section.

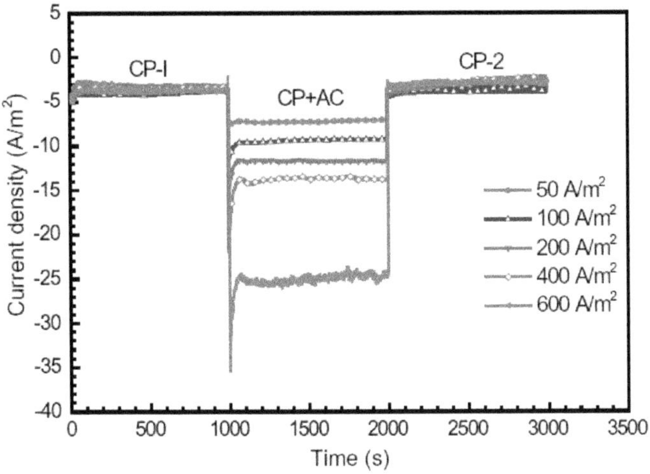


**Figure 6-14.** Current densities recorded on 16Mn steel electrode at various AC current densities under the CP potential of –1,000 mV (SCE) applied potentiostatically in the near-neutral pH bicarbonate solution.[12]

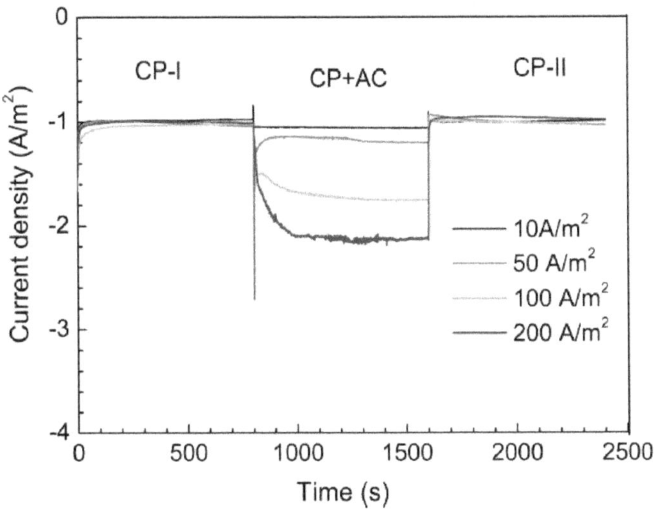


**Figure 6-15.** Current densities recorded on X65 steel electrode at various AC current densities under the CP potential of –1,000 mV (SCE) applied potentiostatically in the near-neutral pH bicarbonate solution.[31]

### 6.3.3. Mechanistic aspect of the effect of AC on CP potential shift

The CP potential applied on pipelines for corrosion protection cannot be maintained at its design value in the presence of AC interference. The "true" DC potential of pipeline steel, which is the summation of the applied CP potential and the potential shift under AC interference, depends on the CP level and the AC current density. Specifically, the DC potential of the steel is shifted negatively by the AC under a CP of –850 mV (CSE) but positively shifted at more negative CP potentials, such as –1,000 mV (CSE).

When the CP is applied galvanostatically, a constant cathodic current is applied on the steel to achieve the required CP potential. If the electric field at the steel–solution interface is disturbed by the AC, the DC potential of the steel would be changed. Thus, the galvanostatic-type CP is not capable of maintaining the steel's cathodic potential in the presence of AC interference owing to the fixed cathodic current output. The dependence of the DC potential of the steel on the CP potential level is attributed to the competition of anodic dissolution of the steel and the cathodic reduction of dissolved oxygen in aerated environments, which are expressed by the Butler–Volmer equation:[35]

$$i_a = i_a^0 \left[ \exp\left( \frac{\alpha_a' nF}{RT} \eta_a \right) - \exp\left( -\frac{\alpha_c' nF}{RT} \eta_c \right) \right] \tag{6-1}$$

$$i_c = i_c^0 \left[ \exp\left( \frac{\alpha_c'' nF}{RT} \eta_c \right) - \exp\left( -\frac{\alpha_a'' nF}{RT} \eta_a \right) \right] \tag{6-2}$$

$$i_{ex} = i_a + i_c \tag{6-3}$$

where $i_a$ and $i_c$ are anodic and cathodic current densities (A/m$^2$), respectively; $i_a^0$ and $i_c^0$ are the anodic and cathodic exchange current densities (A/m$^2$), respectively; $i_{ex}$ is the external current density flowing into the steel electrode; $\alpha_a'$ and $\alpha_c'$ are anodic and cathodic charge-transfer coefficients of the steel, respectively; $\alpha_a''$ and $\alpha_c''$ are anodic and cathodic charge-transfer coefficients of dissolved oxygen, respectively; $\eta_a$ and $\eta_c$ are overpotentials of anodic and cathodic reactions (V), respectively; $n$ is the number of electrons exchanged; $R$ is the ideal gas constant; $T$ is temperature (K) and $F$ is Faraday's constant.

Theoretically, in the absence of the external current source, the anodic dissolution current is equivalent to the cathodic reduction current, indicating that the consumption and supply of electrons are identical in order to maintain the charge-transfer conservation, where the DC potential of the steel reaches a steady-state value. In the presence of external current applied galvanostatically, the current densities $i_{ex}$, $i_a$, and $i_c$ could be affected by AC. While the effect of the AC electric field is applied on the steel instantaneously, the resulting electrochemical responses would be hysteretic.

Two types of electrochemical states could be developed on the steel upon AC application: (a) a non-equilibrium state, where the current equilibrium between the anodic and cathodic reactions is disturbed by AC interference, as illustrated by the inequalities shown below, and (b) an equilibrium state, where the subsequent electrochemical reactions achieve a new equilibrium by altering the cathodic or anodic reaction rates after the application of AC.

1. $|i_a|+|i_{ex}|>|i_c|$; that is, the sum of the supply of free electrons by anodic reaction and the externally applied current is larger than the consumption rate by cathodic reaction. The anodic dissolution of the steel is accelerated, and excessive electrons are generated, resulting in a negative shift of the DC potential. This is equivalent to an increase in the overpotential of the cathodic reaction to maintain the charge-transfer conservation.
2. $|i_a|+|i_{ex}|<|i_c|$; that is, the sum of the supply of free electrons by anodic reaction and the externally applied current is smaller than the consumption rate by cathodic reaction. The AC application causes an acceleration in cathodic reaction, and the electron consumption increases. The DC potential of the steel shifts positively. This is equivalent to an increase in the overpotential of the anodic reaction. Eventually, the supply of electrons is balanced with their consumption on the steel surface.
3. $|i_a|+|i_{ex}|=|i_c|$; that is, the sum of the free electrons by anodic reaction and the externally applied current is equal to the consumption rate by cathodic reaction. As a result, the DC potential of the steel does not change and stays at a steady-state value upon AC application.

Under CP of –850 mV (CSE), the DC potential of the steel shifts negatively when AC is applied, indicating that the AC accelerates the steel dissolution (i.e., $i_a$ increases). The excessive electrons generated cause a negative shift in the DC potential. The higher the $i_{AC}$, the more negative the DC potential. Thus, a CP potential of –850 mV (CSE) is not sufficient to protect the steel from corrosion under the AC interference. However, at a CP of –1,000 mV (CSE), the application of AC results in an increasing cathodic reaction since the consumption rate of electrons by cathodic reduction is much larger than the supply rate of electrons under high cathodic polarization. Therefore, the consumption of electrons decreases to maintain the charge-transfer conservation. Consequently, the DC potential shifts positively. When CP is stopped after 1,000 s, the DC potential of the steel shifts positively, indicating the increasing corrosion tendency in the absence of CP.

The theoretical analysis can be illustrated by a conceptual model developed for the interference of AC on the DC potential of the steel under varied CP potential levels, as shown in Figure 6-16.[31] At the CP potential of –850 mV (CSE), the steel is cathodically polarized and cations are adsorbed on the steel surface, where the double-charge layer serves as a current reservoir for faradic processes. Upon application of AC, excessive ferrous ions are generated owing to the iron oxidation occurring during the positive cycles of the AC. The ferrous ions flow toward the cathode to form corrosion products. The ionic diffusion is accelerated by the generated AC electric field. As a result, the quantity of cations (i.e., ferrous ions) reserved in the double-charge layer reduces, causing the negative shift of the DC potential of the steel. With the increase of the AC current density, more ferrous ions are generated during the positive cycles of AC. The diffusion of the cations toward the cathode causes further negative shifts of the DC potential. When a more negative CP potential of –1,000 mV (CSE) is applied, the steel is sufficiently polarized cathodically to cause ferrous ions, generated owing to AC corrosion, adsorb on the steel surface, resulting in a positive shift of the DC potential.

With an increase of AC current density, the enhanced AC electric field accelerates the diffusion of cations out of the double-charge layer. An equilibrium between the adsorption of cations on the cathodically polarized steel and the diffusion of cations toward the cathode is established, even at high AC current densities. Furthermore, when CP is stopped, the DC potential of the steel is shifted positively by AC. This is primarily attributed to the formation of corrosion products facilitated by AC corrosion and the deposit on the steel surface. With increasing AC current density, the corrosion product layer becomes thick and the DC potential shifts positively as more cations stay in the double-charge layer.

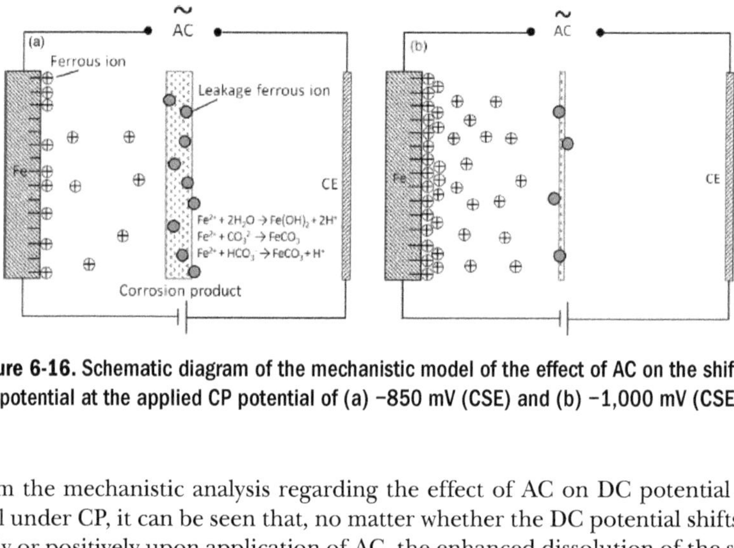


**Figure 6-16. Schematic diagram of the mechanistic model of the effect of AC on the shift of DC potential at the applied CP potential of (a) –850 mV (CSE) and (b) –1,000 mV (CSE).[31]**

From the mechanistic analysis regarding the effect of AC on DC potential of the steel under CP, it can be seen that, no matter whether the DC potential shifts negatively or positively upon application of AC, the enhanced dissolution of the steel by AC and the generation of cations, (i.e., ferrous ions) are the common result. The different shifts at specific CP–AC combinations result from the cathodic polarization of the steel, accelerated corrosion by AC, diffusion of the generated ferrous ions, and the deposit of corrosion products on the steel surface. In other words, the DC potential of the steel would deviate from the applied CP potential either negatively or positively, but the steel suffers from increased corrosion by AC. Therefore, the recorded DC potential in the presence of AC, if it is shifted negatively, could give misleading information that the cathodic polarization of the steel is further increased. This should be paid particular attention when recording and analyzing the steel potential under the mutual effect of AC and CP application.

When CP is applied potentiostatically on the steel, the DC potential of the steel could maintain a constant value (i.e., the CP potential), but the cathodic current would fluctuate in the presence of AC interference. When the CP potential is maintained at -850 mV (CSE), as discussed previously, the AC application results in an additional negative DC potential on the steel that is attributed to the enhanced anodic dissolution of the steel. Thus, the current density becomes anodic in the CP + AC region, indicating that the steel is anodically oxidized and corrosion is enhanced. Under the CP potential of –1,000 mV (CSE), AC application would result in a positive shift of the DC potential owing to the enhanced cathodic reaction and electron consumption. To maintain at –1,000 mV (CSE) potentiostatically, more cathodic current density is required. Since the CP potential is sufficiently negative, the applied AC would not be able to affect the corrosion of the steel at a detectable level.

## 6.4. CP performance under AC interference

Historically, it was thought that AC corrosion could occur on cathodically protected pipelines only when the peak of the positive part of the AC wave was more positive than the protection potential.[36] Thus, a common method to control AC corrosion is to rely on CP application in accordance with industry standards. It has been noted that the "conventional" NACE-recommended CP criterion of -850 mV (CSE) is not adequate to protect buried pipelines in the presence of AC interference.[37]

For example, the 1986 German investigation of the AC corrosion failure detected a high pitting rate despite the CP current density of 1.5 $A/m^2$ to 2.0 $A/m^2$, and the ON potentials of –1.8 V to –2.0 V (CSE) were applied on the buried pipeline.[38] Song et al. found that cathodically protected steel coupons at the cathodic potential of –1.0 V (SCE) still showed significant corrosion rates, which further increased with the high AC current density.[39]

In a field investigation on a leaking gas pipeline that was collocated with several HVAC power lines for about 5 km, Wakelin and Sheldon found that the CP potentials measured directly on top of two nodules associated with coating holidays on the pipeline were –1,080 mV (CSE) and –1,170 mV (CSE).[40] Although the pipe surface at these sites was well protected by CP, corrosion pits still generated under the nodule and coating holidays owing to the induced AC interference. Fu and Cheng demonstrated that, in the absence of AC interference or at relatively low AC current densities (e.g., less than 20 $A/m^2$), the CP potential of -950 mV (CSE) is able to protect the pipeline steel from corrosion attack in simulated soil solutions.[41] However, when the AC current density is higher than 20 $A/m^2$, the CP potential should be more negative to enable full protection. To date, uncertainties still exist on the CP potential interval and the CP current density required to reach acceptable protection of pipelines when AC interference is present.[42]

### 6.4.1. Potentiostatic current density and corrosion rate of CP-applied pipeline steels under AC interference

Figure 6-17 shows the potentiostatic current density curves measured on an X65 steel electrode at cathodic potentials of −850 mV and −950 mV (CSE) in the absence of AC in a concentrated carbonate-bicarbonate solution (0.05 M $Na_2CO_3$ and 0.1 M $NaHCO_3$, pH 9.6).[41] It can be seen that, at −850 mV (CSE), the current density exhibits positive values in the beginning and then drops to negative values with time (Figure 6-17a). At the CP potential of −950 mV (CSE), the current density is negative and reaches a relatively steady value (Figure 6-17b). Thus, the steel is under full cathodic polarization under the potential. Figure 6-18 shows the SEM view of the steel electrode after 48 h of testing. Slight corrosion signs are observed on the electrode at −850 mV (CSE), which is approximately the corrosion potential of the steel. The steel is fully protected from corrosion under the potential of −950 mV (CSE).

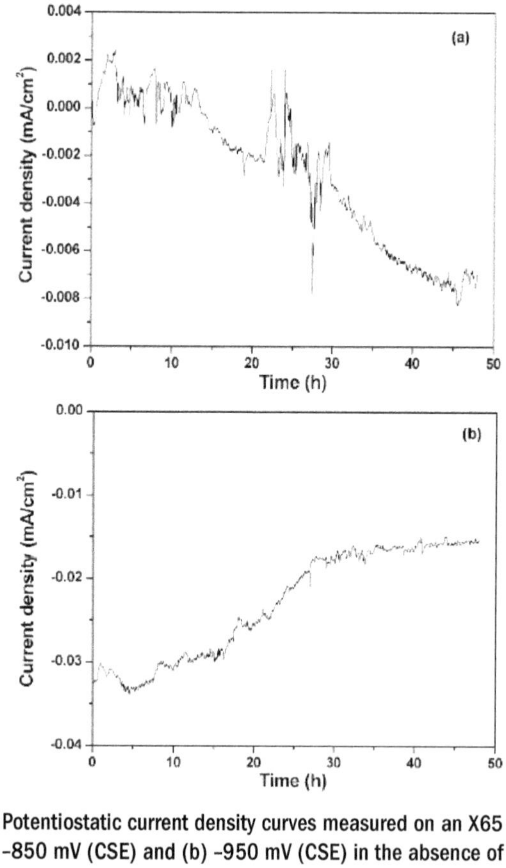


**Figure 6-17.** Potentiostatic current density curves measured on an X65 steel electrode at potentials of (a) -850 mV (CSE) and (b) -950 mV (CSE) in the absence of AC in a concentrated carbonate-bicarbonate solution.[41]

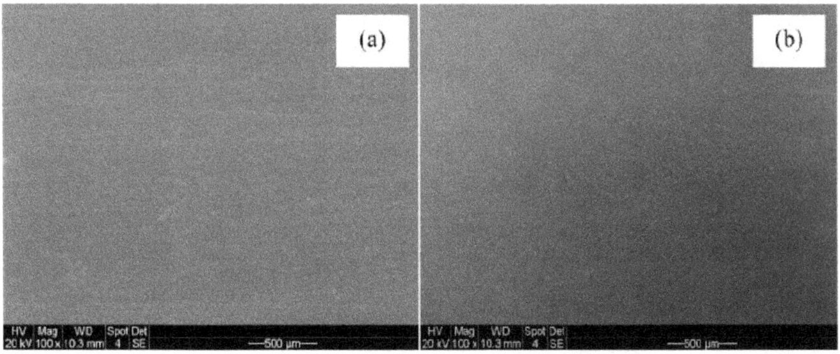


**Figure 6-18.** SEM images of the steel electrode after 48 h of testing in the absence of AC in the carbonate-bicarbonate solution at potentials of (a) -850 mV (CSE) and (b) -950 mV (CSE).[14]

Figure 6-19 shows the potentiostatic current densities measured on the steel electrode at potentials of –850 mV and –950 mV (CSE) in the presence of 20 A/m$^2$ of AC current density. Oscillations of the current density between positive and negative values are observed at the potential of –850 mV (CSE) over the whole time period. The current density is in the negative range when the potential of –950 mV (CSE) is applied. The amplitude of the current density oscillation is approximately 0.01 mA/cm$^2$ for both potentials. The SEM images in Figure 6-20 show that extensive corrosion occurs on the steel electrode at the potential of –850 mV. Thus, the presence of 20 A/m$^2$ of AC current density makes the CP potential of –850 mV (CSE) insufficient to protect the steel from corrosion. Only when the CP potential is shifted to –950 mV (CSE) can the steel corrosion be controlled.

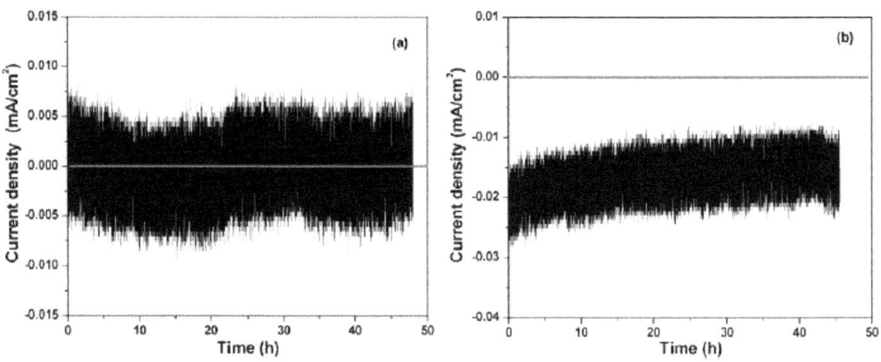


**Figure 6-19.** Potentiostatic current density curves measured on an X65 steel electrode at potentials of (a) -850 mV (CSE) and (b) -950 mV (CSE) in the presence of 20 A/m$^2$ of AC current density in a carbonate-bicarbonate solution.[41]

**Figure 6-20.** SEM images of the steel electrode after 48 h of testing in the presence of 20 A/m$^2$ of AC current density in the carbonate-bicarbonate solution at potentials of (a) -850 mV (CSE) and (b) -950 mV (CSE).[41]

Figure 6-21 shows the potentiostatic current densities measured on the steel electrode at –850 mV, –950 mV, and –1,050 mV (CSE) with 100 A/m$^2$ of AC current density applied. Current density oscillations between positive and negative values are observed at –850 mV (CSE), but the oscillation amplitude increases to 0.05 mA/cm$^2$. At the CP potential of –950 mV (CSE), there are positive values of current density over the time period. When the CP potential is as negative as –1,050 mV (CSE), the current density is completely in the negative value range. Figure 6-22 shows the SEM views of the steel electrode after testing under the conditions in Figure 6-21. At the potentials of –850 mV and –950 mV (CSE), numerous corrosion pits are observed on the steel surface. Only when the CP potential is up to –1,050 mV (CSE) is the steel electrode free of corrosion.

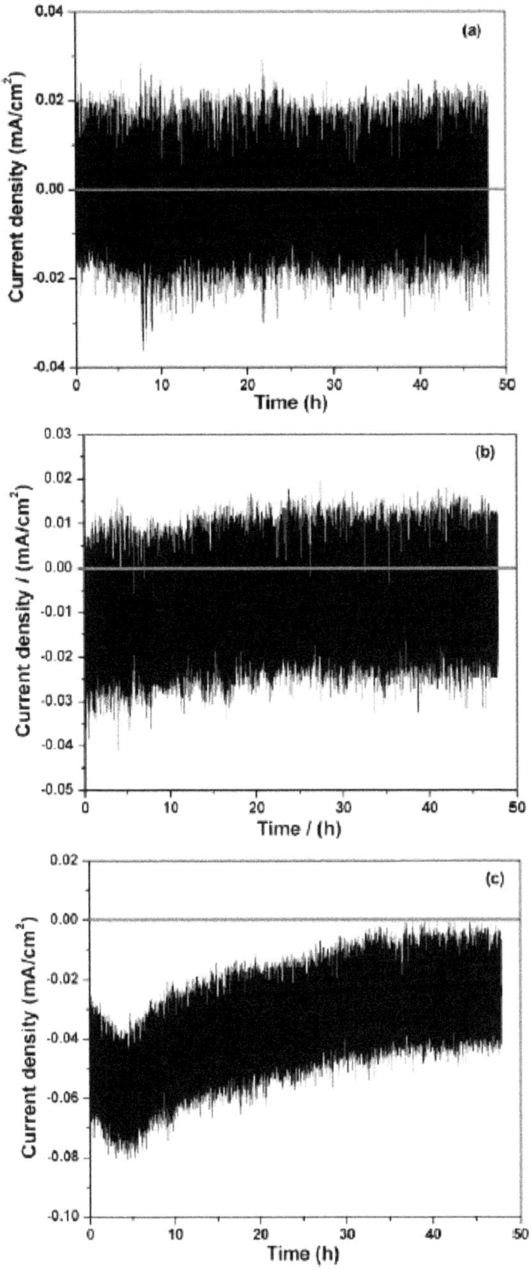


**Figure 6-21.** Potentiostatic current density curves measured on an X65 steel electrode at potentials of (a) -850 mV (CSE), (b) -950 mV (CSE), and (c) -1,050 mV (CSE) in the presence of 100 A/$m^2$ of AC current density in a carbonate-bicarbonate solution.[41]

**Figure 6-22.** SEM images of the steel electrode after 48 h of testing in the presence of 100 A/$m^2$ of AC current density in the carbonate-bicarbonate solution at potentials of (a) -850 mV (CSE), (b) -950 mV (CSE), and (c) -1,050 mV (CSE).[41]

With the further increase of the AC current density to 500 A/m$^2$, the potentiostatic current densities measured on the X65 steel electrode in the carbonate–bicarbonate solution are shown in Figure 6-23. It can be seen that, only when the applied CP potential is shifted to –1,150 mV (CSE) is the potentiostatic current density in the negative range. Figure 6-24 shows the SEM images of the steel electrode at the various CP potentials. At –850 mV (CSE), the steel suffers from extensive pitting corrosion (Figure 6-24a). A few corrosion pits are observed on the electrode surface at the potential of –950 mV (CSE) (Figure 6-24b). When the CP potential is –1,050 mV (CSE), general corrosion occurs on the steel (Figure 6-24c). When the CP potential is as negative as –1,150 mV (CSE), the steel is fully protected and no corrosion occurs, as shown in Figure 6-24d. Clearly, as the AC current density increases, the required CP potential should be more negative to provide full protection to the steel.

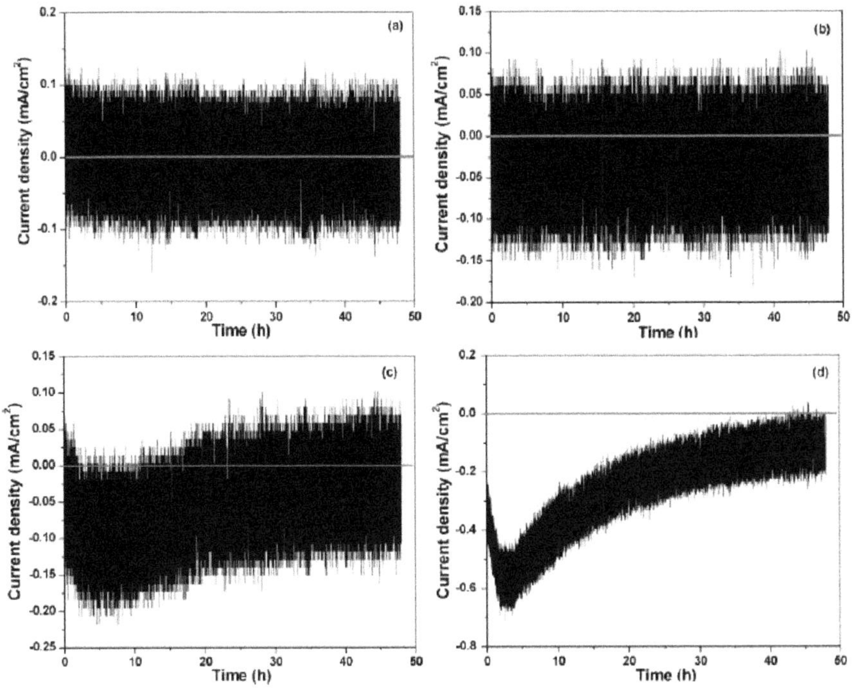


**Figure 6-23.** Potentiostatic current density curves measured on an X65 steel electrode at potentials of (a) -850 mV (CSE), (b) -950 mV (CSE), (c) -1,050 mV (CSE), and (d) -1,150 mV (CSE) in the presence of 500 A/m$^2$ of AC current density in a carbonate-bicarbonate solution.[41]

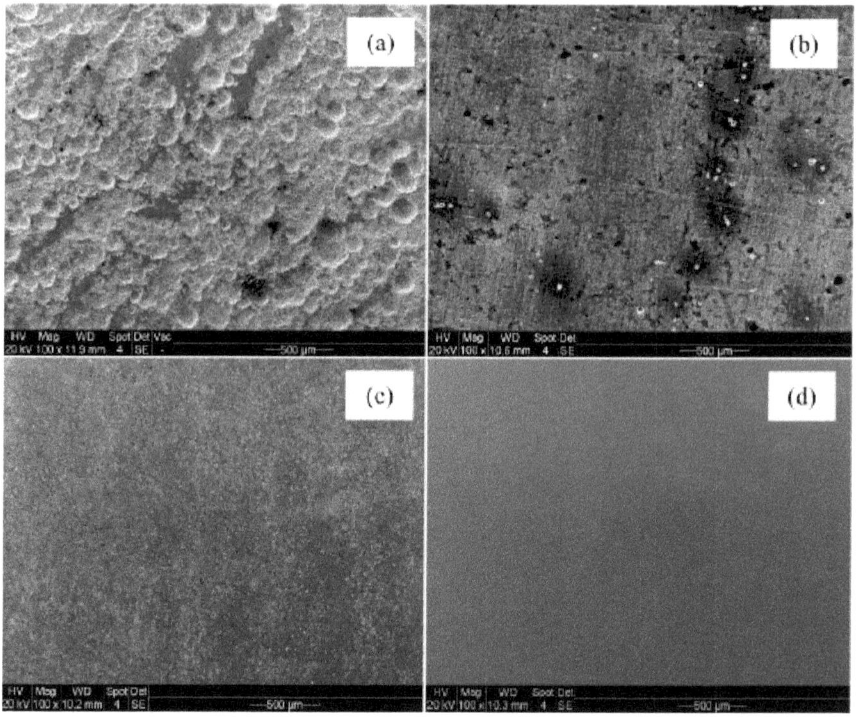


**Figure 6-24.** SEM images of the steel electrode after 48 h of testing in the presence of 100 A/ $m^2$ of AC current density in the carbonate-bicarbonate solution at potentials of (a) -850 mV (CSE), (b) -950 mV (CSE), (c) -1050 mV (CSE), and (d) -1,150 mV (CSE).[41]

The corrosion rates of the steel under various CP potentials were determined from the weight loss measurements in the carbonate-bicarbonate solution at various AC current densities, and the results are shown in Figure 6-25. It can be observed that the corrosion rate of the steel increases with increasing AC current density. Upon CP application, the corrosion rate decreases with the negative shift of the CP potential at specific AC current densities. Moreover, with the increase in AC current density, the ability of the CP potential to fully protect the steel shifts negatively.

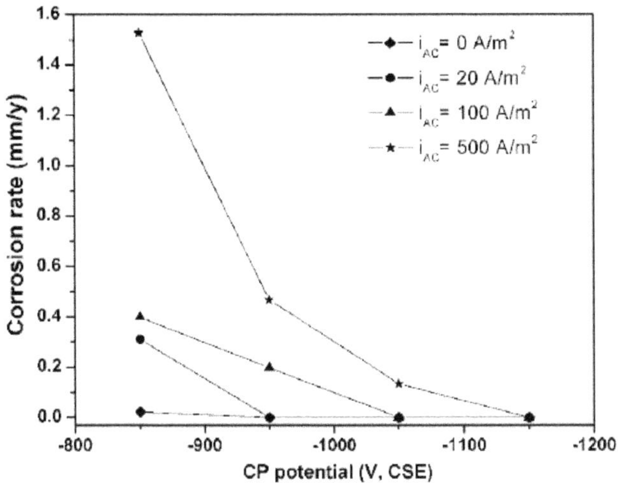


**Figure 6-25.** Corrosion rates of the X65 steel electrode in the carbonate-bicarbonate solution at various AC current densities under varied CP potentials.[41]

Although a negative increase in the CP potential level can reduce or even eliminate the AC corrosion risk on a cathodically protected pipeline, as measured above, AC corrosion could probably remain significant owing to the formation of dissolved dihypoferrite ions ($HFeO_2^-$) at high pH, which may stabilize corrosion at a very high rate, even when the protection potential criterion is being met.[8] Excess hydroxide ions ($OH^-$) and hydrogen can be produced by CP current and the negative cycle of the AC wave, so that overprotection of the pipeline is caused. Thus, for pipelines, especially those made of high-strength steels with a specified minimum yield strength greater than 550 MPa, the applied CP cannot be too negative in order not to cause detrimental effects in the material through hydrogen evolution. Moreover, AC interference can introduce potential threats for enhanced hydrogen evolution and hydrogen entry into the pipeline steels.

## 6.4.2. Effect of AC on CP effectiveness for corrosion protection

It has been acknowledged that AC-induced corrosion is associated with a threshold value of AC current density.[43, 44, 45] For example, Gummow et al. reported that AC corrosion does not occur on steels at an AC current density less than 20 A/m$^2$, while the corrosion is unpredictable at the AC current densities of 20–100 A/m$^2$.[44] Significant corrosion is expected if the AC current densities are greater than 100 A/m$^2$. For CP-applied pipelines, corrosion of the steels depends on both the AC current density and the applied CP potentials. For example, at an AC current density of 20 A/m$^2$, the steel is free of corrosion when a CP potential of –950 mV (CSE) is applied. However, when the applied AC current density increases to 100

$A/m^2$, the steel suffers from extensive pitting corrosion at -950 mV (CSE). Only when the CP potential shifts to –1,050 mV (CSE) is the steel free of corrosion.

However, Tang et al. found that although the increase in CP level could control the AC corrosion of mild steel when the AC current density was less than 300 $A/m^2$, the corrosion of the steel would be accelerated when the CP potentials were too negative, for example exceeding –1,300 mV (CSE).[46] According to the corrosion morphology with superimposed AC and $i_{AC} > 100$ $A/m^2$ under various CP potentials, there is a transition from pitting corrosion to uniform corrosion with a negative shift of CP potential. With consideration of AC corrosion of steels at various AC current densities in the presence and absence of CP, it is believed that the transition of corrosion type is dependent on the magnitude of the positive peak of AC density. Pitting corrosion occurs only when the positive peak current density reaches a critical value. The corrosion pits induced by AC are potential sites for the initiation of stress corrosion cracks. Brenna et al. established that the corrosion rate of carbon steels under CP condition increased as AC and DC current densities increased, (i.e., the corrosion rate depended on both AC interference and CP protection levels).[42] It was found that the critical value of AC current density that may cause a significant corrosion rate on a cathodically protected steel specimen decreased from 500 $A/m^2$ to 10 $A/m^2$ as the cathodic DC current density increased from 0.5 $A/m^2$ to 10 $A/m^2$.

Since the soil chemistry, such as the presence and concentrations of earth-alkaline metal cations ($Ca^{2+}$, $Mg^{2+}$) and alkaline elements ($Na^+$, $K^+$), affects the spread resistance at coating defects and thus the AC corrosion rates,[3, 47] CP performance would be dependent on the chemical composition of the soil. Owing to the CP-driven electrochemical cathodic reactions, which are usually the reduction of dissolved oxygen or water in aerobic or anaerobic soil environments, respectively, the precipitation of nondissolvable hydroxides and carbonates at the coating defects would remarkably affect the spread resistance locally. Du et al. found that the threshold value of AC current density that compromised CP effectiveness depended on the soil chemistry.[48] In a soil solution containing a high content of $Ca^{2+}$ and $Mg^{2+}$, no effective CP potentials were able to reduce the AC corrosion rate of an X70 pipeline steel below 0.1 mm/y when the AC current density was more than 300 $A/m^2$. However, when the soil solution contained a high content of $Na^+$, the threshold AC current density was about 200 $A/m^2$, above which the CP would not be effective for AC corrosion control. Moreover, dense calcareous deposits form at high CP levels (i.e., those more negative than –1,500 mV [CSE]) owing to the strong alkalization of the soil solution; these deposits dramatically increase the spread resistance and thus reduce the AC and DC current densities. However, the breakdown of the deposit could occur by a so-called flake-off mechanism due to hydrogen evolution under negative CP potentials,[49] resulting in reduction of the spread resistance and thus the increase of AC corrosion. The cathodic reactions on the re-exposed steel surface restart and cause solution alkalization, resulting in precipitation of the calcareous deposits. The corrosion stops again with time. Therefore, the soil chemistry plays an important role in AC corrosion and its control at high CP levels, which should be included in the CP criterion design when the soil contains a large amount of earth-alkaline elements. Given the wide variabilities in soil chemistry, investigations conducted under controlled conditions in laboratories are in no way sufficient to determine alternative CP criteria on pipelines.[50] Moreover, laboratory experiments merely screen a series of AC and DC settings (daily). However, in the field, prolonged periods (possibly weeks or months) in a given setting are needed to see the full development of the environment.

## 6.5. Recommendations of CP criteria for pipeline corrosion protection under AC interference

As AC interference adversely affects the effectiveness of CP in pipeline corrosion protection, it has been well accepted that new CP criteria based on field and laboratory studies should be developed for full protection of pipelines from AC corrosion. From a study of the corrosion of steel coupons as a function of AC and DC (CP) current densities, Funk et al. reported that for AC current densities greater than 30 $A/m^2$, the maximum corrosion rate exceeded 0.1 mm/y despite a constant CP current density of 2 $A/m^2$ being applied.[51] In this case, the protection potential criterion was not applicable. When the AC current densities were less than 30 $A/m^2$, there was no AC-induced corrosion at a CP current density of about 1 $A/m^2$. Kajiyama and Nakamura carried out field testing with steel coupons to investigate the AC corrosion risk of 6.6 km of PE-coated pipeline that paralleled a 66 kV, 50 Hz electric power transmission line.[52] It was found that, with a coupon DC current density of 10.8 $A/m^2$ and the polarized potential of −1.12 V (CSE), the coupon AC current density was as high as 184 $A/m^2$. This indicated that AC corrosion at a coating holiday was likely to occur despite the high level of CP. However, overprotection should be avoided because it can cause hydrogen evolution on the steel surface,[53] which may result in hydrogen embrittlement or hydrogen-induced cracking of the steel. Moreover, the elevated solution pH caused by CP-driven electrochemical cathodic reactions can facilitate coating disbondment, resulting in the so-called cathodic disbonding of the coatings. Therefore, for high induced AC voltages, the approach of negatively shifting the CP level is dangerous and should be prohibited. In the case of modest AC voltages, increasing the CP level negatively is probably effective.[8] To evaluate the risk of AC interference on CP performance on metallic structures, three parameters must be considered: the true potential of the steel, the protection (DC) current density, and the AC current density.[54]

There has never been a universal criterion available for protection against pipeline AC corrosion based on CP. Thus, the CP criteria against AC interference, as reviewed herein, are associated with the realistic conditions relevant to pipeline operation. These include (a) CP system selection and CP levels, (b) the risk to CP equipment, (c) the interference of DC from the CP system on HVAC infrastructure, (d) soil resistivity (moisture content, chemicals, salinity, etc.), (e) coating properties and performance condition (pinholes, holidays, disbondment, etc.), (f) HVAC power lines (AC voltage, orientation relative to the buried pipelines, parallel length with the pipeline, etc.), and (g) considerations of underprotection and overprotection.

Figure 6-26 illustrates a CP criterion for buried steel pipelines against AC corrosion.[55] It can be seen that AC corrosion could be mitigated to negligible if the AC current density were below 70 A/m² and the protection current density were below 20 A/m². When the protection current density exceeds 20 A/m², the pipe steel may be overprotected, which could result in significant hydrogen evolution and/or solution alkalinity to induce coating disbondment. Moreover, the underprotection zone (i.e., DC corrosion, as labeled in the figure) defines the minimum requirement of the protection current density. The criterion considers the CP for pipeline corrosion protection in the presence of AC interference using current density, rather than potential (as commonly used in industry). This is reasonable, as the CP performance is subject to the cathodic current flowing to the protected target (i.e., pipelines), which depends on the applied potential and on the resistance of the CP circuit components, such as soil. In other words, the value of the potential is not necessarily indicative of the CP performance, but the value of the DC current density is.

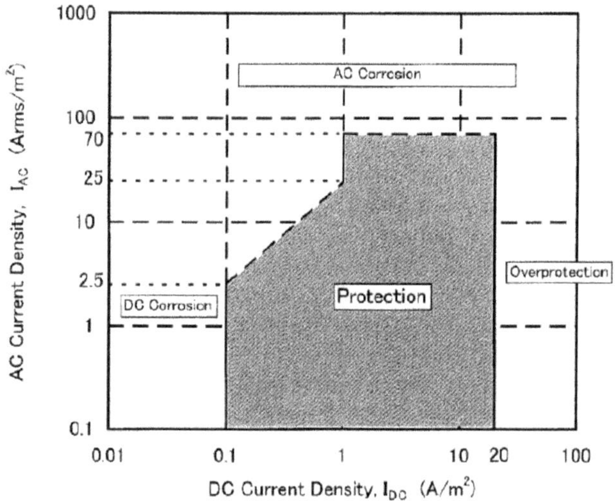


**Figure 6-26.** A new CP criterion against AC corrosion of buried steel pipelines.[55]

In the literature, some published papers proposed new CP criteria based on revised potential values from the industry-accepted standard, such as –850 mV (CSE).[30] For example, Carpentiers et al. stated that under the recommended CP condition steel corrosion can still occur owing to AC interference in the potential ranges that would passivate the steel if AC were absent.[56] Owing to the local alkalization on the steel surface under CP, it was recommended that the protection potential stay below –1.15 V (SCE). Similarly, a new CP level in the range of –1.0 V to –1.2 V (CSE) was considered effective again AC corrosion, and no overprotection conditions were established on the basis of the laboratory results.[54]

Combining both AC current density (i.e., the essential indicator of the AC corrosion rate) and the protection potential (i.e., the frequently used and well-accepted CP parameter), Fu and Cheng proposed a new CP design criterion to protect pipelines from corrosion under AC interference in a carbonate–bicarbonate solution that simulated the soil electrolyte.[41] As shown in Figure 6-27, the pipeline steel can be protected by the 100 mV cathodic polarization relative to the NACE-recommended criterion when the AC current density is smaller than 20 A/m$^2$.

For AC current densities between 20 A/m$^2$ and 500 A/m$^2$, the CP potential should be shifted negatively to protect the steel. The potential along the line ABCD refers to the ideal potential for protection of pipelines from AC corrosion, where the suggested CP potential can provide full corrosion protection to the steel. The area between lines ABCD and EFGH is the protection region, and the area left of line ABCD is the corrosion region. However, the possible detrimental effect should be considered if the AC current density is too high. The area right of line EFGH is the overprotection region. With consideration of the possibly detrimental effect of more negative CP potentials on the pipelines, it is not recommended to use CP to protect coated pipelines if the AC current density is over 500 A/m$^2$. Furthermore, Table 6-1 lists three typical AC current densities that are encountered in the field (20 A/m$^2$, 100 A/m$^2$, and 500 A/m$^2$), ranked as low, moderate, and high AC interference.[41] The recommended protection potentials and their differences from the NACE standard CP potential at specific AC current densities are also provided.

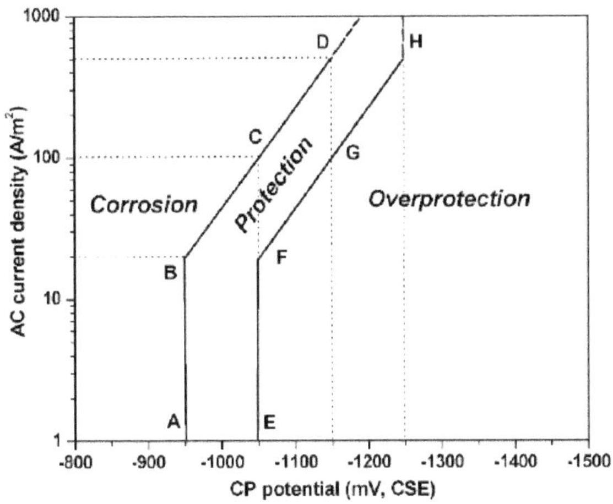


**Figure 6-27.** A new CP design criterion with consideration of AC interference on X65 pipeline steel in concentrated carbonate-bicarbonate solutions.[41]

| AC current density ($A/m^2$) | Level of AC interference | Protection potential (mV, CSE) | Potential difference (mV) |
|---|---|---|---|
| 0 | Zero | -950 | -100 |
| 20 | Low | -950 | -100 |
| 100 | Moderate | -1050 | -200 |
| 500 | High | -1150 | -300 |

**Table 6-1.** Recommended CP potentials and their differences from the NACE standard potential for pipeline corrosion protection at individual AC current densities typically encountered in the field.[41]

Tang et al. proposed a similar CP criterion for the protection of steels from AC corrosion, as shown in Figure 6-28. Assumptions made for the development of the new CP standard included:

- Corrosion of the steel was negligible if the corrosion rate was no more than 0.01 mm/y.
- There was no anomaly of corrosion rate between the selected critical AC current densities.
- The CP potential of –1.3 V (CSE) was defined as the upper limit of protection to avoid overprotection.

In the proposed CP criterion, area I to left of the line ABCDE is the corrosion region. When the CP potential is less negative than the potential along this line, the steel would suffer from AC corrosion. The CP potentials along the line ABCDE are the protection potentials under AC interference. Area II, between lines ABCDE and FHIJ, is the protection region. Area III is the overprotection region, where the applied CP current is too large to provide effective protection to pipelines and prevent AC corrosion. Moreover, hydrogen-induced cracking and cathodic disbondment of the coating might occur in this region.

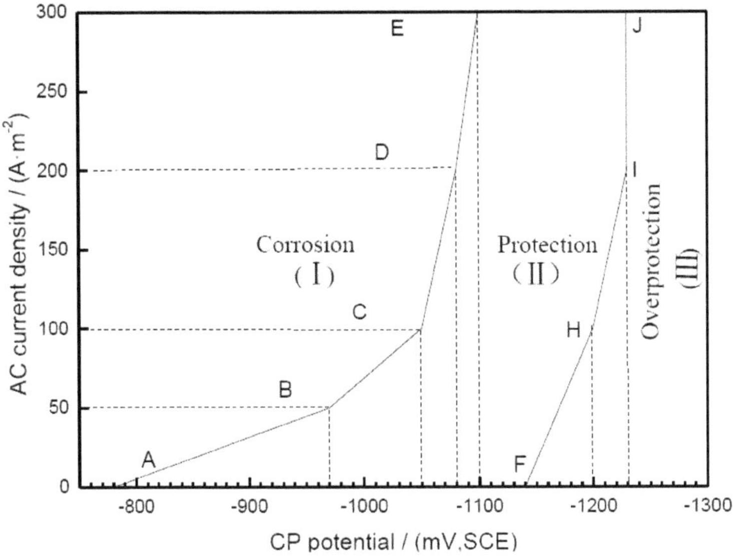


**Figure 6-28.** New CP criterion for mild pipeline steel in the presence of AC interference.[46]

Kajiyama and his team have carried out field observations to assess a new CP potential criterion established by laboratory studies using steel coupons buried for two to three years in various environments under potential AC corrosion risks.[52, 57] They found that, up to an AC current density of 70 A/m$^2$, AC corrosion is reduced to less than 0.01 mm/y by an appropriate level of CP current density. At AC current densities exceeding 70 A/m$^2$, AC corrosion is probable despite a high level of cathodic polarization applied on the pipelines.[55] They proposed a new CP criterion based on coupon DC and AC current densities, listed as follows:

- 0.1 A/m$^2$ $\leqq i_{DC} <$ 1.0 A/m$^2$, $i_{AC} < 25\, i_{DC}$
- 1.0 A/m$^2$ $\leqq i_{DC} \leqq$ 40 A/m$^2$, $i_{AC} <$ 70 A/m$^2$

where $i_{DC}$ and $i_{AC}$ are coupon DC and AC current densities, respectively. Positive values in DC current density indicate the current flowing through the electrolyte to the coupon. The criterion eliminates all corrosion risks—such as AC corrosion, DC stray current corrosion, microbiologically influenced corrosion—and any overprotection risk.

Similarly, Brenna et al. proposed a corrosion risk map that correlated the so-called true potential (i.e., the polarization level on the pipe steel measured by a Luggin capillary in the laboratory, while eliminating the ohmic drop) and the ratio of AC to DC current densities ($i_{AC}/i_{DC}$).[42] It was believed that the consideration of AC current density and CP current density was not sufficient to establish the AC corrosion risk. Instead, the protection potential applied on the pipelines must also be included. The map could be used to assess CP effectiveness in the presence of AC interference through measurements of ohmic drop free potential and current densities. Three corrosion risk levels were identified–low, medium, and high–depending on the corrosion rate of the pipe steel. The medium and high corrosion risks corresponded to the corrosion rate in the ranges of 20 μm/y to 100 μm/y and greater than 100 μm/y, respectively. In the absence of AC interference, pipeline corrosion did not occur if the CP was functional. In the presence of AC interference, at a fixed $i_{AC}/i_{DC}$, the corrosion rate increased as the true potential decreased. At a constant potential, the corrosion rate increased with the increasing $i_{AC}/i_{DC}$ ratio. Overprotection (i.e., cathodic potential more negative than −1.2 V [CSE]) was the most dangerous situation under AC interference. A few amperes per square meter of AC current density (i.e., a ratio of $i_{AC}/i_{DC}$ in the range of 5–20) could cause severe corrosion on the steels. At −0.85 V (CSE), the CP was not effective if the $i_{AC}/i_{DC}$ ratio was higher than 400. In designing a new CP criterion for AC corrosion control, Ormellese et al. found that at CP potentials within −0.85 V to −1.00 V (CSE), the protected carbon steel would corrode in the presence of AC, indicating that the −0.85 V (SCE) criterion was not safe under AC interference.[54, 58] Corrosion was negligible when the potential was in the range of −1.00 V to −1.15 V (CSE), provided that the $i_{AC}/i_{DC}$ ratio was lower than 200 and no overprotection conditions were established. The potential interval could be regarded as a protection potential in the presence of AC interference.

Compared with the adverse effect of AC interference on CP performance in pipeline corrosion prevention, there has been limited attention on the effect of inappropriate CP design on the occurrence of AC corrosion. Lindemuth and Crabtree conducted analysis on four case histories relating to monitoring and mitigating the AC-influenced corrosion of CP-applied pipelines and found that applying excessive levels of CP on a pipeline could worsen the corrosion effect of AC but that modifying the CP level could help control the AC effect.[59] An even distribution of the CP current provided an effective means of controlling AC-influenced corrosion. CP designs should thus be fully integrated with the AC mitigation and monitoring program.

## References

1. NACE, *AC Corrosion State-of-the-Art: Corrosion Rate, Mechanism, and Mitigation Requirements*, publication 35110, NACE, Houston, TX, 2010.
2. Movley, C.M., "Pipeline corrosion from AC: Two UK case histories," CORROSION/2005, paper no. 05132, NACE, Houston, TX, 2005.
3. Büchler, M., "Alternating current corrosion of cathodically protected pipelines: Discussion of the involved processes and their consequences on the critical interference values," *Materials and Corrosion* 63: 1181–1187, 2012.
4. Peabody, A.W., *Peabody's Control of Pipeline Corrosion*, 2nd ed., NACE, Houston, TX, 2001.
5. Revie, R.W., Uhlig, H.H., *Corrosion and Corrosion Control*, 4th ed., John Wiley & Sons, Hoboken, NJ, 2008.

6. British Standards Institution, *Cathodic Protection of Buried or Immersed Metallic Structures: General Principles and Application for Pipelines*, BS EN 12954:2001, London, 2001.
7. Roberge, P.R., *Handbook of Corrosion Engineering*, McGraw-Hill Professional, New York, 1999.
8. Kajiyama, F., "Strategy for eliminating risks of corrosion and overprotection for buried modern pipelines," International Gas Union Research Conference, Seoul, 2011.
9. Kuhn, R. J., "Cathodic protection of underground pipelines from soil corrosion," *Proceedings of the American Petroleum Institute* 14: 153–167, 1933.
10. Von Baeckmann, W., Schwenk, W., Prinz, W, *Handbook of Cathodic Corrosion Protection*, 3rd ed., Gulf Professional Publishing, Houston, TX, 1997.
11. Parker, M.E., Peattie, E.G., *Pipeline Corrosion and Cathodic Protection*, 3rd ed., Gulf Professional Publishing, Houston, TX, 1999.
12. Xu, L.Y., Su, X., Cheng, Y.F., "Effect of alternating current on cathodic protection on pipelines," *Corrosion Science* 66: 263–268, 2013.
13. Cheng, Y.F., Norsworthy, R., *Pipeline Coatings*, NACE, Houston, TX, 2017.
14. Pikas, J.L., "Case histories of external microbiologically influenced corrosion underneath disbonded coatings," CORROSION/1996, paper no. 96198, NACE, Houston, TX., 1996.
15. Jack, T., King, F., Cheng, Y.F., Worthingham, R., "Permeable coatings and CP compatibility," Proc. 4th International Pipeline Conference, Calgary, Alberta, 1889–1893, 2002.
16. Norsworthy, R., "Coatings used in conjunction with cathodic protection-shielding vs. non-shielding pipeline coatings," 17th International Corrosion Congress, paper no. 4017, Las Vegas, NE, 2008.
17. Floyd, R. (2004), "Testing and mitigation of AC corrosion on 8" line: A field study," CORROSION/2004, paper no. 04210, NACE, Houston, TX, 2004.
18. Office of the Federal Register, U.S. Code of Federal Regulations, Title 49 "Transportation," Part 192, Washington, DC, 2012.
19. Office of the Federal Register, U.S. Code of Federal Regulations, Title 49 "Transportation," Part 195, Washington, DC, 2012.
20. Munger, C.J., "Causes and prevention of paint failure," in *Good Painting Practice - Steel Structures Paint Manual*, Vol. 1, J.D. Keane, D. Berger, H. Hower, B.R. Appleman, eds., Steel Structures Painting Council, Pittsburgh, PA, 1994.
21. Cheng, Y.F., "Controversy contained: In-depth look at CP shielding," *World Pipelines* 13: 50–54, 2013.
22. Kuang, D., Cheng, Y.F., "Probing potential and solution pH under disbonded coating on pipelines," *MP* 54: 40–45, 2015.
23. Kuang, D., Cheng, Y.F., "Effect of alternating current interference on coating disbondment and cathodic protection shielding on pipelines," NACE Northern Area Western Conference, Calgary, Alberta, 2015.
24. Ormellese, M., Goidanich, S., Lazzari, L., "Effect of AC interference on cathodic protection monitoring," *Corrosion Engineering, Science and Technology* 46: 618–623, 2011.
25. Camitz, G., Johansson, C., Marbe, A., "Alternating current corrosion on cathodically protected steel in soil: A long-term field investigation," CeoCor

International Congress, European Committee for the Study of Corrosion and Protection of Pipes and Pipeline Systems, Brussels. 2000.
26. Gummow, R.A., *Cathodic Corrosion Considerations for Pipelines with AC Mitigation Facilities*, Pipeline Research Council International, Arlington, VA, 1999.
27. Kouloumbi, N., Batis, G., Kioupis, N., Asteridis, P., "Study of the effect of AC-interference on the cathodic protection of a gas pipeline," *Anti-Corrosion Methods and Materials* 49: 335–345, 2002.
28. George, J., "Alternating current interference on cathodically protected pipeline from railway electrification systems," CORROSION/2018, paper no. 10723, NACE, Houston, TX, 2018.
29. Ames, D.W., *Cathodic Protection Theory and Practice*, V. Ashworth and C. Googan, eds., Ellis Horwood, New York, 241–251, 1986.
30. NACE, *Control of External Corrosion on Underground or Submerged Metallic Piping Systems*, SP0169-2013, Houston, TX, 2013.
31. Kuang, D., *Pipeline Corrosion and Coating Failure under Alternating Current Interference*, PhD thesis, University of Calgary, Alberta, 2016.
32. Du, Y., Lu, M., Tang, D., Chen, S., "Researches on the effects of AC interference on CP parameters and AC corrosion risk assessment for cathodic protected carbon steel," CORROSION/2017, paper no. 8962, NACE, Houston, TX, 2017.
33. Ouadah, M., Touhami, O., Ibtiouen, R., "Diagnosis of the AC current densities effect on the cathodic protection performance of the steel X70 for a buried pipeline due to electromagnetic interference caused by HVPTL," *Progress in Electromagnetics Research* 45: 163–171, 2016
34. Brenna, A., *A Proposal of AC Corrosion Mechanism of Carbon Steel in Cathodic Protection Condition*, PhD thesis, Politecnico di Milano, Italy, 2012.
35. Bard, A., Faulkner, L., *Electrochemical Methods. Fundamentals and Applications*, 2nd ed., John Wiley and Sons, New York, 2001.
36. Büchler, M., Voute, C., Stalder, F., "Characteristics of potential measurements in the field of AC corrosion," CeoCor 6th International Congress, European Committee for the Study of Corrosion and Protection of Pipes and Pipeline Systems, Brussels, 2003.
37. Goidanich, S., Lazzari, L., Ormellese, M., Pedeferri, M.P., "Effect of AC on cathodic protection of carbon steel in simulated soil conditions," EUROCORR 2006, paper no. 279, Maastricht; European Federation of Corrosion, London, 2006.
38. Gummow, R.A., Wakelin, R.G., Segall, S.M., "AC corrosion: A new threat to pipeline integrity?" Proc. 1st International Pipeline Conference, Vol. 1, ASME, New York, 1996.
39. Song, H.S., Kho, Y.T., Kim, Y.G., Lee, S.M., Park, Y.S., "Competition of AC and DC current in AC corrosion under cathodic protection," CORROSION/2002, paper no. 02117, NACE, Houston, TX, 2002.
40. Wakelin, R.G., Sheldon, C., "Investigation and mitigation of AC corrosion on a 300 mm natural gas pipeline," CORROSION/2004, paper no. 04205, NACE, Houston, TX, 2004.
41. Fu, A.Q., Cheng, Y.F., "Effect of alternating current on corrosion and effectiveness of cathodic protection of pipelines," *Canadian Metallurgy Quarterly* 51: 81–90, 2012.

42. Brenna, A., Lazzari, L., Pedeferri, M., Ormellese, M., "Cathodic protection condition in the presence of AC interference," *La Metallurgia Italiana* 6: 29–34, 2014.
43. Prinz, W., Schöneich, H.G., "Alternating current corrosion of cathodically protected pipelines," Proc. International Gas Research Conference, Orlando, FL; Gas Research Institute, Des Plaines, IL, 502–511, 1992.
44. Gummow, R.A., Wakelin, R.G., Segall, S.M., "AC corrosion: A new challenge to pipeline integrity," CORROSION/98, paper no. 566, NACE, Houston, TX, 1998.
45. Goidanich, S., *Influence of Alternating Current on Metal Corrosion*, PhD thesis, Politecnico di Milano, Italy, 2005.
46. Tang, D., Du, Y., Lu, M., Jiang, Z., Dong, L., "Study on CP criteria for mild steel in the presence of AC interference," CORROSION/2014, paper no. 3802, NACE, Houston, TX, 2014.
47. Nielsen, L.V., "Role of alkalization in AC Induced corrosion of pipelines and consequences hereof in relation to CP requirements," CORROSION/2005, paper no. 188, NACE, Houston, TX, 2005.
48. Du, Y., Xiao, Y., Xie, S., Lu, M., "Research on the effects of environmental parameters on AC corrosion behavior," CORROSION/2018, paper no. 10676, NACE, Houston, TX, 2018.
49. Junker, A., Nielsen, L.V., Møller, P., "Effect of chemical environment and pH on AC corrosion of cathodically protected structures," CORROSION/2017, paper no. 9352, NACE, Houston, TX, 2017.
50. Junker, A., Nielsen, L.V., Heinrich, C., Møller, P., "Laboratory and field investigation of the effect of the chemical environment on AC corrosion," CORROSION/2018, paper no. 10844, NACE, Houston, TX, 2018.
51. Funk, D., Prinz, W., Schöneich, H.G., "Investigations of AC corrosion in cathodically protected pipes," *3R International* 31: 336–34, 1992.
52. Kajiyama, F., Nakamura, Y., "Effect of induced alternating current voltage on cathodically protected pipelines paralleling electric power transmission lines," *Corrosion* 55: 200–205, 1999.
53. Cheng, Y.F., Niu, L., "Mechanism for hydrogen evolution reaction on pipeline steel in near-neutral pH solution," *Electrochemistry Communications* 9: 558–562, 2007.
54. Ormellese, M., Lazzari, L., Brenna, A., Trombetta, A., "Proposal of CP criterion in the presence of AC-interference," CORROSION/2010, paper no. 10032, NACE, Houston, TX, 2010.
55. Hosokawa, Y., Kajiyama, F., Nakamura, Y., "New CP criteria for elimination of the risks of AC corrosion and overprotection on cathodically protected pipelines," CORROSION/2002, paper no. 02111, NACE, Houston, TX, 2002.
56. Carpentiers, P., Gregoor, R., Pourbaix, A., "AC corrosion: detection, investigations and mechanisms," EUROCORR 2003, paper no. 307, Budapest; European Federation of Corrosion, London, 2003.
57. Kajiyama, F., Okamura, K., "Evaluating cathodic protection reliability on steel pipe in microbially active soils," *Corrosion* 55: 74–80, 1999.
58. Ormellese, M., Lazzari, L., Goidanich, S., Sesia, V., "CP criteria assessment in the presence of AC interference," CORROSION/2008, paper no. 8064, NACE, Houston, TX, 2008.

59. Lindemuth, D., Crabtree, D., "AC corrosion control: When too much cathodic protection might just be a bad thing," CORROSION/2018, paper no. 11271, NACE, Houston, TX, 2018

7

# Inspection and Monitoring Techniques for AC Corrosion of Pipelines

The effectiveness of the AC interference mitigation system and CP performance for pipeline corrosion protection is usually evaluated by continuous monitoring of AC interference (i.e., AC voltage and/or AC current density) on pipelines collocated with HVAC power lines or AC-powered rail systems. Before selection and implementation of an appropriate inspection or monitoring technique, it is crucial to identify the areas on the pipelines where AC corrosion is likely to occur. In general, the following factors and associated parameters and conditions can guide the choice of the inspection techniques to be used in the field:

- **HVAC power lines.** HVAC power lines paralleling buried pipelines provide the primary source of AC interference to the pipelines, resulting in AC corrosion. All parameters relevant to the operation of the power lines should be included for consideration.
- **Other pipelines.** Other pipelines in the vicinity, whether they are parallel or crossing, may contain fault current, which can be picked up from nearby facilities, causing AC corrosion of the target pipelines.
- **Road crossings.** Roads crossing the buried pipelines do not directly induce AC interference. However, chemical salts, microorganisms, and other factors contributing to increased conductivity of the medium close to the roads can decrease the resistivity of the soil through which the AC current flows. This would elevate the AC level and thus accelerate AC corrosion.
- **Voltage peak locations.** There are locations where AC voltage peaks are expected to measure. These include stretches where pipelines deviate away from the paralleling HVAC power lines, horizontal directional drilling entry or exit points, and the presence of aboveground facilities that are electrically isolated from the buried pipelines with isolation kits.

It is critical to determine the AC interference by measuring or monitoring AC voltages or AC current densities at these locations. The methodology for measuring and monitoring the induced AC interference was not sufficiently mature until recently.[1] Generally, continuous monitoring is required over a minimum of seven days so that the variations of AC voltage or AC current density over a typical day during a typical week are accurately measured. These measurements provide information regarding the variation pattern and the amount of demand on the power grid by consumers throughout the day and the week. Furthermore, AC corrosion is more related to the time-weighted average AC and DC effects, rather than the maximum conditions. Thus, a reasonable monitoring and inspection program should consider the variability that can occur in the field and the associated pipeline integrity risks.[2]

## 7.1. Conventional measurement techniques

While AC corrosion occurs on the external surface of buried pipelines, the techniques enabling the inspection and monitoring of external corrosion can provide information about AC corrosion if the AC interference sources are adjacent to the pipelines. However, not all techniques are specifically applicable to AC corrosion. In fact, most techniques measure the corrosion rate and the corrosion-induced consequences–such as metal loss, defects, and cracks–rather than the reasons causing pipeline corrosion, such as AC interference. Commonly used pipeline inspection programs, such as ILI tools, are not always able to detect the initiation or progression of external defects such as coating failures, which are considered the major target of AC-induced corrosion. Since extensive literature resources describe the techniques and methods for online pipeline corrosion inspection and monitoring, this chapter includes those that are designed and used for AC corrosion only.

The most commonly acceptable techniques for the detection and monitoring of the AC corrosion risk are through the measurement and monitoring of AC voltage and AC current density using specialized test coupons and probes. Depending on the type of installation of the coupons and probes, the monitoring can be done manually or using remote monitoring equipment, providing real-time data inputs. Manual monitoring, such as periodic field visits to the test station or scheduled pipeline CP surveys, provides only instant information about the pipeline conditions at the location and time the measurement is conducted. For remote monitoring, continuous data collection, transmission, and storage are accomplished, along with alarm notification of abnormal AC voltage or AC current levels and fault conditions.[3]

### 7.1.1. Buried coupon test station

One of the most widely accepted methods for monitoring the AC corrosion of buried pipelines is to measure the AC current density, rather than just AC voltages, using coupon test stations installed adjacent to the pipelines. Both AC and DC coupons are used and are typically installed at the test stations. The instant-OFF (polarized) potential and DC current density are measured and monitored using the DC coupons, and the AC potentials (or voltages), AC current densities, and DC current densities are measured and monitored using the AC coupons.[4] The measurement results are then analyzed and compared with the level that would result in pipeline corrosion when exceeded. Sometimes, electric resistance probes are installed to monitor the corrosion rate of the pipelines under AC interference, providing validation to the coupon measurements. Since the AC current density at a coating holiday on pipelines cannot be measured directly, the AC current density is measured on the steel coupon that is buried at the pipe depth when the coupon is connected to the steel pipe. The coupon and a reference electrode should be placed as close to the pipeline as possible to minimize the ohmic drop across the soil. Thus, the coupon DC potential can be considered as the polarized potential.[5] In general, the recommendation is that a 1 $cm^2$ coupon is installed at the pipe depth to measure the AC current density. Figure 7-1 shows the installation of a steel coupon adjacent to a buried pipeline that is under AC interference.[4] The coupon and the pipe are electrically connected via a conducting wire, providing the real-time data. Both the coupon and the pipe are adjacent and under the same environmental conditions, including soil chemistry, CP current, and AC interference. Coupon currents can be measured by the voltage drop across an internal shunt resistor between the coupon and the pipe. Although the coupon DC and AC potentials are not aimed at determining the CP level, these values are also useful in understanding the CP performance condition of the buried pipelines. The coupon AC potentials are measured relative to the reference electrode, instead of remote earth, and used to roughly determine the levels of AC interference.

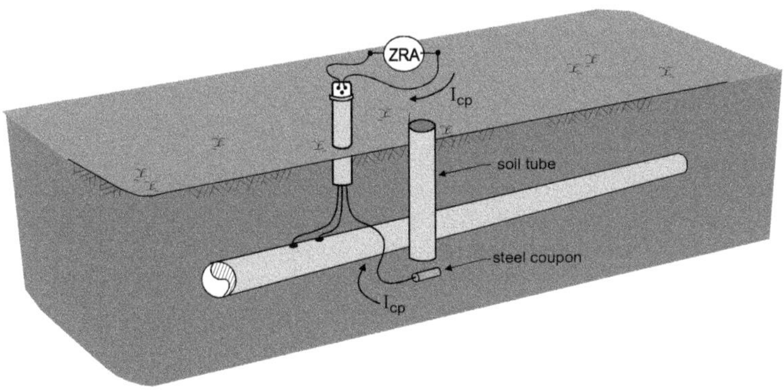


**Figure 7-1.** Schematic diagram of the installation of a steel coupon electrically connected with a buried pipeline that is under AC interference.[4]

Generally, the following considerations are critical to the coupon installation:

- A good, stable, and durable electrical contact is established between the coupon and the steel pipe.
- The same soil condition is present where the coupon is installed and at the pipeline.
- The coupon is positioned correctly relative to the reference electrode.
- The influence of neighboring structures on the measurement of the coupon is mitigated or eliminated.
- A single coupon is buried at any one location.

Moreover, the following requirements must be met before measurements are conducted on the buried coupons:

- All CP test stations and drainage are under normal operating conditions.
- All groundings, such as AC discharging devices and capacitors, are connected.
- The reference electrode is located at the ground surface, sufficiently close to the pipeline.

In addition to the AC current density, the parameters measured and calculated using the buried coupons also include AC potential, DC potential, and DC current density to enable complete understanding of the AC corrosion process and its mitigation at the test site. Büchler and Schöneich noted that although the AC current density determined on the coupons provides information on the corrosion state, the AC potential measured on pipelines serves as an indicator rather than a criterion.[6]

**AC current density.** AC current density is typically measured using a steel coupon with an area of 1 $cm^2$. It can be used to directly evaluate the AC corrosion risk. Generally, if the root mean square of the AC current density is lower than 30 A/$m^2$, the AC corrosion risk is considered low.[7] Furthermore, the coulometric development provides a technique to determine the amount of AC corrosion on the coupons in the field.[8] Characterization and analysis of the steel coupons subjected to the AC interference allow evaluation of the coulometric oxidation of the steel to determine the weight loss resulting from AC corrosion on the coupons.

**Coupon-to-soil potential.** The pipeline is generally regarded to be protected from AC corrosion if the coupon-to-soil OFF potential is more negative than the NACE-recommended protection potential range. For iron or steels in aerobic environments, the protection potential recommended by NACE is –850 mV (CSE), and in anaerobic soils containing sulfate-reducing bacteria, the recommended potential is –950 mV (CSE).[9] The DC potentials are measured with 10 $cm^2$ coupons.

**Ratio of AC current density to DC current density.** In addition to AC current density, the DC current density between the coupon and the pipeline must be measured simultaneously. For the buried coupon, if the ratio of the root mean squares of AC current density and DC current density is:

- Lower than 3, the AC corrosion risk is low.
- Between 3 and 10, the AC corrosion risk is medium, and further investigation is advised.
- Higher than 10, the AC corrosion risk is high, and immediate mitigation measures should be taken.

**Case analysis.** AC corrosion was claimed to have caused significant metal loss on a buried oil pipeline made of API 5L X42 steel, which was constructed in the 1990s and primarily coated with coal tar, with several short replacement segments coated with FBE. The segment of the pipeline containing the corrosion features was collocated with one 345 kV and one 69 kV HVAC transmission line.[10] Field assessment and laboratory analysis were conducted on six corrosion features, and it was confirmed that AC corrosion occurred on the pipeline. The AC current densities calculated from the AC voltages measured at all features exceeded 30 A/m$^2$, and even exceeded 100 A/m$^2$ at three features, with a maximum current density of 160 A/m$^2$. At the same time, online AC/DC voltage and current density measurements were conducted on a coupon test station, which was approximately 46–134 m from the six corrosion features assessed and analyzed. The steel coupon had a surface area of 1 cm$^2$. The reported AC voltage had a maximum reported value of 2.03 V, with an average of 1.11 V. The AC current density was calculated using the measured current on the coupon, divided by its area of 1 cm$^2$. The maximum value of AC current density was 35 A/m$^2$, compared with the maximum AC current density of 160 A/m$^2$ measured on the six corrosion features on the pipeline. Obviously, there was a big difference between the measured AC current density on the test coupons and the calculated value. However, the AC voltage measurements for both the coupon and the pipeline were comparable. Analysis showed that the discrepancy between the test coupon measurements and the calculated AC current density for the corrosion features on the pipeline was primarily attributable to the variation in soil resistivity. Generally, the AC current density at a coating holiday is controlled by the local soil resistivity near to the holiday and increases with a decrease in the soil resistivity. When a test coupon station is located far away from the buried pipeline, the local soil resistivity might be different, reflected in an appreciable difference in the AC current density. For this reason, the coupon measurements were believed not to be representative of the AC corrosion on the pipeline. Thus, test coupons should be buried sufficiently proximate to the pipeline so that they experience identical soil chemistry and resistivity.

**CP monitor.** Kajiyama and Nakamura developed an advanced instrumentation, named a "CP monitor," that was based on the buried coupon method to measure and assess the AC corrosion of pipelines; they also modified the CP potential criterion of pipelines in the presence of AC interference.[11] Conical coupons with a surface area of 10 cm$^2$ were used in the CP monitor to ensure good contact between the coupons and the electrolyte (i.e., the soil solution). Simultaneous measurements of coupon AC and DC current densities ($i_{AC}$ and $i_{DC}$) and the ON potentials with a sampling rate of 0.1 ms were conducted during a period of 24 h. A low pass filter with a cut-off frequency of 73 Hz was used to avoid abnormal electrical spikes and harmonic currents.

The criterion based on the measured coupon DC and AC current densities was established for an effective control of pipeline corrosion, including AC corrosion, DC corrosion, and microbiologically influenced corrosion, as shown below:[11]

$$0.1\ A/m^2 \le i_{DC} < 1.0\ A/m^2,\ i_{AC} < 25 i_{DC}$$

$$1.0\ A/m^2 \le i_{DC} < 40\ A/m^2,\ i_{AC} < 70 i_{DC}$$

Positive values of $i_{DC}$ indicate the current flowing through the electrolyte to the coupon. It was noted that, for pipelines buried in proximity to HVAC power lines, AC-powered rail transit systems, DC-powered rail systems, and foreign cathodically protected metallic structures, both DC and AC coupons should be installed to assess the pipeline AC and DC corrosion and evaluate the CP performance.[12]

## 7.1.2. Ultrasonic-based coupon testing method

For the coupon testing method used for AC corrosion measurements, the metal coupon serves as a simulation of the corrosion behavior of the underground structure that is being protected, such as pipelines. An electric resistance probe is used to measure changes in electrical resistance or potential in the vicinity of the underground structure. However, the changes in electrical potential or resistance are usually "noisy" and are not always consistent.[1] Moreover, the measured signals can be affected by many variables, such as temperature variations, the presence of stray current, instrumentation calibration issues, and operational error. It is thus expected that the coupon testing would not be affected by environmental or operational conditions and that data are provided with sufficient precision.

Warner et al. reported a method using a traditional metal coupon imbedded with ultrasound transducers, enabling direct measurements of the coupon metal loss under AC interference with improved accuracy (Figure 7-2). In this method, ultrasound is used to directly measure the coupon thickness as a function of time. Moreover, the ultrasound-based coupon testing can be integrated into CP test stations. The obtained results include both AC and DC current densities, which are combined with metal loss, enabling more accurate and reliable evaluation of the integrity of buried pipelines and CP performance in the presence of AC interference. The ultrasound transducers are positioned directly underneath the coupon's surface, and they measure the thickness of the coupon, which changes along with the corrosion of underground pipelines in the presence of AC interference. Thus, the obtained information provides a direct rather than an inferred measure of the corrosion rate of steel coupons.[1]

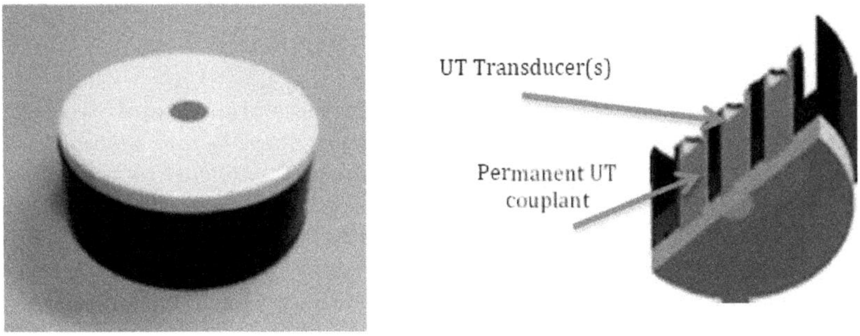


**Figure 7-2.** Ultrasound coupon for AC monitoring and its cross-sectional view.[1]

There are several benefits to using ultrasonic coupon testing in AC corrosion measurements. First, it responds to regulatory and NACE guidelines that "external corrosion rates should be determined by directly comparing measured wall thickness changes."[13] Second, real-time precise corrosion rate information is provided without physical access to the buried coupon, which is usually inconvenient and costly. Third, remote acquisition of corrosion data is achieved, with improved data accuracy and reliability.

### 7.1.3. AC voltage modulation technique

Normally, test stations for CP monitoring can experience AC voltage peaks due to the adjacency between the pipeline and the HVAC power lines.[7] Electric multimeters and CP measurement instruments are capable of measuring both DC and AC voltages. Measurement of voltage between a pipeline and a reference electrode is probably the most frequent method used to determine pipeline corrosion and CP performance. Therefore, when a DC pipe-to-soil potential is measured at a CP test station, it is convenient to switch the instrument from DC to AC poles to measure the AC voltage of the pipeline. The AC voltage should be measured with respect to remote earth.[14]

When the AC voltage measurement is conducted in proximity to a grounding electrode or some other AC mitigation facilities, the AC current discharging to the ground would create a potential gradient in the soil owing to the soil resistivity. This can result in the measured voltage being less than the actual voltage to remote earth. If it is greater than 10 V (CSE), the AC voltage should be monitored for 24 h to determine whether the AC interference varies over the day. Empirical evidence suggests that further investigation is required when the AC voltage (mV, CSE) is greater than the soil resistivity (Ω·cm) at the pipeline depth, where $i_{AC}$ would be greater than 20 A/m$^2$ on a 1 cm$^2$ coating holiday.[7] Moreover, both AC and DC pipe-to-soil potentials should be routinely monitored to determine if the AC voltage remains below 15 V, a level that would result in an acceptable AC corrosion current density based on a maximum $i_{AC}$ of 50 mA/cm$^2$ and a ratio less than 5 for $i_{AC}$ to $i_{DC}$. It is suggested that, for a more conservative mitigation program, the AC corrosion current criteria could be set at 20–30 mA/cm$^2$.[7]

If the AC corrosion risk can be evaluated on the basis of induced AC voltage, a handheld voltmeter can simply and promptly detect the locations where AC corrosion is likely. Therefore, although the assessment of AC corrosion likelihood based on AC voltages can be misleading, it is convenient and effective to measure AC voltage to roughly evaluate the likelihood of AC corrosion.[5] However, it is AC current density ($i_{AC}$, A/m$^2$), rather than AC voltage ($V_{AC}$, V), that is directly related to AC corrosion. Thus, the AC voltage measured on pipelines can be used to calculate the AC current density if the local soil resistivity and spread resistance at the coating defect are estimated. The AC voltage measurement on buried pipelines is usually not used by itself to evaluate the tendency of AC corrosion.

Furthermore, the infrastructure in CP test stations is often complicated and can comprise grounding systems, junction boxes, anodes, and other equipment. AC voltages measured locally are considered as the touch voltages.[15] However, the AC voltage on the pipeline, which would typically be higher, should be measured relative to remote earth, while the touch voltages outside the local area would be close to the AC voltage to remote earth for a well-coated pipeline. Incorrectly recorded AC voltages could cause an underestimation of the electrical shock hazards and AC corrosion risks.

The phase of most power lines varies seasonally in summer and winter. Peak powerline loading is recorded in the summer during hot weather, and daily between 6 pm and 7 pm, when people started using appliances and air conditioning.[14] Moreover, the soil moisture typically varies from wet in the spring to dry in the fall, which affects the resistance of the soil medium for current flow. The variations in the environmental conditions mean that AC-induced voltages and the applied AC mitigation program should be monitored frequently. One annual measurement is not a reliable indicator of the AC interference risk to the pipeline. For example, it is unlikely that the peak AC voltage would be captured by one AC measurement, as peak or maximum power line load conditions only occur at certain periods. Thus, to ensure that the level of AC interference on a pipeline is correctly assessed, fluctuations in power line loading need to be considered and recorded seasonally, monthly, weekly, and even hourly.

In addition to being collocated along HVAC power lines, pipelines could be buried adjacent to AC-powered railway systems. As a result, the pipelines could be subject to additional AC corrosion risk from the rail systems. In a field test, AC pipe-to-soil voltages were measured on a pipeline running parallel to a rail line for about 150 m.[16] The induced AC voltages for the pipeline fluctuated between 4 $V_{rms}$ and 33 $V_{rms}$. The AC voltage could vary, especially when multiple electric-powered trains traveled simultaneously through adjacent rail tracks. Since the recorded AC voltages exceeded the NACE-recommended control threshold of 15 $V_{rms}$,[15] it was strongly recommended to mitigate the AC interference through suitable techniques.

**AC voltage close interval survey.** The AC voltages along a pipeline have been proposed as a primary indicator of electrical safety hazards and the AC corrosion risk under the AC interference of collocated power lines.[14] Since the locations of the CP test stations and AC voltage peaks on pipelines do not always coincide, it may be possible to determine the AC voltage profile by conducting AC close interval survey (CIS) along the pipeline.

To date, there has been no indication in the literature that an AC CIS on buried pipelines is not theoretically possible.[14] Performing a CIS of AC voltages on pipelines has been attempted.[17] Research has also been conducted to investigate the use of AC CIS in evaluating AC corrosion risks.[18, 19] Some also regard AC CIS as a primary means of characterizing the AC level on pipelines.[20] However, in Lindemuth et al. an attempt to conduct an AC CIS failed owing to the interference of power lines on the instrument and measurements of AC pipe-to-soil potential.[17] Although various methods can be used to minimize the interference effects, there is a fundamental reason why the data obtained from an AC CIS are not valid, as explained below.[14]

Unlike DC voltages, the AC voltages induced on buried pipelines are complex numbers. An AC voltage–measuring circuit should consist of impedance elements, which are in series or parallel connections of resistance, capacitance, and inductance (Figure 7-3). For a well-coated pipeline on which the coating is intact, the soil voltage gradient due to the discharge of AC current is negligible owing to the high resistance of the coating. The AC potential of the soil immediately above the pipe is the same as the potential at remote earth. As a result, an AC CIS conducted on the pipeline with a voltmeter that is connected to the CP test station would produce a series of identical measurements, all being equivalent to the AC voltage at the test station with respect to remote earth.[14] When the external coating degrades and extensive coating defects exist, AC voltage gradients with varied magnitudes would exist in the soil because the resistance at the coating defects would be comparable to the soil resistance. The measured AC CIS results would be different, depending on the reference electrode position. However, these measurements are composed of the voltage at the test station, with the addition of an unknown amount of voltage drop in the soil. Thus, the AC voltage CIS is invalid, and the AC voltage relative to earth at any point on a pipeline can only be measured by connecting the voltmeter directly to that point.[14]

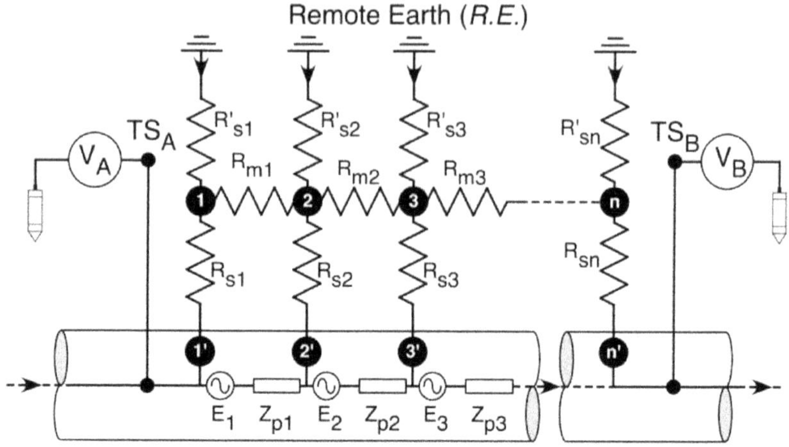


**Figure 7-3. Circuit diagram for a CIS of induced AC pipeline voltages.**[14]

## 7.2. An online AC corrosion monitoring and assessment technique

An online AC corrosion monitoring and assessment technique was developed, where the AC/DC real-time signal measurements were performed through a National Instruments USB-6009 multifunction DAQ card and a National Instruments DAQ data-logging software package that was written and coded on the LabVIEW 2011 64 bit programming platform.[21] The input analog voltage range of the hardware is ±20 V, with an input resolution of 14 bits. The sampling rate is set at 10 kS/s. The software includes three modules: (a) a data analysis module for analyses of the recorded AC/DC mixed signals obtained by the DAQ card as well as calculation of the amplitude, frequency, phase, and power spectrum; (b) a data operation and display interface for the input of settings and the display of real-time potential-time plots; and (c) a data storage module to save the analyzed data as a text file. The technique can separate the recorded mixed potential signals that contain both DC and AC voltage components.

While Figure 3-9 shows the real-time DC potentials that are separated by the DAQ software from the recorded mixed potential signals on a 16Mn steel electrode at various AC current densities in a simulated soil solution, Figure 7-4 shows the DAQ-separated AC potentials recorded on the steel electrode in the same time domain under various $i_{AC}$ in the solution. It can be seen that all curves show an ideal sine wave shape and have the same frequency of 60 Hz (i.e., the measuring frequency). Moreover, the magnitude of the AC potential increases with the AC current density.

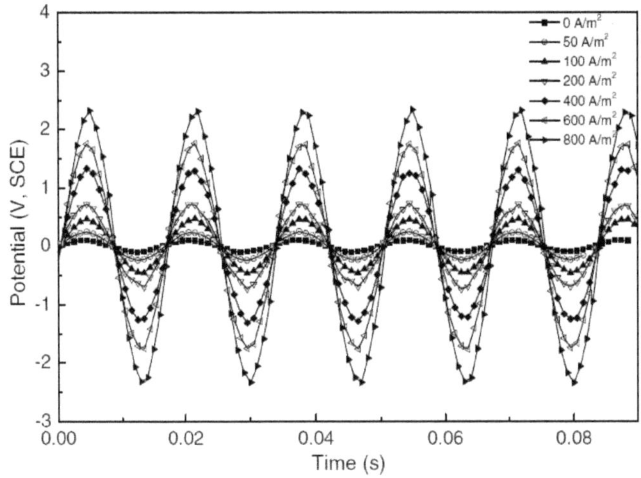


**Figure 7-4.** DAQ-separated AC potentials recorded on a 16Mn steel electrode in the same time domain as Figure 3-9 under various $i_{AC}$ in a simulated soil solution.[21]

Figure 7-5 shows the AC peak-to-peak potential ($V_{PP}$) under various $i_{AC}$ in the test time period shown in Figure 3-9. In stage i (i.e., from 0 s to 1,000 s) and stage iii (i.e., after 2,000 s), no AC is applied. In stage ii, the $V_{PP}$ is maintained at a constant value at individual AC current densities up to 400 A/m², and the $V_{PP}$ increases at higher $i_{AC}$ values. However, when $i_{AC}$ is equal to 600 A/m² or 800 A/m², $V_{PP}$ increases gradually to a steady-state value of 5.2 V (SCE) and 5.7 V (SCE), respectively, which are much more positive than those measured under an $i_{AC}$ lower than 400 A/m².

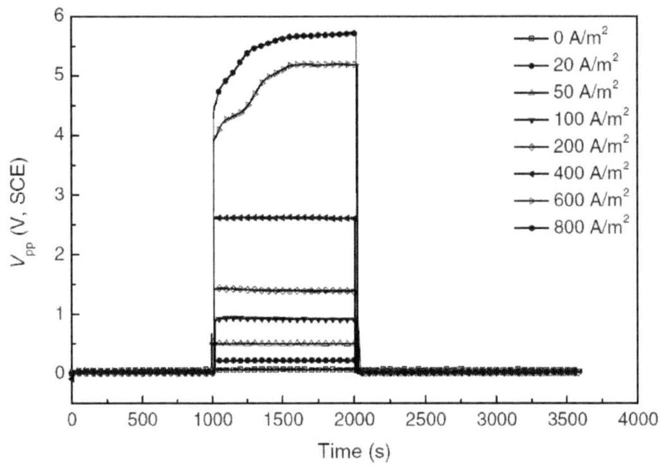


**Figure 7-5.** AC peak-to-peak potential ($V_{PP}$) at various AC current densities in the time period shown in Figure 3-9: (a) corrosion potential before 1,000 s; (b) AC-ON potential from 1,000 s to 2,000 s; (c) AC-OFF potential after 2,000 s.[21]

Figure 7-6 shows the most positive potential ($V_{P+}$) separated by DAQ under various AC current densities. The stages from 0 s to 1,000 s and after 2,000 s are not under AC application. When the $i_{AC}$ is 400 A/m² or less, $V_{P+}$ increases immediately to a steady-state value. However, at an $i_{AC}$ of 600 A/m² or 800 A/m², the $V_{P+}$ increases gradually to reach steady-state values of 1.95 V (SCE) and 2.25 V (SCE), respectively. The steady-state value of $V_{P+}$ increases with increasing $i_{AC}$.

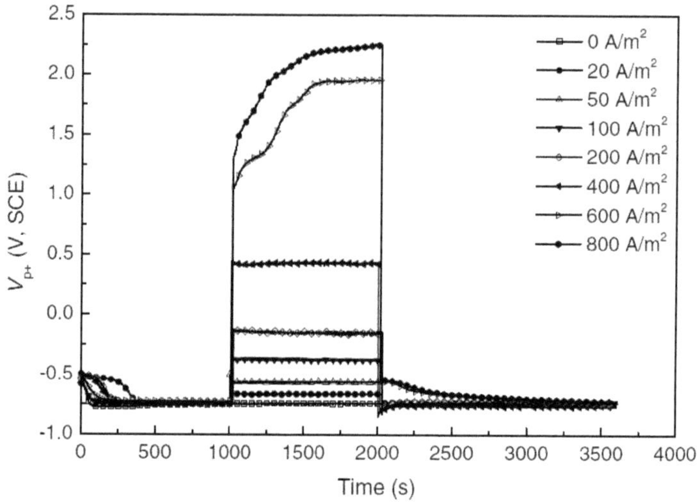


**Figure 7-6. The most positive potential ($V_{P+}$) of the steel under various AC current densities.**[21]

The most negative potential ($V_{P-}$) on the steel electrode was also separated by DAQ, and the results are shown in Figure 7-7. The stages from 0 s to 1,000 s and after 2,000 s are not under AC application. It can be seen that the change of $V_{P-}$ is identical to that of $V_{P+}$ in Figure 7-6, but in a different direction. The higher the AC current density applied, the more negative the steady-state value of $V_{P-}$.

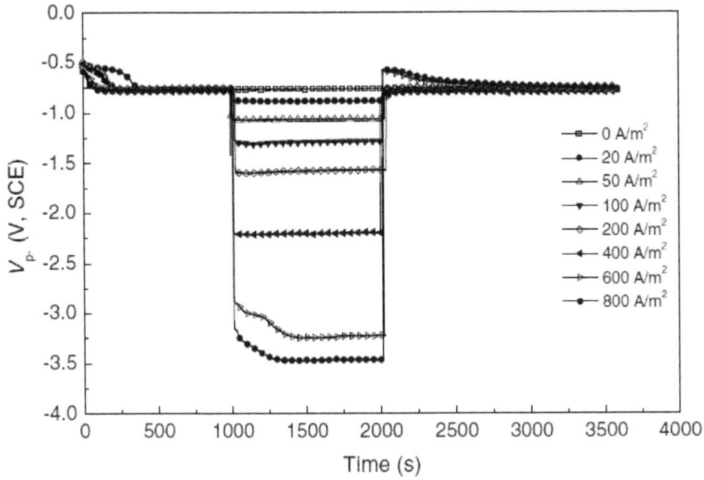


**Figure 7-7. The most negative potential ($V_{P-}$) of the steel under various AC current densities.**[21]

Figure 7-8 shows the dependence of the AC peak-to-peak potential ($V_{PP}$), the most positive potential ($V_{P+}$), and the most negative potential ($V_{P-}$) on AC current density ($i_{AC}$). Linear relationships are observed for the three potentials as a function of the AC current density, and the following empirical equations can be derived:

$$V_{PP} = 0.0074 i_{AC} + 0.059 \quad (7\text{-}1)$$

$$V_{P+} = 0.0039 i_{AC} - 0.795 \quad (7\text{-}2)$$

$$V_{P-} = -0.0035 i_{AC} - 0.860 \quad (7\text{-}3)$$

These equations can be used to calculate $i_{AC}$ by measuring and separating the AC voltages from the recorded mixed signals in the field. Although AC-induced corrosion is a serious problem in pipeline industry, it is usually easy to measure AC potential, but very difficult to measure AC current density directly owing to the lack of information about the spread resistance of the coating defect and thus the AC current flowing through it. However, it is feasible to measure AC voltages at coating defects using the developed DAQ technique. It is expected that the AC current density can then be determined from the derived relationships. The empirical equations shown above are derived under specific testing conditions. Further validation of the model and equations is required for generalization in determining AC current density in the field. The technique provides a promising alternative for monitoring and assessing the AC corrosion of buried pipelines in the field.

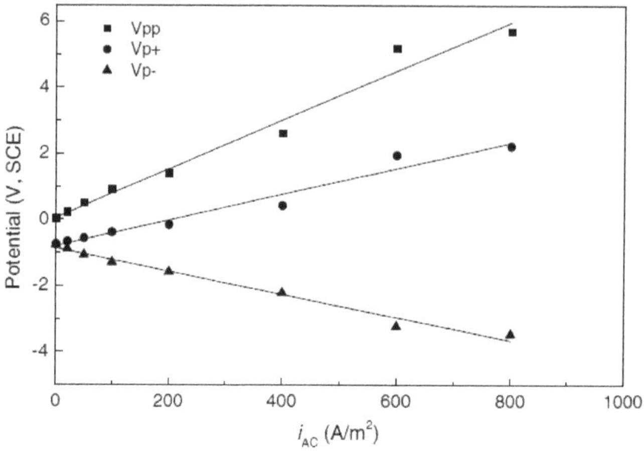


**Figure 7-8.** Dependence of AC peak-to-peak potential ($V_{PP}$), the most positive potential ($V_{P+}$), and the most negative potential ($V_{P-}$) on the AC current density ($i_{AC}$).[21]

## 7.3. A graphical user interface-based AC corrosion monitoring technique

A graphical user interface (GUI)-based technique was developed and operated through the deployment tool in MATLAB for monitoring AC-induced corrosion and conducting AC corrosion risk assessments on pipelines.[22] With initial inputs about the pipelines (e.g., pipe diameter, pipe wall thickness, burial depth, type and properties of the coatings, geometry of the coating defects, length of the target segment, coating thickness), the HVAC power lines (e.g., frequency, operating voltage, peak load current per conductor), and the soil (e.g., soil resistivity, soil relative permeability, aeration, chemical composition, microorganisms), the GUI provides the results, including the induced AC voltage on the pipeline, the pipe-to-soil potential, the relative risk value of AC corrosion, and the AC current density through the coating defect on the pipe. While various conventional methods are used for measuring and monitoring the pipeline, coating, soil, and CP data as inputs, the GUI serves as a central processing unit to deliver relevant results based on the AC corrosion assessment model. The AC corrosion risk assessment criteria are monitored using the GUI to ensure that a threshold value above which there is a high likelihood of AC corrosion damage to the pipeline is not exceeded. If the threshold is exceeded, the GUI will alert the pipeline operator to take adequate measures, such as installation of the AC mitigation system or other AC reduction strategies.

The basic procedure for the GUI technique is as follows. First, the data–including the pipeline, soil, and power line data described above–must be available as inputs. Second, a program computes the AC corrosion risk assessment criteria using the following equation:[22]

$$\text{Relative risk} = V_{p/s} / \rho_{soil} \tag{7-4}$$

where $V_{p/s}$ is the pipe-to-soil potential along the pipeline section at any point, and $\rho_{soil}$ is the soil resistivity. The AC current density through a coating defect on the pipeline is used as the most important parameter for assessing and predicting corrosion induced by AC interference. The threshold value of AC current density is selected for the pipeline corrosion risk assessment. For estimation of AC current density, $i_{AC}$, the spread resistance at a coating defect, $R_{spread}$, to remote earth is expressed as follows:[22]

$$R_{\text{spread}} = \frac{\rho_{\text{soil}}}{2d}\left[1 + \frac{8t_c}{d}\right] \tag{7-5}$$

where $d$ is the diameter of the coating defect, and $t_c$ is the coating thickness. The current, $I$, is then obtained by Ohm's law:

$$I = \frac{2d^2 V_{p/s}}{\rho_{\text{soil}}(d+8t_c)} \tag{7-6}$$

Assuming a circular defect, the current density, $i_{AC}$, through the defect at a point $x$ along the pipeline section is expressed as:

$$i_{\text{AC}} = \frac{8V_{p/s}(x)}{\pi\rho_{\text{soil}}(d+8t_c)} \tag{7-7}$$

If the estimated AC current density at the coating defect is greater than 100 A/m$^2$, the GUI gives information on the AC corrosion condition (i.e., that corrosion damage is expected) on the pipeline. Recommendations are given regarding the installation of an AC corrosion mitigation system. If the AC current density is smaller than 30 A/m$^2$, the GUI would show no corrosion expected. When the AC current density is between 30 A/m$^2$ and 100 A/m$^2$, a "low corrosion likelihood" message is shown and a mitigation system may be installed. The field input data may be varied to provide different outputs for the analysis of pipeline AC corrosion under different conditions.

# References


1. Warner, J., Leary, R., King, D., "The use of ultrasound for monitoring the impact of AC-induced corrosion on underground structures," CORROSION/2017, paper no. 9722, NACE, Houston, TX, 2017.
2. Lindemuth, D., Crabtree, D., "AC corrosion control: When too much cathodic protection might just be a bad thing," CORROSION/2018, paper no. 11271, NACE, Houston, TX, 2018.
3. Hilleary, J., DeWitt, J., "Continuous verification monitoring at AC mitigation stations," CORROSION/2011, paper no. 11310, NACE, Houston, TX, 2011.
4. Fieltsch, W., Shahinas, F., "CP considerations for pipelines subjected to AC interference," NACE Northern Area West Conference, Calgary, Alberta, 2015.
5. Kajiyama, F., "Strategy for eliminating risks of corrosion and overprotection for buried modern pipelines," International Gas Union Research Conference, Seoul, 2011.

6. Büchler, M., Schöneich, H., "Investigation of alternating current corrosion of cathodically protected pipelines: Development of a detection method, mitigation measures, and a model for the mechanism," *Corrosion* 65: 578–586, 2009.
7. Gummow, R.A., *A/C Interference Guideline Final Report*, Canadian Energy Pipeline Association, Calgary, Alberta, 2014.
8. Büchler, M., Voute, C.H., Schöneich, H.G., "Effect of cathodic protection levels and defect geometry on the AC corrosion on pipelines," CeoCor International Congress, European Committee for the Study of Corrosion and Protection of Pipes and Pipeline Systems, Brussels, 2007.
9. NACE, *Control of External Corrosion on Underground or Submerged Metallic Piping Systems*, SP0169-2013, Houston, TX, 2013.
10. Finneran, S., Norfleet, D., Buckingham, L., "Field examination of AC accelerated corrosion featured in monitored locations," CORROSION/2018, paper no. 10911, NACE, Houston, TX, 2018.
11. Kajiyama, F., Nakamura, Y., "Effect of induced alternating current voltage on cathodically protected pipelines paralleling electric power transmission lines," *Corrosion* 55: 200–205, 1999.
12. Hosokawa, Y., Kajiyama, F., "Case studies on the assessment of AC and DC interference using steel coupons with respect to current density CP criteria," CORROSION/2006, paper no. 06161, NACE, Houston, TX, 2006.
13. NACE, *The Use of Coupons for Cathodic Protection Monitoring Applications*, RP0104-2004, Houston, TX, 2004.
14. Fieltsch, W., Wakelin, R., "The AC close interval survey and other common AC measurement errors," CORROSION/2017, paper no. 9455, NACE, Houston, TX, 2017.
15. NACE, *Alternating Current Corrosion on Cathodically Protected Pipelines: Standard Practice for Risk Assessment, Mitigation, and Monitoring*, Proposed NACE Standard Practice, Draft 1a, Houston, TX, 2015.
16. George, J., "Alternating current interference on cathodically protected pipeline from railway electrification systems," CORROSION/2018, paper no. 10723, NACE, Houston, TX, 2018.
17. Lindemuth, D., Creel, J., Hernandez, H., Neeley, M., "AC mitigation and cathodic protection for a long regulated pipeline," CORROSION/2010, paper no. 10105, NACE, Houston, TX, 2010.
18. Yunovich, M., Waters, S., *External Corrosion Direct Assessment for Unique Threats to Underground Pipelines*, Pipeline and Hazardous Materials Safety Administration, U.S. Department of Transportation, Washington, DC, 2007.
19. Simmonds, D., Orton, S., "Development of over-line survey systems for monitoring pipelines within an alternating electric field," 23rd Colloquium on Safety and Reliability of Gas Pipelines, Prague, 2014.
20. NACE, *AC Corrosion State-of-the-Art: Corrosion Rate, Mechanism, and Mitigation Requirements*, Publication 35110, Houston, TX, 2010.
21. Xu, L.Y., Su, X., Yin, Z.X., Tang, Y.H., Cheng, Y.F., "Development of a real-time AC/DC data acquisition technique for studies of AC corrosion of pipelines," *Corrosion Science* 61: 215–223, 2012.
22. Adedeji, K.B., Ponnle, A.A., Abe, B.T., Jimoh, A.A., Abu-Mahfouz, A.M., Hamam, Y., "GUI-based AC induced corrosion monitoring for buried pipelines near HVTLs," *Engineering Letters* 26: 489–497, 2018.

# 8 Modeling for Evaluation of AC Interference and the Risk of AC Corrosion of Pipelines

Modeling and evaluation of AC interference and its adverse impact on buried pipelines from adjacent AC sources is complex and usually difficult. In recent decades, the development of advanced numerical methods and computer-based tools for the location of pipeline segments potentially exposed to AC interference, simulations of the level of AC interference and its effect on pipelines, evaluations of the risk of AC corrosion of the affected pipelines, and the design of AC interference mitigation programs have been paid much attention. The main functions of the computer-based modeling include (a) the determination and evaluation of AC interference from HVAC power lines or similar sources and the resulting impact on the collocated pipelines and (b) the design of appropriate mitigation programs to reduce and control AC interference and pipeline corrosion. Moreover, the models can analyze the interacting contributions of multiple variables and factors that affect the AC interference, providing technical recommendations to pipeline and power line operators for optimal design of the operating conditions and ROW selection to avoid the problem. The modeling development also contributes to improved understanding of the AC corrosion mechanism.

## 8.1. Historical background

Even as late as the mid-1970s, analytical techniques for calculating AC interference levels were not accurate. Carson's equations, which were originally developed to predict the coupling effect between HVAC power lines and the adjacent aboveground communications circuits, were applied to underground pipelines, where the resulting errors were up to an order of magnitude (10 times) or more.[1] Obviously, this method was not applicable for buried pipelines because the pipelines exhibited a continuously distributed resistance to earth. The induced voltage was continuously drained to earth, making the induced AC voltage not a problem to the pipelines.[2]

In 1976, the American Gas Association and the Electric Power Research Institute co-initiated a project to develop a manually calculating program to predict the inductively induced voltages on pipelines from the parallel HVAC power lines with an accuracy on the order of 10%.[3] This was the first software to predict steady-state AC-induced voltages on buried pipelines. Further work was conducted to simulate realistic situations, in which the pipelines and HVAC power lines are not always parallel. In the 1980s, the Electromagnetic and Conductive Coupling Analysis from Powerlines to Pipelines program was developed to account for both inductive and conductive coupling effects during fault conditions on HVAC power lines.[4] The early-stage AC interference modeling focused on a first-order approximation of the interaction between HVAC power lines and pipelines.[5, 6, 7, 8, 9] It was assumed that the power lines had a determining influence on the voltage induced on the pipeline and that the pipeline had a negligible impact on the voltage associated with the power lines. The approximation is acceptable in most cases, but there are situations in which this could cause significant errors, either underdesigning or overdesigning the protection for the pipeline.[10] This is particularly true during a phase-to-ground fault on an HVAC power line tower that has a strong through-earth coupling with a collocated pipeline.

Nowadays, with the increased complexity of the pipeline–power line collocation, predictive models have become more complicated. Advanced computing capabilities are required to reflect the actual status in AC interference modeling, such as power line superpositions and phase conductor bundling, an unrestricted number of bonds between pipelines and the mitigation system, the nonparallel collocation of pipelines and power lines, and HVAC power lines under fault conditions.[11] Two software packages developed by Safe Engineering Services & Technologies were claimed to possess capabilities to model the AC interference effect at various complexity levels.[10] The first one accounted for all possible interactions between the power lines and collocated pipelines and was thus highly accurate. By solving Maxwell's electromagnetic field equations, the method allowed computation of the current distribution in a network consisting of both aboveground and buried conductors with arbitrary orientations, where the effect of a uniform or layered earth of arbitrary resistivity, permittivity, and permeability is completely considered.[12] The second software package made the approximation that the pipeline had a negligible effect on the power lines. This method calculated the power line parameters for the conductor system and then built a circuit model including these parameters and ground impedances. The software computed the voltage and current distributions in the power line system and pipeline. Finally, the power line ground injection currents and the induced electromotive force values on the pipeline from parallel overhead conductors were calculated to determine the interactions between the buried conductors. Although this method is adequate in many applications, it can result in overdesign or underdesign.[10]

In addition to AC interference from the HVAC power lines, transient rail systems were also modeled, and a computer program was developed that was capable of predicting the steady-state induced AC voltages on pipelines and railway installations, under sponsorship of the Electric Power Research Institute, the Association of American Railroads, and the American Railway Engineering and Maintenance-of-Way Association.[13]

# 8.2. Modeling for evaluation of AC interference

## 8.2.1. Principles for modeling of AC interference

The applicability of the modeling approach on any physical processes or phenomena relies on correct initial assumptions of the processes occurring in the specific environment under certain conditions, the input of correct initial parameters and magnitudes, and a proper mathematical modeling tool. For assessment of the AC interference on buried pipelines from HVAC power lines, most models used today were initially developed for AC interference from a single power line. Moreover, a relatively small number of segmentations of the pipelines were targeted and modeled. Owing to uncertainty in the analysis and difficulty in the collection of various input conditions, the use of complex models–although proposed and developed by industry and academic communities–has been limited in practice.

The unavailability and uncertainty in critical input variables can be caused by many factors, such as the HVAC load current and phasing, the exact distribution of soil resistivity along the pipeline and power line ROW, the horizontal directional drilling installations at road crossings, natural and artificially made obstructions, and discontinuities in horizontal mitigation conductors. All of them can cause complex models to lose accuracy in the assessment of AC interference.[11] Yet, as the collocation complexity increases in terms of the amount of collocated infrastructure and its geometric routing, the limitations of the basic modeling methods and software should be modified and improved.

The accurate simulation of AC interference and the associated impact on buried pipelines requires a thorough understanding and knowledge of the involved variables and parameters, as well as their interactions, including the current load in the power lines; phase and frequency; tower design; bulk and local soil resistivity; and pipeline parameters such as pipe geometry, burial depth, diameter, electrical connections or isolations, coating and its performance condition, and CP. They all significantly affect the AC interference and the modeling associated with it. Frequently, the input data are collected from multiple sources, such as field measurements, power line and pipeline operators, internal reports, or open publications. Although multiple models have been developed and used in practice, their applicable limits should be considered. Moreover, verification of the accuracy of the modeling results depends on the expertise of analysts and the reliability of the data input.

The software package commonly used for AC interference modeling is the Transmission Line Model, which calculates the longitudinal electric field using established fundamental Carson or Maxwell equations for electromagnetic fields.[14] The geometry and routing of the collocated pipeline and the HVAC power line are used as inputs to the model, which also has the function of simulating the AC interference for multiple pipelines and power lines, with multiple input variables and conditions. The modeling software can calculate the longitudinal electric field for each section and solve the fundamental equations to calculate the AC voltage and the AC current density for a given collocation.

While most models and developed software are for simulation of the electromagnetic field developed on the affected pipelines, as described in detail below, there is another model based on the double-charge layer at the interface between the coated pipeline steel and the soil electrolyte in the presence of AC interference and CP application.[15] The model considered the equivalent electric circuit, illustrating the steel–soil solution interface, containing both AC source (induced) and DC source (CP) circuits. The impedance of the interface was constituted by the spread resistance; an interfacial double-layer capacitor; some diode-like components representing the anodic and cathodic Butler-Volmer kinetics of all included electrochemical processes, and their related equilibrium potentials; and the diffusion properties of the involved chemical species in terms of Warburg impedances. Particularly, the spread resistance considers the summative effect of the geometry and size of coating defects, the response of the chemical environment to alkalization, and the soil resistivity; the double-layer capacitance is affected by the AC current flowing through the interface. The electrochemical reaction thermodynamics and electrode process kinetics, as well as the mass transfer of corrosive species toward the steel surface for corrosion reactions, are used to quantify the rates of individual reactions occurring at the steel–soil solution interface. Since the model requires knowledge of electrochemistry and electric double-layer structure, as well as the electrode process kinetics, it is appropriate to be used in laboratory research to investigate mechanistically the interaction of multiple affecting factors on the AC corrosion of steel. Practically, the model seems too complicated for field application, especially if the users do not have a background in corrosion science and electrochemistry.

### 8.2.2. Modeling of affecting factors and variables

AC interference on pipelines from collocated HVAC power lines depends on numerous affecting factors that interact with one another. It is thus challenging to separate the effect of a single variable from others in the collocation scenario. As a result, it is difficult, if not impossible, to determine the threshold level for individual variables, outside of which (mostly, below which), the AC interference becomes negligible. Therefore, considering the most important affecting variables, rather than all of them, is a viable approach for AC interference modeling.

**HVAC sources.** The primary factor influencing the level of AC interference (i.e., induced AC voltages) on a pipeline collocated with HVAC power lines is the line voltage and the circuit conductor configuration of the towers. In general, with the increased power line voltage, the phase conductor load current increases, which is the primary driver of the electromagnetic field, and thus the AC corrosion of pipelines. Figure 8-1 shows the predicted maximum induced AC voltages on a pipeline at various HVAC line voltages for both horizontal and vertical tower configurations.[16] It can be seen that the induced AC voltage on the pipeline increases dramatically as the power line voltage increases. In addition, the vertical tower configurations enhance the induced voltage on the pipeline by approximately 50% compared with the horizontal tower configurations. Similar simulation results were also obtained in which, although the induced AC voltage was not uniformly distributed, the vertical configuration of HVAC power lines induced a higher AC voltage maximum value than the horizontal configuration under identical operating conditions.[17] Further supporting evidence was obtained by modeling the induced AC voltages on a gas pipeline that was paralleled with power lines carrying different AC voltages, where the 115 kV power line contributed less to the induced AC interference on the pipeline than did the 345 kV power line.[18]

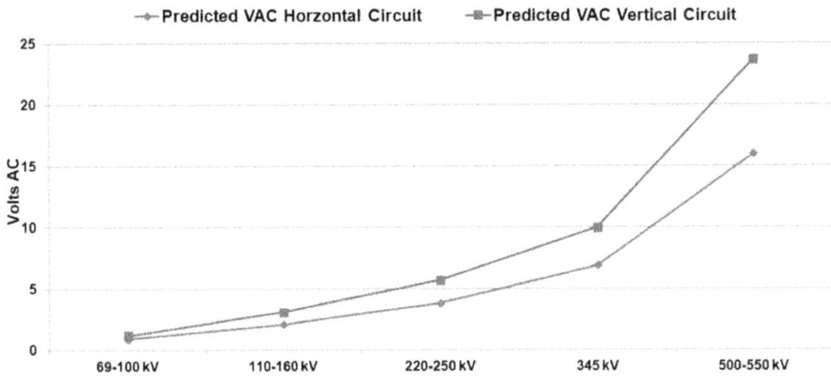


**Figure 8-1.** Predicted maximum induced AC voltages as a function of the power line voltages and tower configurations.[16]

In addition to the power line voltages, the current imbalance also affects the pipe-to-soil potential and the AC current density. Figure 8-2 shows the variability of the pipe potential and the AC current density when the currents carried by the power lines change randomly by ±5% and their phase changes by ±2 degrees.[11] As a comparison, the induced AC voltage and AC current density on the pipeline were also calculated with the assumption that the power line currents were in a balanced condition with an equal magnitude and displaced in phase by 120°. It can be seen that the unbalanced currents can induce AC voltages and AC current densities on the pipeline that are either larger or smaller than those calculated for the balanced condition.[19]

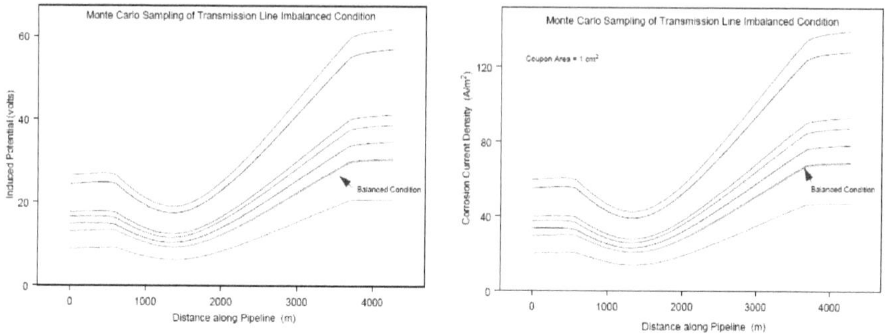


**Figure 8-2. Induced AC voltages and AC current densities on the pipeline for imbalanced phase currents carried by HVAC power lines.[11]**

Moreover, the height of the HVAC power lines affects the induced AC voltage on the buried pipelines. Generally, when the height of the power lines increases, the induced AC voltage decreases owing to the reduced electromagnetic field applied on the pipeline.[17]

**Soil resistivity.** The resistivity of the soil affects the AC voltage distribution and the AC current density along the pipeline that is under AC interference. In addition to the moisture and chemical contents, the water-freezing of the soil in the winter could result in an increase in the soil resistivity by several orders of magnitude, compared with that in the summer. This is particularly true in Canada and the northern area of the U.S., where the frost penetration can vary from centimeters up to a meter or more in depth.[20] Generally, frozen soil conditions can greatly affect the resistive coupling effect under the power line faulting as well as the effectiveness of the AC interference mitigation system. However, the seasonal change of soil conditions is usually negligible in pipeline integrity management. Modeling was conducted for the assessment of AC interference on a gas pipeline under steady-state and fault conditions by two HVAC power lines (315 kV and 735 kV of system voltages) for both summer and winter soil conditions.[20] Figure 8-3 shows the modeling results of the AC voltages and AC current densities under the steady-state condition of the power lines in summer and winter with the installation of zinc ribbon mitigation systems.[20] It can be seen that there is little change in the AC voltage and AC current density between summer and winter when the zinc ribbons are installed below the frost line (i.e., Winter Deep in the figure). If the zinc ribbons are installed within the frost (i.e., Winter Shallow in the figure), the resistance of the mitigation increases in the winter, resulting in an increase in AC voltage and AC current density, and thus an increased risk of AC corrosion with the AC current densities above the 50 $A/m^2$ limit.

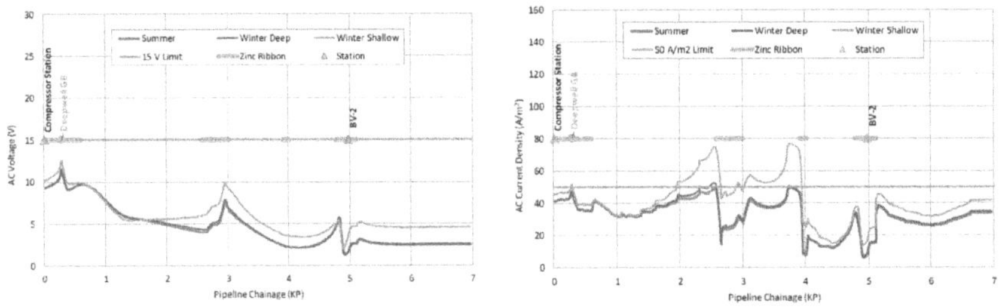


**Figure 8-3. Modeling results of steady-state AC voltages and AC current densities in summer and winter with the installation of zinc ribbon mitigation systems.**[20]

Generally, both average and specific soil resistivities should be considered in AC interference assessment. While the average soil resistivity at the pipeline burial depth is one of the controlling factors in the analysis of induced AC voltage, the local resistivity of the soil layer specific to the pipe surface affects the electrochemical corrosion activity at a coating defect, contributing directly to AC corrosion. The soil resistivity combined with the coating resistance determines the AC current density, which is related to the AC corrosion rate. Thus, a single layer soil resistivity model is not capable of providing the required information for computations of AC interference on pipelines. Two-layer models for soil resistivity have been investigated for many years, but they are not sufficiently accurate in many situations.[21, 22] It is proposed that at a minimum a three-layer model is appropriate for the determination of soil resistivity.[23] Figure 8-4 shows a typical three-layer soil model derived from Wenner four-pin resistivity data, where the regions in which to calculate the step and touch potentials, the pipe current density, and the pipe potentials are labeled.[11] It can be seen that the step and touch potentials and pipe-to-soil potential calculations are affected by the soil resistivity of more than one layer. For small defects on the coated pipelines, the local soil resistivity dominates the magnitude of the AC current density on the pipelines.

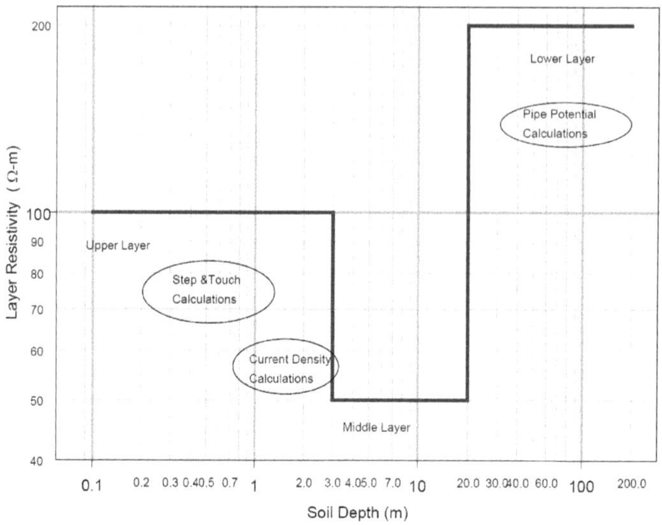


**Figure 8-4.** A three-layer soil resistivity model for AC interference assessment.[11]

Through the use of coupons and other monitoring equipment, the AC current density, along with the AC voltage, can be measured directly. If the size of the coating defect on the coupons is known, the soil resistivity can be calculated through Equation (3-7). The calculation results can be validated by actual measurements over a period of time. A correlation is then derived between the calculated soil resistivity and the measurements as a function of the seasonal change. As the AC current density is related to the soil resistivity under a certain AC-induced voltage, the changes in soil resistivity are then used to indicate the variations of AC current density. Therefore, calculations of the soil resistivity provide a useful tool in evaluating the risk of AC corrosion of buried pipelines. In areas where the AC interference is induced and the soil resistivity changes seasonally, continuous monitoring of the soil resistivity can help identify the scenarios in which the AC corrosion risks should be periodically evaluated for further testing and mitigation.[24]

**Separation distance between the pipeline and HVAC power lines.** The separation distance between the pipeline and power lines is an important variable controlling the induced AC voltage on the pipeline. Generally, as the distance increases, the induction on the pipeline decreases, which is attributed to the increased distance from the electromagnetic field decreasing the coupling effects. Figure 8-5 shows the electrical field under a power line with three phase conductors with horizontal and vertical arrangements (12 m of separation between each phase and a 1,000 A load current) as a function of the distance to the power line. It can be seen that the magnitude of the electrical field increases rapidly as the distance decreases. Moreover, as the distance to the power line reduces to about 50 m or smaller, a slightly higher electrical field is produced from the horizontal arrangement of the power line. As the distance increases up to 200 m, there is negligible difference in the electric fields between horizontal and vertical arrangements.

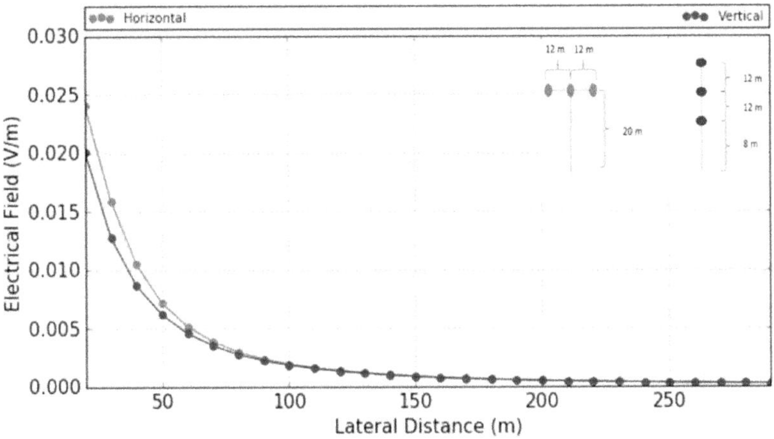


**Figure 8-5. Electrical field under a power line with three phase conductors with horizontal and vertical arrangements as a function of the distance to the power line.**[25]

**Collocation length of the pipeline and power lines.** The collocation length of the pipeline and the power lines also affects the magnitude and distribution of induced AC voltages along the pipeline. The collocation length is defined as the distance along the ROW that a pipeline parallels or crosses the power lines within a certain separation distance and angle that allow for the inductive coupling effect to occur. Generally, the collocation length determines the length of the pipeline exposed to the electromagnetic field. As the collocation length increases, the induced AC voltage on the pipeline increases. Figure 8-6 shows the predicted maximum induced AC voltage as a function of the collocation length. It can be seen that the AC voltage increases linearly with the increasing collocation length up to 10,000 feet (3,048 m). Beyond that, the predicted $V_{AC}$ does not increase owing to the diminishing effect of the electromagnetic coupling. Similar results were also confirmed in other simulations.[17]

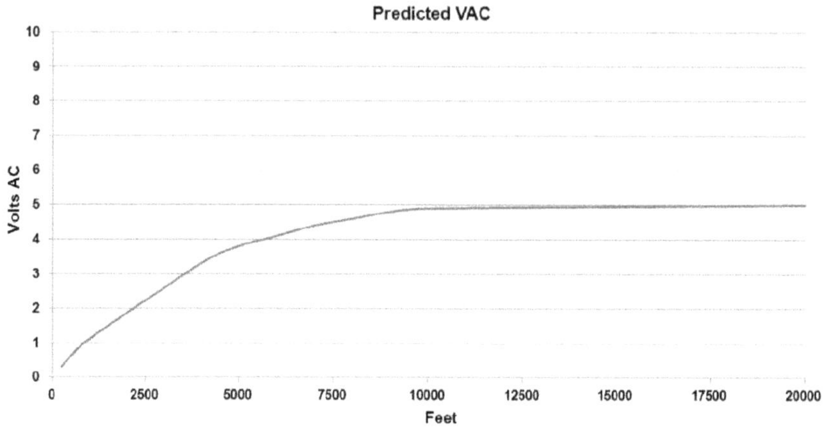


**Figure 8-6.** Predicted induced AC voltage as a function of the collocation length between a pipeline and an HVAC power line.[16]

**Orientation between the pipeline and power lines.** The angle at which a pipeline and HVAC power lines cross remarkably affects the magnitude of inductive coupling on the pipeline at the crossing. Generally, as the angle increases, the magnitude of the induction decreases. Particularly, for a perpendicular crossing (i.e., the pipeline crosses at or near 90° to the power line), the induction on the pipeline is minimized. Empirically, crossing angles greater than 60° result in a low or even negligible induction on the adjacent pipelines.[25] However, along with the increasing current-carrying capacity of power lines, AC interference may result in high AC voltages on the pipeline even at relatively wide angle crossings.

**Coating resistance.** The resistance of pipeline coatings is critical to the level of induced AC voltage on the pipelines. Generally, the coating resistance depends on the coating type, performance condition, and thickness, and on the degradation modes, all of which may be varied along the pipeline length. A poorly coated pipeline or a degraded coating with a low resistance allows multiple paths for AC voltage to dissipate to the ground, reducing the buildup of induction. The result is low AC voltage and a low AC current density discharge at the coating holiday. If a pipeline is well coated with high dielectric strength coatings, the resistance to ground along the pipeline is high, resulting in a great inductive coupling effect building up on the pipeline. This was detected on a newer FBE-coated pipeline, compared with an older CTE-coated pipeline, in proximity to a collocated HVAC power line.[26] Shwehdi and Johar also reported that the FBE coatings mainly used in the U.S. and the 3LPE coatings used in Europe have both made the problem of AC interference on pipelines more severe than earlier coatings.[27]

Figure 8-7 shows the modeling of AC voltages induced on an FBE-coated (Pipe 1) and a CTE-coated (Pipe 2) pipeline.[26] Although the induced AC voltages on the two pipelines fluctuated, the AC voltages would shift 20–30 V when the power company switched substations. It was apparent that the older CTE-coated pipeline had a lower AC voltage and thus did not require the same level of mitigation as the newer FBE-coated pipeline. Owing to the high resistance of FBE coating and the relatively few grounding paths, the induced AC voltage can build on an FBE-coated pipeline, which is hazardous from a safety standpoint to the personnel. Moreover, the corrosion risk is high, as the AC current can discharge from just a few holidays. Similar results (i.e., a higher coating resistance in higher AC voltages on pipelines) were also obtained in other modeling.[25]

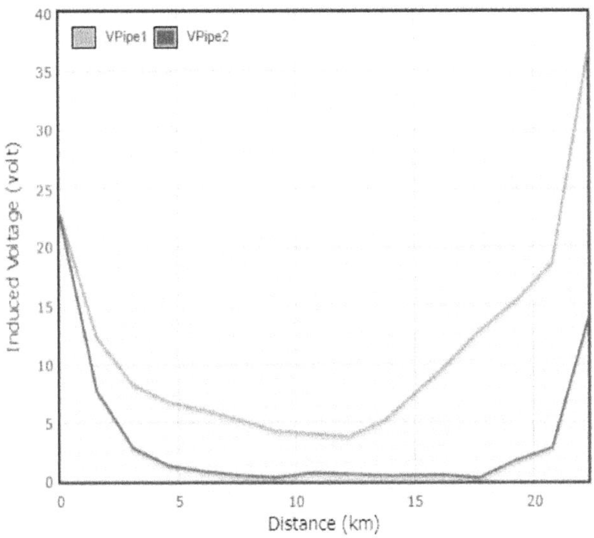


**Figure 8-7. Modeling of AC voltages induced on an FBE-coated (Pipe 1) and a CTE-coated (Pipe 2) pipeline.[26]**

**Pipeline diameter.** The diameter of the pipeline collocated with an HVAC power line affects the level of induced AC voltages on the pipeline. However, experience has shown that the effect is relatively small, compared with the influences resulting from other factors mentioned in this section. Generally, as the pipe diameter decreases, the surface area exposed to the longitudinal electric field decreases. For a smaller diameter pipeline, the longitudinal electric field influences a smaller surface area, resulting in a greater AC voltage being induced. For example, an increase in pipeline diameter from 150 mm to 1,220 mm can result in a 20% decrease in the induced AC voltage on the pipeline, regardless of the interfering current. Figure 8-8 shows the induced AC voltage profiles on a pipeline with varied diameters for the parallel collocating lengths of 1 km and 10 km. It can be seen that, for the shorter parallel distance of 1 km, the effect of the pipeline diameter on the induced voltage is negligible. When the parallel distance increases to 10 km, the effect of the pipeline diameter on the induced voltage is remarkable. As the pipe diameter becomes smaller and the parallel distance longer, the induced voltage along the pipeline increases.

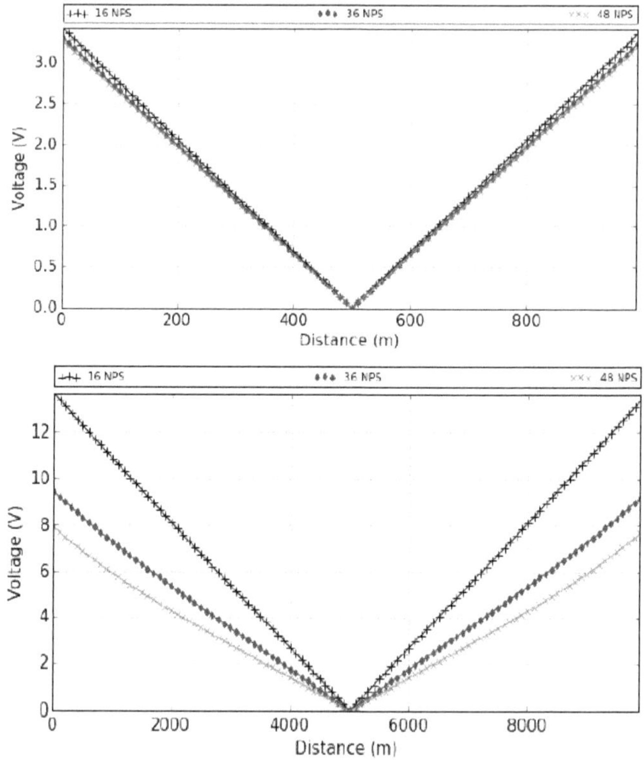


**Figure 8-8. Induced AC voltage profiles on a pipeline with varied diameters for the parallel lengths of (a) 1 km and (b) 10 km.**[25]

**Pipeline burial depth.** Generally, the burial depth of pipelines has a minor influence on the induced AC voltage on the pipeline. The level of AC interference decreases slightly with the increasing burial depth of the pipelines. For example, it was modeled that there was little change in the predicted induced AC voltages for a typical pipeline with buried at 1.22–1.83 m.[16]

## 8.2.3. Modeling of AC harmonic interference

AC currents can contain considerable harmonics of the fundamental AC frequency, which have been observed on pipelines.[28] Field measurements have demonstrated that these harmonics could produce significant interference on pipe-to-soil potentials.[29] Generally, the harmonics produced on a pipeline depend on the amplitude of harmonic currents in the collocated power lines and the phase relationship between the currents in the different phase conductors.

The electric current in power lines is usually not purely sinusoidal. Instead, it shows considerable distortions in the waveform. For a power system operating at 60 Hz, in addition to the fundamental component (60 Hz) of the current, there are also components at the even and odd harmonics (120 Hz, 180 Hz, etc.). A pipeline might have a significant response to the harmonic currents of 180 Hz, 360 Hz, 540 Hz, and so on, carried in an adjacent power line.[29] The harmonic-associated electric field induced on the pipeline produces variations in the pipe-to-soil potential. Figure 8-9 shows the variations of the pipe-to-soil potential along a pipeline collocated with a power line for the fundamental (60 Hz), second harmonic (120 Hz), and third harmonic (180 Hz) components. It can be seen that the 180 Hz amplitude causes a larger variation of pipe-to-soil potential on the pipeline than the 60 Hz amplitude, although the 60 Hz current in the power line is 100 times larger than the 180 Hz power line current. At the end of the pipeline, the amplitude of pipe-to-soil potential can be as high as 7 V for the third harmonic.

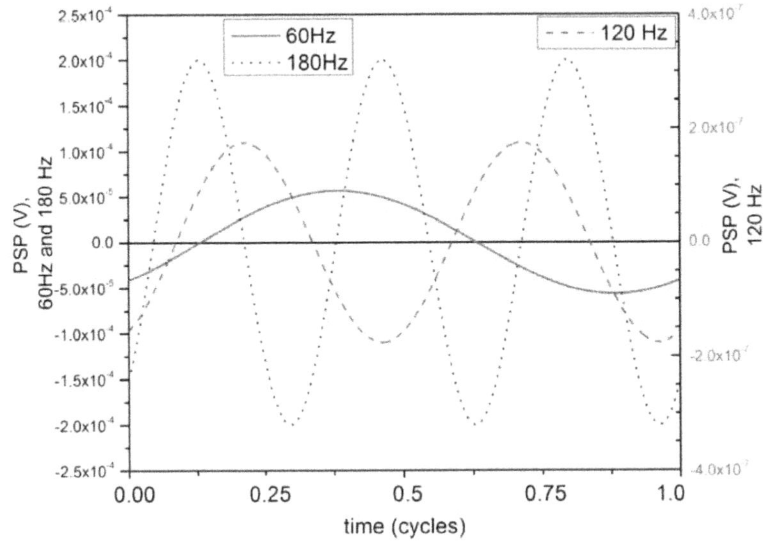


**Figure 8-9.** Variations of pipe-to-soil potential along a pipeline collocated with a power line due to harmonics for the fundamental (60 Hz), second harmonic (120 Hz), and third harmonic (180 Hz) components.[29]

# 8.3. Application of computational models for assessment of AC interference on pipelines

Nowadays, multiple software packages are available to help assess the AC interference in pipeline–power line shared corridors and to design mitigation programs for reduced electrical shock hazards and risk of AC corrosion of pipelines. The modeling results contribute to an optimized design of the physical layout of the utilities and reasonable recommendations for mitigative actions. In many cases, the assessment models, in combination with experimental testing, also develop a mechanistic understanding of the AC corrosion processes, contributing to a strong knowledge base in the area of pipeline AC corrosion.

## 8.3.1. Optimization of power line design for minimization of AC interference

One of the primary functions of modeling AC interference is to help optimize the HVAC power line design to minimize interference on pipelines. Kemp et al. reported that two existing 115 kV power lines in a shared utility corridor collocated with two petroleum pipelines for about 32 km in western Washington state were to be upgraded to operate at 230 kV.[31]

In addition to a number of field measurements, sensitivity studies, and design iterations, the Elysca IRIS (Inductive and Resistive Interference Simulator) software package was selected, after a comparison of the AC voltages modeled with the field measurements, to develop a detailed model of the shared utility corridor. Several power line structures and the corresponding designations were evaluated in an effort to minimize the impact of AC interference on the collocated pipelines. Moreover, two primary routes were considered for the assessment. The AC interference modeling results at four loading scenarios and various combinations of tower configurations along the two routes obtained the maximum induced AC voltages on the pipelines, which ranged from 3 V to 24 V, which corresponded to the maximum theoretical current densities of between 1 A/m$^2$ and 83 A/m$^2$. A further study of the predicted AC voltages and theoretical AC current densities on the optimized phasing, routing, and tower configurations determined the current load scenario with the minimum AC interference. The induced AC voltages on the pipelines were below the 15 V shock hazard safety threshold, and the AC current density did not exceed the threshold of 20 A/m$^2$. Thus, the likelihood of AC corrosion of the pipelines was low.[30]

## 8.3.2. Modeling of AC voltages and AC current densities on pipelines

The induced AC voltage and AC current density along the pipeline are primary indicators of the electrical safety hazards and the AC corrosion risk, respectively, under interference from collocated HVAC power lines. At a test station installed for a gas pipeline collocated with a 500 kV HVAC power line and an electric power substation, the AC voltage was always higher than 15 V during testing, and the maximum value was 47.5 V.[32] A computing model was established for AC interference analysis, after validation by comparing the modeled AC voltages and the AC current densities with the measured values (Figure 8-10)

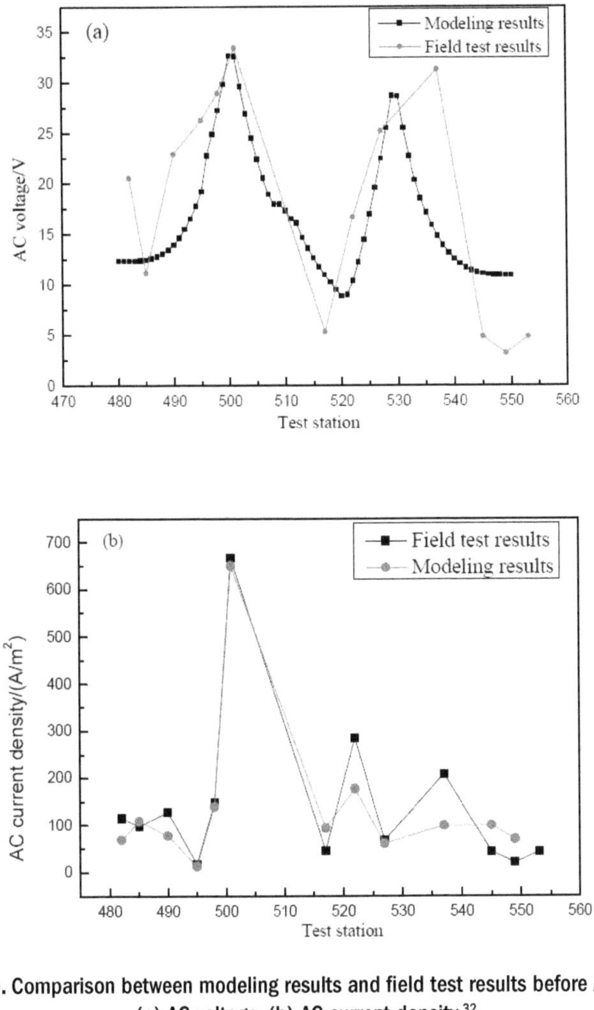


**Figure 8-10. Comparison between modeling results and field test results before AC mitigation: (a) AC voltage, (b) AC current density.**[32]

The AC interference mitigation involved the selection of zinc ribbons as the ground bed material to avoid galvanic corrosion. Solid-state DC decouplers, which allowed only the AC current to pass, were installed between the pipeline and the zinc ribbons to prevent the CP current flowing away. The zinc ribbon configuration, including the length, burial depth, and the length parallel with the pipeline, was determined and optimized by the modeled AC voltages and AC current densities (Figure 8-11). It can be seen that the overall mitigation effect is satisfactory. The AC voltage on the pipeline decreased remarkably after mitigation and was under 15 V. The AC current density was under the 30 A/m$^2$ threshold, except at one test station, where further mitigation measures would be taken. Therefore, numerical modeling provides an effective tool in AC interference evaluation and mitigation design. Numerical modeling can be used in the design stage to evaluate the possible effect of AC interference on pipelines, recommending the optimal design of the power line configuration and service conditions. For existing AC interference, the modeling approach can be used to optimize the mitigation program.

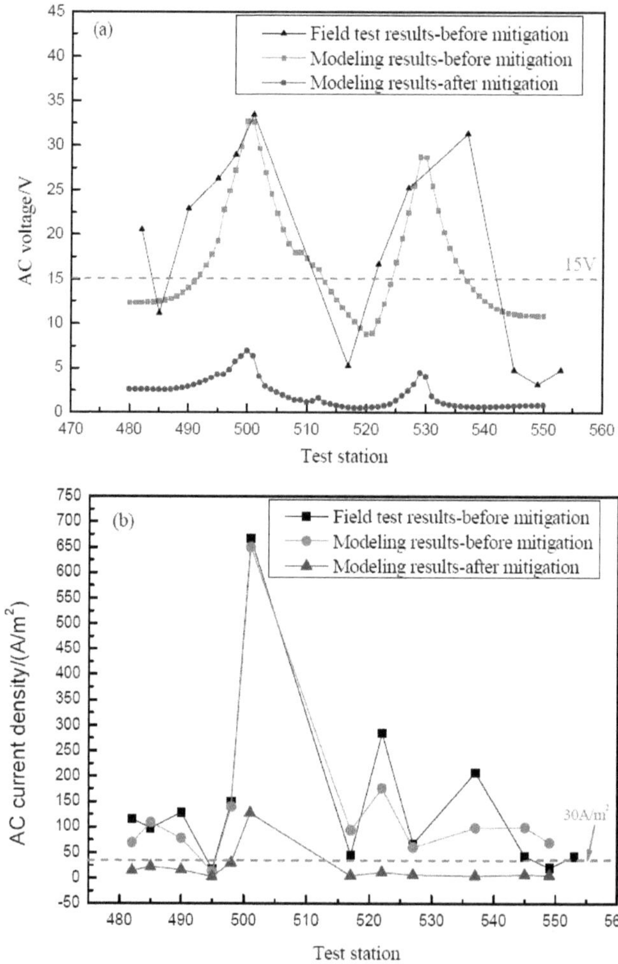


**Figure 8-11. Modeling results of AC interference on a pipeline upon mitigation installation: (a) AC voltage, (b) AC current density.**[32]

**Modeling of AC voltage CIS on pipelines.** Since CP potential CIS has been commonly used in direct assessment of CP performance and coating condition, an AC voltage CIS along a pipeline could provide a possible method to predict the risk of AC corrosion. Although it was discussed in Chapter 7 that AC voltage CIS is not theoretically viable, attempts have been made to use CDEGS software to model the AC voltage CIS for a pipeline–power line configuration, where a coated pipeline (with a coating resistance of 100 kΩ/m$^2$) was collocated with a 115 kV power line with a load current of 832 A for 7.8 km.[33] The soil resistivity was 10 Ω·m. As a comparison, a pipeline with a degraded coating (coating resistance of 5 kΩ/m$^2$) was also modeled. The modeling and measurement results demonstrated that the AC voltage measured at the end of an unshielded survey wire was affected by the voltage induced on the wire by the parallel power line and by voltage gradients in the soil. Although it was possible that the AC voltage measured at the end of the survey wire was similar to that measured at a local CP test station, the measurement results obtained between the two connection points could be completely different from the actual AC voltages on the pipeline. When the voltage gradient in the soil is negligible (i.e., the coating is well intact), the measured AC voltage at the end of the survey wire is the sum of the AC voltage on the pipeline at the connection point and the AC voltage on the survey wire. When the coating is in poor condition, the voltage gradient of the soil is included in the voltage summation. In either case, the measurement does not represent the AC voltage of the pipeline at locations remote from the connection point.[33] Therefore, the AC CIS modeling approach is not valid for mapping the AC voltages on a buried pipeline collocated with HVAC power lines.

### 8.3.3. Modeling of fault currents

When a power line is faulted, both inductive and conductive effects must be considered. The magnitude of the current flowing into the faulted tower is required. If this information is not available, fault analysis should be conducted to determine the injected current at the faulted tower. As shown in Figure 8-12a, except for the regions near to the terminal stations, the available current at a faulted tower is quite small. Depending on the location of the buried pipeline relative to the terminal stations, mitigation requirements for the fault case may not be so complex.

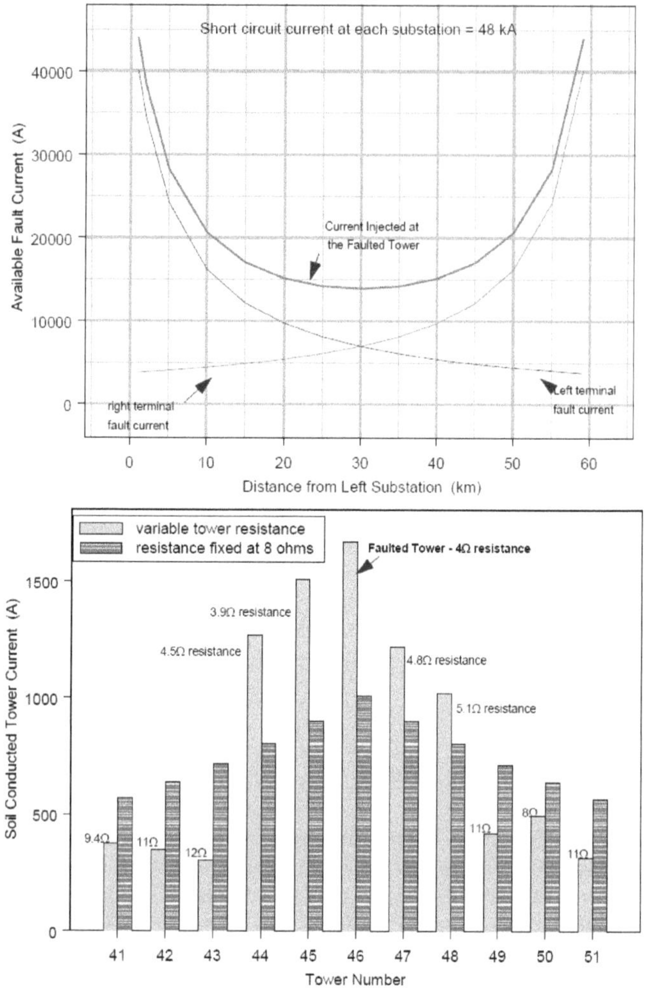


**Figure 8-12.** (a) Modeling of the current injected at a faulted tower, (b) modeling results, compared with actual measurements, of the tower soil resistance and the calculated tower current.[11]

At the faulted tower, the input current is separated into two components: the inductive current returning to the substations via the sky wires and the conductive current leaving the tower flowing into the soil. An accurate determination of the split factor is critical, since the former reduces the inductive coupling to the pipeline, and the latter increases the soil conductive current.[11] Generally, the calculation of the current split factor is subject to the assumption that all tower footings have the same resistance. Since the resistance at individual tower footings is usually unknown, the modeling approach is essential to the determination of the fault current. Figure 8-12b shows the modeling results, compared with the actual measurements, of the soil resistance at the towers and the calculated tower current. Since the induced potential on the pipeline is proportional to the conductive current, the modeling of individual tower resistances in the analysis can generate improved accuracy in fault current modeling.

### 8.3.4. Modeling of the mutual coupling effect between a pipeline and grounding anodes

Buried horizontal anode conductors that are installed parallel to and bonded to the pipelines is a common technique for AC interference mitigation. When computing the mitigated pipe-to-soil potentials, it is possible to disregard the mutual coupling effect between the pipeline and the anode conductors without incurring excessive errors (Figure 8-13a). However, computations of the pipe current density require that the mutual coupling between the pipeline and the mitigation anode conductors be considered simultaneously. If not, large errors in the calculation of the current density can be caused (Figure 8-13b).

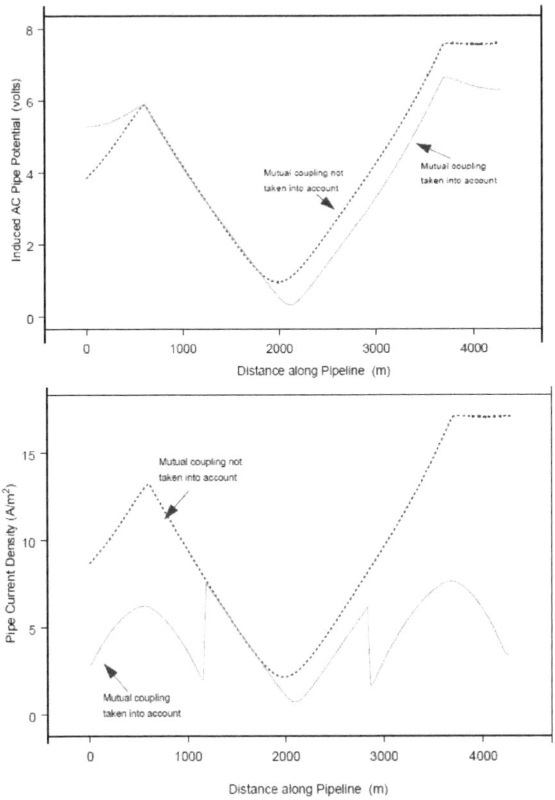


**Figure 8-13.** Modeling results of the (a) pipe-to-soil potential and (b) pipe current density with and without consideration of the mutual coupling effect between the pipeline and the mitigation anode conductors.[11]

## 8.4. Reliability analysis of the AC interference modeling

Even if modeling calculations have been conducted with established numerical methods, the field measurements usually show lower AC voltage levels than the calculated values. For example, Wahl and Schmautzer stated that calculation results were often up to seven times higher than the measurements conducted on pipelines.[34] The discrepancies between measurements and modeling calculations are mostly due to unknown or assumed parameters used in the models. The factors that can reduce the induced AC voltages on the pipelines at varying degrees and cause a discrepancy between calculated and measured AC voltages include use of the load current rather than the maximum operational currents in modeling, the reduction effect of global earthing systems, the reduction effect of pipelines running in parallel, the reduction effect of parallel HVAC power systems with grounding conductors, the reduction effect of local earthing systems, incorrect or inadequate pipeline coating parameters, and the influence of specific soil resistivity.[34]

Generally, knowledge of the specific soil resistivity and local coating resistance (i.e., spread resistance at the coating defects) is critical to AC interference modeling since both can affect the induced AC voltage at the measuring location. The load current of the power lines during the measurement period must be known. Moreover, the presence of any conductive materials within the AC interference area can act as a reduction factor, decreasing the induced AC voltages on the pipelines. These materials also produce influencing voltages, which may even increase the AC voltages.[34] Therefore, field measurements must be conducted while the modeling is performed to calculate the AC interference parameters. The measurement results can validate the calculations and improve the modeling accuracy and reliability. With well-calibrated models, the measurement workload can be reduced, and some unnecessary measurements can even be avoided. The necessary measurements (e.g., the local soil resistivity) can be conducted more effectively and efficiently.

# References


1. Bonds, R.W., *The Effect of Overhead AC Power Lines Paralleling Ductile Iron Pipelines*, Ductile Iron Pipe Research Association, Golden, CO, 2017.
2. Dabkowski, J., "A probabilistic approach to calculate ac induction levels on power line collocated pipelines," CORROSION/1995, paper no. 358, NACE, Houston, TX, 1995.
3. Dabkowski, J., Taflove, A., *Mutual Design Considerations for Overhead AC Transmission Lines and Gas Pipelines*, EPRI Report EL-904, Electric Power Research Institute, Palo Alto, CA, 1978.
4. Dawalibi, F., Southey, R.D., Malric, Y., Tavcar, W., *Power Line Fault Current Coupling to Nearby Natural Gas Pipelines: Volume 1, Analytic Methods and Graphical Techniques, Electromagnetic and Conductive Coupling Analysis of Powerlines and Pipelines (ECCAPP)*, Electric Power Research Institute, Palo Alto, CA, 1987.

5. Dabkowski, J., Taflove, A., "Prediction method for buried pipeline voltages due to 60 Hz AC inductive coupling, Part I: Field test verification," *IEEE Transactions on Power Apparatus and Systems* PAS-98: 788–794, 1979.
6. Dabkowski, J., Taflove, A., "Prediction method for buried pipeline voltages due to 60 Hz AC inductive coupling, Part II: Pipeline grounding methods," *IEEE Transactions on Power Apparatus and Systems* PAS-98: 1814–1823, 1979.
7. Frazier, M.J., *Utility Corridor Design: Transmission Lines, Railroads, and Pipelines*, EPRI Report EL-4147, Electric Power Research Institute, Palo Alto, CA, 1985.
8. Dawalibi, F., Southey, R.D., "Analysis of electrical interference from power lines to gas pipelines, Part I: Computation methods," *IEEE Transactions on Power Delivery* 4: 1840–1846, 1989.
9. Dawalibi, F., Southey, R.D., "Analysis of electrical interference from power lines to gas pipelines, Part II: Parametric analysis," *IEEE Transactions on Power Delivery* 5: 415–421, 1990.
10. Southey, R.D., Dawalibi, F.P., Li, Y., Ruan, W., "Increasing the cost-effectiveness of ac interference mitigation designs with integrated electromagnetic field modeling," CORROSION/2005, NACE, Houston, TX, 2005.
11. Dabkowski, J., "The evolution of AC predictive and mitigation software," CORROSION/2017, paper no. 8912, NACE, Houston, TX, 2017.
12. Dawalibi, F.P., Donoso, F., "Integrated analysis software for grounding, EMF, and EMI," *IEEE Computer Applications in Power* 6: 19–24, 1993.
13. EPRI (Electric Power Research Institute), *Power System and Railroad Electromagnetic Compatibility Handbook: Revised First Edition*, Palo Alto, CA, 2006.
14. INGAA (Interstate Natural Gas Association of America), *Criteria for Pipelines Co-Existing with Electric Power Lines*, Final Report No. 2015-04, INGAA Foundation, Washington, DC, 2015.
15. Santana-Diaz, E., Nielsen, L.V., Junker-Holst, A., "A critical review of parameters for meaningful AC corrosion modelling," CORROSION/2018, paper no. 10847, NACE, Houston, TX, 2018.
16. Simon, P.D., Lopez-Garrity, M., Wagner, D., Garrity, K., "Identification and prioritization of potential HVAC interference locations on a 6,000 mile crude and product transmission pipeline system," CORROSION/2017, paper no. 9260, NACE, Houston, TX, 2017.
17. Ouadah, M., Touhami, O., Ibtiouen, R., "Diagnosis of the AC current densities effect on the cathodic protection performance of the steel X70 for a buried pipeline due to electromagnetic interference caused by HVPTL," *Progress in Electromagnetics Research* 45: 163–171, 2016
18. Wakelin, R.G., Sheldon, C. (2004), "Investigation and mitigation of AC corrosion on a 300 mm natural gas pipeline," CORROSION/2004, paper no. 04205, NACE, Houston, TX.
19. Stet, D., Micu, D.D., Czumbil, L., Darabant, L., Ceclan, A., "Simulation of interferences between power lines and gas pipelines in unbalanced phase currents state," *International Journal for Computation and Mathematics in Electrical and Electronic Engineering* 31: 1178–1184, 2012.
20. Fieltsch, W., Gudino, E., Shahinas, F., Wymenga, T., "Effects of seasonal variation on AC interference and mitigation design," CORROSION/2018, paper no. 11231, NACE, Houston, TX, 2018.

21. Seedher, H.R., Arora, J.K., "Estimation of two layer soil parameters using finite Wenner resistivity expressions," *IEEE Transactions on Power Delivery* 7: 1213–1219, 1992.
22. Del Alamo, J.L., "A comparison among eight different techniques to achieve an optimum estimation of electrical grounding parameters in two-layered earth," *IEEE Transactions on Power Delivery* 8: 1890–1898, 1993.
23. Yang, H., Yuan, J., Zong, W., "Determination of three-layer earth model from Wenner four-probe test data," *IEEE Transactions on Magnetics* 37: 3684–3892, 2001.
24. Hilleary, J., "Calculating and tracking soil resistivity change in high AC corridors," CORROSION/2017, paper no. 8984, NACE, Houston, TX, 2017.
25. Garcia, A., Krissa, L.J., DeWitt, J., "Effect of transmission pipeline properties on alternated induced voltage," CORROSION/2017, paper no. 9786, NACE, Houston, TX, 2017.
26. Pikas, J., "Looking back and understanding a case history of AC corrosion," CORROSION/2018, paper no. 10603, NACE, Houston, TX, 2018.
27. Shwehdi, M.H., Johar, U.M., "Transmission line EMF interference with buried pipeline: Essential & cautions," Proc. International Conference on Non-Ionizing Radiation at UNITEN, Kuala Lumpur, 2003.
28. Boteler, D.H., Croall, S., Nicholson, P., "Measurements of higher harmonics in AC interference on pipelines," CORROSION/2010, paper no.14755, NACE, Houston, TX, 2010.
29. Trichtchenko, L., Boteler, D.H., "Frequency dependence in AC harmonic interference to pipelines," CORROSION/2018, paper no. 11004, NACE, Houston, TX, 2018.
30. NACE, *Mitigation of Alternating Current and Lightning Effects on Metallic Structures and Corrosion Control Systems*, SP0177-2014, Houston, TX, 2014.
31. Kemp, D., Krebs, B., Finneran, S., "Minimizing HVAC interference on pipelines through transmission line design optimization," CORROSION/2018, paper no. 11285, NACE, Houston, TX, 2018.
32. Jiang, Y., Bi, W., Chen, K., Liu, M., Zhao, J., "Mitigating pipeline AC interference using numerical modeling and continuous AC voltage monitoring," CORROSION/2018, paper no. 11659, NACE, Houston, TX, 2018
33. Fieltsch, W., Wakelin, R., "The AC close interval survey and other common AC measurement errors," CORROSION/2017, paper no. 9455, NACE, Houston, TX, 2017.
34. Wahl, C., Schmautzer, E., "Pipeline voltage," *Pipeline International Journal* 1: 23–31, 2015.

# 9 Mitigation and Control of AC Corrosion of Pipelines

## 9.1. General principles for mitigation and control of AC interference

The key strategy in designing an AC mitigation system is to keep a safe distance between the buried pipeline and HVAC power lines to mitigate and minimize the AC interference, including the arcing risk during a power line fault. The target is to make the pipeline electrically grounded to earth to dissipate stray current and prevent hazardous voltages under either fault or steady-state conditions, minimizing or even eliminating the risk of AC corrosion.[1] The main purposes of an AC interference mitigation program include the prevention of voltage spikes during fault conditions, reduction of AC current density to protect against AC-induced corrosion, and maintenance of AC step and touch potentials below a threshold AC voltage (e.g., 15 V) to protect personnel from shock hazards.[2]

A mitigation system should be installed for pipelines under AC interference when collocated with HVAC power lines or AC-powered rail transit systems.[3] Although the mitigation system design may vary across the industry, in general, all techniques involve a grounding system to pass the interfering AC to the earth. Typically, the mitigation system can be either a surface or deep grounding design.

The primary principles of an AC mitigation design include:

- Keep a safe pipe-to-power line separation distance to minimize the risk of arcing to the pipeline during a power line fault.
- Install gradient control mats at pipeline appurtenances to protect personnel working on the pipeline during a power line fault.

- Electrically ground the pipeline to earth to prevent accumulation of hazardous voltages under both fault and steady-state operating conditions and to minimize the risk of AC corrosion.

Two methods are usually suggested for the design of the AC interference mitigation system: (a) field measurements and design and (b) theoretical modeling design.[4] The three general steps for mitigation of AC interference on pipelines based on field measurements and design are:

- **Collect and analyze data.** All data about the power lines and pipelines should be gathered, including actual AC potentials, soil properties and conditions, alignment of the buried pipeline to power lines, coating and its performance condition, and AC current densities. All aboveground pipeline appurtenances should be documented, and the parallel lengths and separation distances between the power lines and the pipelines should be analyzed. Potential areas where peak AC voltages can be experienced should be identified.
- **Design the AC mitigation system.** The power lines, pipeline, and soil data should be combined into a comprehensive design program, which should produce designs that protect both personnel and the pipeline. Conventional methods, such as DC decoupling devices, connection to available natural drains, additional grounding, and installation of test coupons, are usually integrated into a combined mitigation program. This program should be considered part of the overall design of the pipeline system prior to construction if a risk of AC interference is identified.
- **Test and monitor the results.** The pipeline should be monitored online to record all parameters relevant to AC interference. The mitigation system should be adjusted in response to operational changes in electric transmission and distribution or to degradations of the coating or CP performance during pipeline operation.

The alternative method for design of the AC interference mitigation program is theoretical modeling based on computation and simulation of the interference level and its impact on pipelines. Prior to design and installation of an actual AC mitigation system, AC modeling is usually conducted to provide information regarding the optimal mitigation methodology, and the quantity and location of the mitigation required based on numerous input factors, such as soil resistivity, physical characteristics of the pipelines, coatings and CP, operating parameters of the HVAC transmission system, and the spatial alignment between the pipeline and the HVAC power lines. The software program would calculate the conditions required to mitigate the predicted AC voltage and AC current density, as well as the residual AC voltage and AC current density; these conditions can be used in the design of the AC mitigation system. With extensive experience and sufficient input conditions, theoretical modeling is usually effective in designing the AC mitigation system. The accuracy of the approach is subject to the developed model and the availability of the required data.

Mitigation and control of AC interference on pipelines is often achieved through a three-tiered approach comprising primary, secondary, and tertiary mitigative controls.[5] The primary mitigative control is to minimize the source of the AC interference. For example, the electromagnetic field generated by HVAC power lines can be reduced by optimizing the phase wire arrangement, phasing, load unbalance, and so on. This is the most effective and convenient way to mitigate AC induction on a pipeline, in which the AC interference is minimized at the production source. Secondary mitigation is aimed at reduction and control of the AC interference that would otherwise reach the pipeline, but nothing is done on the source side. The techniques used at this stage can intercept or block the AC interference before it reaches the pipeline. A typical example is the placement of elevated conductors between the pipeline and the power lines to sacrificially induce a potential from the electromagnetic field, reducing the induced AC voltages on the pipeline. Tertiary mitigation reduces the AC interference that has reached the pipeline. A technique that has been widely employed in industry is to install grounding systems to dissipate the AC current that flows toward the pipeline. Although the primary mitigative control is the most effective, it may not always be feasible.[5] Consideration should be given to potential AC interference problems whenever a new pipeline or HVAC transmission line project is initiated in a shared utility corridor.

In addition to their effect during normal operation, AC interference and the induced pipeline corrosion can also be significant during construction of a pipeline project. The situation is even worse when the CP system is not in place. Thus, implementation of AC corrosion mitigation measures by evaluating the AC corrosion susceptibility of the pipelines should be considered in the pipeline design. For example, a permanent AC mitigation system, such as grounding facilities, can be installed during pipeline construction. Frequent information communications should be maintained between the pipeline and the AC power line operators, and an agreement on the collocation plan should be reached. When a new HVAC power line is planned in the vicinity of an existing pipeline, mitigation measures must be in place before energization. Modeling computation and analysis should be performed to determine the factors affecting the generation and level of AC interference. Based on discussions between the power line and pipeline operators, it may be possible to achieve optimal arrangements to minimize the AC corrosion risk on the pipeline. For example, the selection of appropriate shield wires on the power line can reduce the electromagnetic induced voltages on the pipeline and protect the power lines from lightning.[1] However, this method does not apply to pipelines being installed in the vicinity of existing HVAC power lines.

## 9.2. Surface and deep grounding

Grounding, as an equivalent of a reduction in coating resistance of pipelines, has been reported to be an effective means of AC interference mitigation and is used extensively in the field.[6] No matter the type of grounding system, all AC mitigation systems involve the installation of a low-impedance grounding system to the affected pipelines to allow AC-induced current and fault current to be quickly discharged and carried away from the pipelines.[2]

## 9.2.1. Surface grounding

Surface grounding generally refers to the installation of mitigation grounding facilities and assembly units at or near the ground surface or the pipe depth. Figure 9-1 shows the installation of surface grounding by metal wires. Bare copper cables, zinc ribbons, or other units are buried parallel to the pipe path in the same ROW and are connected to the pipeline, shielding the pipeline and dissipating harmful stray currents away from the pipeline by providing a low-resistance path to earth. The mitigation system installed at specific locations can effectively reduce the inductive AC interference along the pipeline. Moreover, the metal ribbons also serve as a shielding layer to protect the pipeline from damage resulting from a power line fault generating a rapid discharge of fault current near the pipeline and a direct current arcing in the soil.

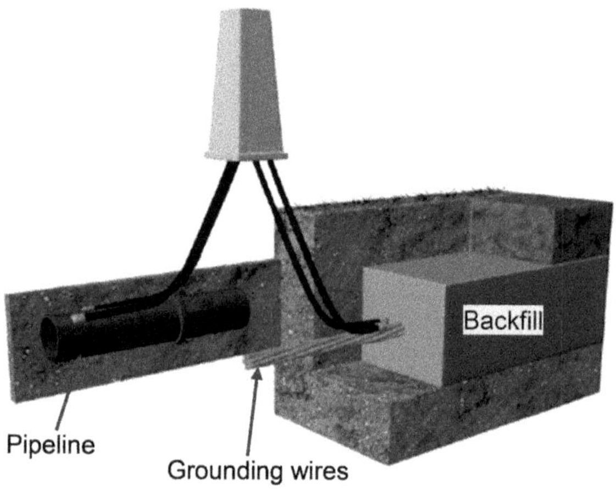


**Figure 9-1.** Schematic diagram of the installation of surface grounding by metal wires.[7]

Surface grounding, compared with other AC interference reduction methods, provides a cost-effective and convenient method for installing the mitigation system along a significant length to reduce the AC interference on the adjacent pipelines. In particular, during new construction of pipelines, surface grounding systems can be installed directly in the pipe trench or laid parallel to the pipe at ROW in an adjacent trench. Additional mitigation ribbons can be connected in parallel, contributing to the enhanced reduction of the grounding resistance.

## 9.2.2. Point drain grounding

In a point drain grounding system, grounding conductors are drained into the soil locally to achieve a low-resistance grounding. The conductors are connected to the pipeline at specific locations (preselected by either modeling or historical pipeline failure experience) to reduce the AC voltage level on the pipeline.

## 9.2.3. Deep grounding

Deep drilled ground wells offer another form of AC interference mitigation grounding, where one or more anodes are drilled and installed vertically into the ground to achieve low ground resistance locally (Figure 9-2). The grounding system runs perpendicular to the pipeline, deep into the ground, generally up to 200 ft (60 m) or more.[8] With a greater capacity for mitigating AC interference, deep well grounding is generally selected when one or both of the following criteria are satisfied:

- The soil resistivity at the surface is much greater than (at least 20 times) the soil resistivity at certain depths.
- Surface grounding is not feasible owing to construction obstacles such as roads, railways, or ROW access.

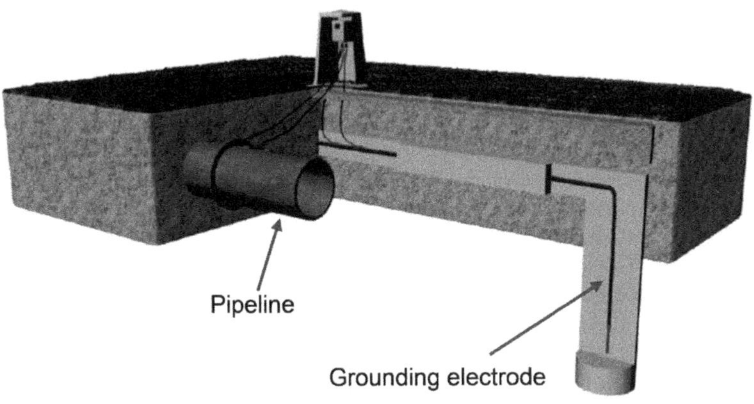


**Figure 9-2.** Schematic diagram of the installation of a deep grounding electrode.[7]

When both parallel ribbon installation and deep grounding are feasible for AC interference mitigation, the surface parallel ribbons are more efficient and economical. This approach can achieve a low resistance to ground at a low cost, making it the most common form of AC interference mitigation.[2]

## 9.2.4. Grounding materials

Grounding materials selected for AC mitigation should enable the grounding system to possess a longevity in the specific environment, a low total installation cost, and an effective resistance reduction at the installation site. As the grounding systems are usually decoupled from DC, the CP current is not available for corrosion protection. Thus, the life expectancy is based on the corrosion rate of the grounding materials in the environment.[9] Some common materials that can be used for AC interference mitigation by grounding include zinc ribbons or wires, galvanized steel wires, and bare copper wires. All selections have their unique benefits and detriments. Consideration of the performance, cost, and constructability is required.

**Zinc materials.** Zinc grounding systems are a traditional selection for AC interference mitigation. However, the systems can experience problems such as passivation and corrosion of the zinc materials.[1] It is generally believed that the passivation results in an increase in the impedance of the zinc, reducing the effectiveness of the zinc grounding in AC mitigation. Previously, although specific soil environments were identified under which zinc could be passivated, few data were found to confirm a measured increase in impedance.[10] In addition, rapid corrosion of zinc in soil environments was believed to affect its service life. Gummow stated that accelerated corrosion of zinc ribbons was expected and, thus, should be accounted for in the design of AC mitigation systems, despite the lack of information on the magnitude of the accelerating effect.[1] It was reported that AC could accelerate corrosion of zinc dramatically and that the corrosion rate increased with the AC current density.[11] It was also noted that this work was conducted at AC current densities that were several orders of magnitude greater than what was expected in service. Thus, the performance and service life of grounding zinc ribbons in the presence of AC is an issue that should be considered during installation of mitigation and control systems. An AC current density of 20–300 $A/m^2$, as tested in the laboratory, is related to a current discharge from a 1 $cm^2$ coating holiday, not to the current density levels discharging from a bare grounding conductor, typically ranging from over 100 to a few thousand meters in length.[9] Electrochemical studies conducted recently showed that, although zinc was passivated in carbonate solutions at room temperature, for the frequencies of concern typical of HVAC power lines in a range of 50–60 Hz, the measured impedance was nearly identical between the initial and the 24 h electrochemical impedance spectroscopy (EIS) scans.[9] This indicated that zinc passivation had no adverse effect on the total impedance. Moreover, as the time increased to 96 h, the impedance of the zinc sample decreased slightly from the initial tests prior to passivation. When a high content of sulfur is contained in the soil, such as the areas containing agricultural fertilizers, zinc sulfate can form on the zinc surface, causing resistivity of the zinc materials up to 6,000,000 Ω·cm, making the zinc material insulate to pass current. Moreover, a direct connection of zinc grounding materials with the pipelines can increase the demand on CP current output.

**Copper materials.** Bare stranded copper wires have increasingly been selected for grounding in recent years. While the performance of copper wires is excellent for AC mitigation, the issues of corrosion and theft affect their use. Bare copper, which is more corrosion resistant than steels and zinc in soil environments, can also experience corrosion, especially in the presence of chlorine ions and sulfur compounds. Moreover, the copper cable could be affected by stray current interference from nearby ICCP systems.[7] Thus, protective measures must be taken on the copper materials, such as sacrificial anodes or conductive cement backfill, to achieve a reasonable service life, while maintaining their ability to pass current to the soil.[1] Although copper materials possess a low impedance, the total impedance of the system can be limited by the solution resistance of the soil.[9] As a result, the advantage of the low impedance of the copper material cannot be fully used, causing a minimal variation of the total impedance when using copper, as compared with zinc material. At the same time, warnings and mandatory protections should be implemented to avoid theft. Luk et al. reported that installation of numerous interconnected copper ground rods was effective to mitigate the electrical shock hazard and AC corrosion risks to an acceptable level on pipelines collocated with a 240 kV power line.[12] For example, prior to mitigation, the pipeline AC voltages under steady-state conditions were above the 15 V safety limit, reaching a maximum value of 25.1 V. After installation of the ground copper rods, the pipeline AC voltages were below the 15 V safety limit, reaching a maximum value of 12.7 V only.

**Steel materials.** Steel materials possess a higher resistance for grounding than copper systems. The limited current-carrying ability of the steels requires the selection of large-diameter steel wires in the mitigation system. However, the steels are normally cheap, and the theft issue is not relevant. Generally, the steels suffer from more corrosion in a wide variety of environments, such as soils, than copper or zinc do. The steel grounding system must be protected for as long as its ability to pass current is not degraded.

**New grounding electrode.** A new electrode material used for grounding–a prepackaged copper wire electrode–was reported.[7] The electrode consisted of copper wire in a cotton sock, filled with conductive carbon backfill material, and was supplied in lengths up to 150 m. The backfill material included either a corrosion inhibitor or a special conductive cement backfill that did not allow corrosive chemicals to touch the copper wire. The design ensured a high ground performance to dissipate the stray current to earth and a sufficient service life expectation.

### 9.2.5. Backfill materials

During grounding, it is common to install the grounding conductors in a special backfill material. The purpose of the backfill can vary, depending on the conductor materials chosen and the type of backfill used. Generally, the benefits of the AC mitigation backfill include the following:

- **Enhanced surface area.** Conductive backfills such as carbon or conductive concrete can effectively increase the surface area of the grounding conductors, reducing the overall resistance to earth.
- **Corrosion or passivation protection.** Some backfills are designed to protect the grounding conductors from corrosion or passivation, avoiding adverse effects on the conductors' service life and degradation of their performance.
- **Hydroscopicity.** Some hygroscopic backfills readily retain water from the environment, helping to maintain a low uniform resistance around the grounding conductors.

Commonly used backfills include native soil, bentonite, or conductive concretes. For a typical conductive concrete over copper AC mitigation electrodes, a large copper core is encapsulated in a conductive concrete backfill, which provides sufficient strength to resist external loads such as soil and surface load. Upon assembly, the AC mitigation electrode becomes a single, solid conductor with a large surface area that can dissipate a large amount of current. The electrode system can protect the copper core from corrosion and deterioration, thus extending the service lifetime of the system.

## 9.3. Overvoltage arresters

While grounding mitigation can be effective for controlling AC interference, researchers have identified that the AC interference from HVAC power lines on the collocated buried pipelines could not be adequately mitigated below recommended limits for AC voltages or current densities even if the grounding strategy was implemented.[5] This may be attributed to elevated AC current carried by the power lines, unfavorable soil resistivity conditions, or physical restrictions that limit the level of grounding that can be practically installed. Thus, additional mitigation techniques should also be assessed and used to further reduce the induced AC interference where necessary. Semiconducting overvoltage arresters have been used to mitigate and control AC corrosion of buried pipelines.

In principle, the arresters consist of a set of anti-parallel thyristors that are activated by an AC voltage across pipe and ground wires. When the arrester is activated, the AC is conducted by the arrester and flows from the pipeline to the ground though the arrester. As a result, the AC voltage drops. Empirically, the activation voltage for arresters is in the range of 22–25 V.[13] When the voltage is below this range, the arrester acts as an open switch. To avoid interference of AC on CP potential measurements, AC filters are usually installed to give a direct reading of DC voltage. Field measurements demonstrated that the DC current flowing through the energized arrester is on the order of few milliamperes. It could not be effectively attenuated by adding more diodes in the positive current path.

The installation of overvoltage arrests was demonstrated to be effective at mitigating AC interference on buried pipelines.[13] In a demonstration test that was conducted on a 47.3 km gas pipeline under AC interference, a large potential difference was experienced at the end of the pipeline and the highest long-term AC-induced voltage was recorded. The results showed that the arresters were active for 33% of the entire time monitored. This indicated that, during the inactive period, the pipeline experienced AC voltages ranging between 12 V and 25 V, whereas during active arrester periods, the AC voltage was less than 10 V. Thus, the overvoltage arrester allowed the pipeline to attain a high AC-induced voltage level (i.e., up to 25 V) for long time intervals. Moreover, during the activation period of the arresters, the DC voltage between the pipe and the earthing wire shifted toward zero. This suggested that the DC was flowing from the grounding wire to the pipe through the arresters. Thus, the potential of the pipeline and of the earthing wire tended to reach the identical value.

Field-testing results confirmed that the installation of overvoltage arrests can reduce AC voltage on pipelines without affecting the CP performance.[13] During activation of overvoltage arresters, the AC voltage is lowered while the DC is carried by the arresters. Although the AC voltage is reduced by the arresters, the remaining AC voltage on the pipeline may still be sufficiently high to cause AC-induced corrosion. Thus, the overvoltage arrestors serve as a complementary method, in addition to grounding materials, for AC interference control.

**Alternating voltage arrester in continuous function.** Although overvoltage arresters can block AC voltage on the pipelines, Kioupis et al. noted that overvoltage arresters did not satisfy the desired functionality for the CP system.[14] They reported on the use of an electronic device, which can conduct AC but block DC, to reduce the AC level on the pipe but without degrading the CP performance. The device was connected in parallel with each overvoltage arrester across the pipeline and grounding wires. The device can achieve a maximum steady-state (long-term) AC voltage of 30 $V_{rms}$, a maximum steady-state DC voltage of 15 V, a maximum steady-state AC discharge current of 10 A, a steady-state AC discharge current at 30 $V_{AC}$ in a test circuit with a 20 Ω resistor of 1.5 A, and a steady-state DC leakage current at 15 $V_{DC}$ in a test circuit with a 20 Ω resistor of 10 μA. It has an extremely high DC blocking capability (a DC leakage current on the order of few microamperes) so that the DC flow to and from earth is totally prevented. Consequently, the CP function, line current, and DC potential measurements are not severely affected. Moreover, the DC leakage current does not increase with the DC voltage, facilitating investigation of CP problems or coating defect surveys with an improved accuracy of the measurements.

Field measurements were conducted on a gas pipeline, 45 km in length, under AC interference originating from electric power transmission lines (eight 150 kV and two 380 kV lines) to test the effectiveness of the device.[14] The pipeline had already been equipped with semiconducting overvoltage arresters to reduce AC voltage on the pipeline whenever exceeding a certain safety level, in this case 25 $V_{rms}$. Those arresters were insufficient to prevent AC corrosion, but they could interfere with the CP of the pipeline. It was found that, after installation of the device, AC voltage levels were minimized, and DC voltage variations were diminished. Therefore, the AC corrosion risk was lowered and the possibility of DC potential deviations from the protection potential was decreased. The CP measurements were generally not deviated. Moreover, the "ON–OFF" potential values were not considerably altered. The total protection current was decreased, indicating improved CP operation.

## 9.4. Transformer use for AC interference mitigation

Land et al. reported that transformers can be added to the grounding path to create a resonant condition, referred to as a "circuit" consisting of a resistor, an inductor, and a capacitor, from the known frequency of the induced AC, which will remove energy from the circuit in the form of a magnetic field or electric field.[8] Additionally, a transformer can be designed to transfer energy from the AC mitigation system (i.e., primary circuit) to the secondary circuit, effectively draining energy from the AC mitigation system. This further reduces the induced AC voltage on a pipeline. Figure 9-3a shows the typical design condition, where ground ribbons are connected to the pipeline through a capacitor. When a circuit is tuned to its resonant condition, the power angle can be reduced to zero, thus removing unnecessary impedance due only to capacitance. Figure 9-3b shows the effect of resonance with the additional inductance added and the circuit tuned to resonance.

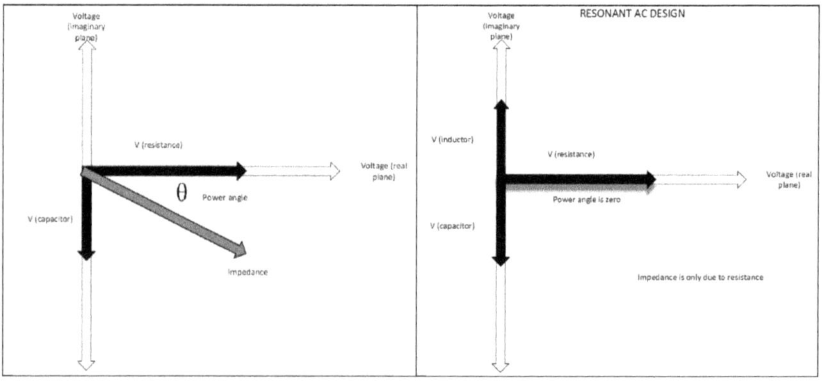


**Figure 9-3.** Complex voltage plane representation of AC mitigation using (a) just a capacitor and (b) a capacitor in series with an inductor where the circuit is operating at resonance.[8]

Field tests were conducted on a pipeline under 10 A of AC current, with zinc ribbons connected in parallel to the pipe.[8] Although the zinc grounding could reduce the induced AC voltage, when the transformer was connected, the AC voltage on the pipe decreased to a lower level. The lowest AC voltage was measured when the ground zinc ribbon was connected in parallel with the primary circuit of the transformer. The transformer was set to its highest ratio from the primary circuit to the secondary circuit, and the resistance of the secondary circuit was minimized. It was demonstrated that the transformer was successful in removing energy induced on the pipeline. As a comparison, AC mitigation by grounding only transmits the induced AC current to ground.

# 9.5. DC decoupling technique

For AC interference mitigation systems for buried pipelines, the detrimental effect of the systems on the CP applied on the pipelines must be considered. In principle, the AC mitigation systems effectively dissipate the current in the soil, and they do not compromise the CP performance. However, the materials and structures in grounding and shield wires used in AC interference mitigation can change the operation of the CP system. A direct effect is to change the surface area on pipelines to be protected by the CP. Moreover, it is not unusual that the grounding structures can shield the CP current from reaching the pipeline surface, decreasing its effectiveness for corrosion protection. DC decouplers are usually used between the pipeline and grounding materials. When used in conjunction with the AC mitigation systems, DC decouplers can effectively mitigate and avoid the adverse effect on CP. Furthermore, DC decouplers prevent galvanic coupling between the buried pipeline and the mitigation electrodes.

The operating principle of the decouplers integrated into the circuit is that, while they allow AC current to pass to ground, all DC current (i.e., CP current) is blocked. This can isolate the pipeline's CP system from the AC mitigation system, preventing the grounding conductors from taking CP current from the pipeline. Figure 9-4 shows a typical installation of a decoupler with the pipeline through copper wires, and Figure 9-5 shows the electrical principle of the use of a solid-state DC decoupling device. Steady-state AC current passes through the capacitor, while AC fault current flows through the thyristors. The surge protector passes lightning current, and the inductor prevents the lightning current from damaging the capacitor and thyristors.

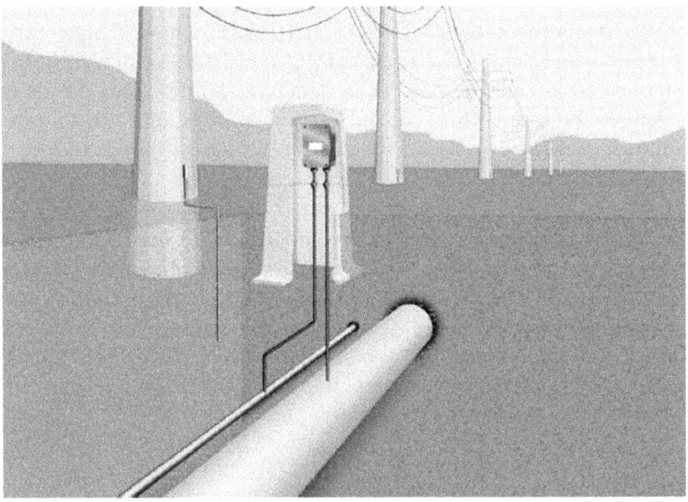

**Figure 9-4.** Schematic diagram illustrating the connection of a DC decoupler with the pipeline through copper wires.[15]

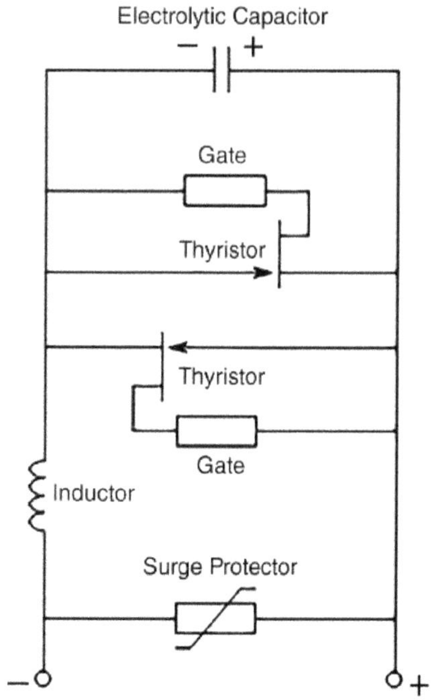


**Figure 9-5. Electrical principle of the use of a solid-state DC decoupling device.**[11]

The DC decouplers are also used to block DC current at aboveground appurtenances, such as block valves, metering stations, and launcher or receiver stations. Moreover, the decouplers installed across electrical isolation flanges can prevent "burning over," which can occur when an AC fault current or lightning surge is large enough to arc over the gap between flange faces or exceeds the rating of the isolation flanges. Presently, there are various types, sizes, and ratings of commercial decouplers, and their selection should depend on the predicted AC interference and mitigation design.

Generally, DC decouplers are used in combination with grounding anodes such as zinc ribbons or copper wires. As copper is more noble than steels and zinc and does not corrode rapidly in the soil, grounding copper anodes offer a low impedance to ground to discharge AC current. Decoupling devices are connected between the copper cable and the protected pipeline. As a result, the AC interference can be reduced to such a level that the AC voltage and current do not constitute a threat to humans or the assets.

## 9.6. Gradient control facilities

When high levels of AC voltage are present on a pipeline either during a fault condition or as the result of inductive coupling during the normal operation of HVAC power lines, personnel near and/or touching any aboveground appurtenance are at risk of electrical shock step or touch hazards. Gradient control mats are primarily used in a local environment to mitigate hazardous AC touch or step potentials. The ground mat provides buried bare conductors, which are typically galvanized steel, copper, or zinc. When connected to the pipeline, the mat provides local touch and step voltage protection by creating an equipotential area around the appurtenance.

## 9.7. Industry standards and practical recommendations

Industry standards have been proposed and developed to mitigate and control pipeline corrosion under AC interference. The primary factors for the evaluation and control of AC corrosion likelihood are AC voltage and AC current density. The AC voltage on the pipeline is measured relative to remote earth. While the voltage measurement is easy to conduct, it is not sufficiently accurate, since the measured value depends heavily on the soil resistivity. To date, there has been no agreement on the threshold value of AC voltage below which the AC interference is negligible.[16] Instead, the AC current density is directly associated with the AC corrosion risk. However, AC current density measurements are not easy to perform, and proper probes are required to be installed near the pipeline to measure the AC current density and true potential.[17] Since numerous factors affect the AC corrosion phenomenon, threshold values of the affecting parameters can only be considered from a probabilistic viewpoint. No deterministic relationship has been found between the parameters and the AC corrosion rate.[18]

### 9.7.1. NACE International criteria

According to NACE standard practice SP0177, the AC voltage should be mitigated below 15 V with respect to local earth for HVAC power lines under steady-state operating conditions where personnel could readily make contact with the pipeline or appurtenance.[3] This voltage requirement does not necessarily eliminate the likelihood of AC corrosion of pipelines. For example, in extensive field investigations, corrosion attack was found on steel samples that had an AC voltage less than 5 V.[19] The Canadian standard CAN/CSA-C22.3 No. 6-M91 also recommends that the AC voltage on a pipeline be reduced to less than 15 V.[20]

In North America, the commonly accepted criterion of AC current density below which AC corrosion is not expected to occur is 20 A/m$^2$, specifically:[21]

- The likelihood of AC-induced corrosion is low at AC current densities less than 20 A/m$^2$.
- The likelihood of AC-induced corrosion is unpredictable for AC current densities between 20 A/m$^2$ and 100 A/m$^2$.
- The likelihood of AC-induced corrosion is high at AC current densities greater than 100 A/m$^2$.

The level of AC current density is a function of the soil resistivity, the induced AC voltage, and the size of a holiday in the coating. The average bulk soil resistivity to the pipeline depth is one of the controlling factors affecting the magnitude of induced AC. However, the specific resistivity of the soil layer directly next to the pipe surface is directly related to the corrosion activity. Moreover, it is suggested that, when a pipeline is above a minimum AC current density, the typically accepted levels of CP applied on the pipeline would not be able to control AC corrosion. Wakelin and Sheldon reported that, when soils were contaminated with chloride ions, the AC-induced voltages should be reduced to lower values than the NACE-recommended value in order to effectively control AC corrosion.[22]

In 2018, NACE issued SP21424-2018, providing guidelines and procedures for risk assessment, mitigation, and monitoring of corrosion on underground, cathodically protected steel piping systems caused by proximity to AC power supply systems.[23] The AC current density of 30 $A/m^2$ was recommended as the upper limit, which was used in conjunction with the DC current density (i.e., the $i_{AC}/i_{DC}$ ratio). This combined criterion is also recommended in other industry reports by the INGAA Foundation[24] and CEPA.[1]

### 9.7.2. European Committee for Standardization

To protect the personnel readily in contact with the pipeline or appurtenance, in Germany, the AC voltage should not exceed 65 V with respect to local earth for steady-state conditions.[25] The European Committee for Standardization issued the standard CEN/TS-15280, where, to control the likelihood of AC corrosion on a buried pipeline, the induced AC voltage does not at any time exceed 10 V when the surrounding soil resistivity along the pipeline is greater than 25 Ω·m, and 4 V when the local soil resistivity measured along the pipeline is less than 25 Ω·m.[26] It is further stated that, if the AC current density is larger than 30 $A/m^2$, AC corrosion may happen, and if the AC current density exceeds 100 $A/m^2$, AC corrosion will happen. The corrosion rate of steels is greater than 1 mm/y in the presence of AC current densities higher than 500 $A/m^2$.[16] Both CEN/TS-15280[26] and ISO 18086[27] recommend the CP criteria for pipelines subject to AC interference to be:

- AC current density: $i_{AC} < 30\ A/m^2$, or
- DC current density: $i_{DC} < 1\ A/m^2$, or
- $i_{AC}/i_{DC} < 3$.

## 9.8. Development of AC interference management program for pipelines

The objectives of an effective AC interference mitigation and control program for a cathodically protected pipeline include (a) assessment of the pipeline segments under AC interference risk, (b) prevention of the pipeline from suffering from AC accelerated corrosion and pitting corrosion, (c) protection of the external coatings from damage or disbondment accelerated by fault currents or exceeding the dielectric strength of the coating, and (d) avoidance of deviation of the CP potential from the applied value, which would decrease its effectiveness for corrosion protection. An effective AC interference mitigation and control program also aims to prevent damage to pipeline components such as isolating fittings, transformer-rectifiers, resistance bonds, galvanic anodes, and electrical monitoring equipment. Furthermore, the program will prevent hazardous step and touch voltages that could harm personnel and the public who might make contact the pipeline and/or its appurtenances. The mitigation strategy usually includes the following units, which are integrated into an interrelated program: AC voltage field survey, AC interference assessment, AC mitigation design, AC mitigation field application, AC mitigation effectiveness post-assessment, and AC voltage and AC current density online monitoring. Therefore, at collocations of buried pipelines with HVAC power lines, it is critical for pipeline operators to investigate the possibility of AC interference, to evaluate the magnitude of the interference, to design a mitigation program, to install mitigation facilities and systems, and to monitor the performance of the mitigation system and the CP system applied on the pipelines.[1] All of these are essential tasks to an effective AC interference mitigation and control program for pipelines.

The development of an appropriate and effective AC interference management program for pipelines usually applies in the situations as listed below, but may also apply in others:

- An existing pipeline collocated with an existing HVAC power line
- A new pipeline intended to be constructed with collocation with an existing HVAC power line
- An HVAC power line intended to be constructed with collocation with an existing pipeline
- An extension to a pipeline that is on or near an existing HVAC power line and within 5 km of the entrance or departure point between a power line and a collocated pipeline.[1]

An effective AC interference management program should be integral to the integrity management program for pipelines. The initial step is to review each segment of the pipeline over its entire length and record the crossings and collocations (parallel) with HVAC power lines located within 300 m lateral separation from the pipeline. Substations should also be identified within a distance of 300 m from each pipeline.[28] A database should be developed containing each HVAC crossing, collocation, and substation, as well as such information as external pipeline coatings (types, history, performance, etc.), the nearest CP test stations (AC pipe-to-soil potentials), the distance between the pipeline and the nearest HVAC tower footing, the length of paralleling between the pipeline and the power lines, the maximum AC voltage recorded at the nearest test point, the AC tower type (number of circuits, vertical or horizontal arrangement of HVAC lines, tower design).

Since the soil resistivity is critical to an AC interference assessment, the national soil database, which is usually open to public, could be used to obtain relevant soil information at the pipeline segments. The soil database could include the soil types within the area of interest, the percentage of the area covered by each soil type, and–if available–the soil conductivity of each soil type. In summary, a complete, accurate database is key to the development of a reliable AC interference and pipeline AC corrosion risk assessment program. Simon et al. reported that a HVAC transmission line geographic information system database was developed and used to acquire the voltage of each power line at identified locations in each pipeline system.[28]

Three basic risks that result from AC interference on buried pipelines at crossings and collocations between the pipelines and HVAC power lines are usually modeled and assessed: personnel safety issues due to exposure to AC voltages induced during normal operation and/or fault conditions, AC-influenced (usually accelerated) corrosion during normal operation, and the coating or pipe wall damage during a fault.[28]. Each risk is influenced by specific factors at the individual sites. The factors dominating the safety risk by induced and fault AC voltages include the geophysical relationship of the pipeline–HVAC power line collocation, soil resistivity, coating resistance, HVAC power line load, HVAC tower geometry (phasing, conductor layout, height), and the HVAC fault current and duration.[28] The affecting factors for pipeline AC corrosion include the geophysical relationship of the pipeline–power line collocation, soil resistivity, estimated coating holiday size (coating type and performance condition), coating resistance, HVAC power line load, and HVAC tower geometry (phasing, conductor layout, height). The affecting factors for fault (arcing) include the separation between the pipeline and the closest tower grounding, soil resistivity, soil type (gravel, loam, clay), and HVAC line voltage. With the information listed above, a software package can be used to calculate the predicted AC interference. If some factors are unknown, the importance of the factors should be weighed (i.e., sensitivity analysis of each factor should be performed). In addition to the three basic risks, it has been proposed that using the ratio of the induced AC voltage to the soil resistivity to evaluate the relative risk of AC corrosion at any point along the length of a buried pipeline.[29]

The effects of individual factors on AC interference, usually AC voltages induced on the pipelines, can be modeled by the selected software. As described in Chapter 8, all affecting factors can be modeled to quantify their contributions to the AC interference that generates from HVAC power lines to the collocated pipelines.

In addition to modeling analysis, periodic inspection and online monitoring of the CP performance and corrosion occurrence on pipelines are important for AC corrosion evaluation. For example, a visual inspection conducted in conjunction with an external corrosion direct assessment dig should inspect whether a telltale tubercle (i.e., a hard deposit containing soil and corrosion products) is observed at the coating defect. Such tubercles are often associated with a distinct morphology characterizing AC corrosion.[4] Additionally, ILI data indicating significant external wall loss on a pipeline that is shared with power lines in an energy corridor should be investigated further, including an AC interference analysis. Some newly developed ILI tools have been reported to show promising results for detecting and measuring AC current.[4] Furthermore, the risks induced by AC interference to the CP equipment should be considered as well. Rectifiers and electronics should be protected with surge protectors and lightning arrestors. Shunts, switches, and resistors are usually susceptible to electric transients, and selection of reliable parts should be made. Cables should be sized for the inclusion of AC current, in addition to the design DC current.

The management program should also include proper monitoring and control of DC interference from the CP system on HVAC infrastructure such as power line structure anchors, grounding, and foundations.[30] Generally, power lines with shield wires are a preferred low-resistance path and susceptible to DC interference. Corrosion would occur where the DC current discharges. DC interference testing should be performed on structures close to CP ground beds. Mitigation measures usually include the installation of sacrificial anodes on the structure legs.

For an improved understanding of the AC interference management program and its implementation, two successful experiences are cited and described herein. The first case involved a comprehensive AC interference analysis and the implementation of an extensive AC corrosion mitigation and monitoring system conducted on a 328 km portion of a regulated pipeline, which was collocated with multiple HVAC power lines and countless adjacent pipelines with CP.[31] After detailed understanding and analysis of the collocating situation between the target pipeline and the power lines and adjacent pipelines, the field study was conducted to define the significance of AC interference on the pipeline and establish a sound basis for the required mitigation. The measurements included soil resistivity, instantaneous AC and DC pipe-to-soil potentials, and coupon current densities. The AC pipe-to-soil potential was monitored as a function of time to understand the variations in power line loading and the electromagnetic interaction between adjacent power lines and between the power lines and the pipeline. Since there was a strong interaction between the AC and DC electrical signals imposed on a pipeline, the relationship between coupon AC and DC current densities was derived. The performance of the pipeline coating was also assessed on the basis of measurements and the evaluation of CP data and AC current attenuation measurements along a section of the pipeline. All the field work and information collection provided a firm foundation for the AC interference evaluation conducted in the next stage.

Computer simulations were performed to evaluate the pipeline AC interference conditions under normal steady-state operations and during a possible ground fault of the power lines. Once validated by comparing with the measured field data, the unmitigated simulation results served as the reference for comparing different mitigation approaches. The modeling results included the unmitigated AC current density and AC pipe-to-soil potential profiles under the normal operating conditions of the power lines. Generally, AC corrosion from normal steady-state power line operations is the most prevalent pipeline integrity threat. To date, most of the modeling analysis has thus focused on the steady-state conditions. The modeling was conducted when the planned mitigation was applied on the pipeline to obtain AC current density and AC pipe-to-soil potential profiles on the mitigated pipeline. In this program, the mitigation included numerous deep anode grounds, which were connected to the pipeline through DC decoupling devices. Additionally, bare copper cables were selected for the parallel mitigation conductors and high-silicon cast iron anodes in coke breeze backfill were used for the deep anode grounds. As a result, the maximum modeled AC current density for the target pipeline was 30 A/m$^2$, and the modeled AC potentials were less than 5 V.

Fault analysis was also performed to obtain the modeled AC pipe-to-soil potential profiles resulting from the power line to ground fault. These included the profiles for the unmitigated condition and the profiles for the mitigation scenarios devised for steady-state AC interference and for faults. From the fault analysis results, the mitigation system for faults was proposed and connected to the pipeline through electronic DC decoupling devices.

Finally, the planned AC interference mitigation was installed with quality assurance tests to determine and document specification compliance, including various test station measurements, DC decoupler current measurements, parallel mitigation conductor electrical continuity measurements, and resistance-to-earth measurements for each deep anode ground. Monitoring was performed on the AC mitigated pipeline, and optimization was probably performed on the pipeline CP level to reduce its effect on the AC interference. Remote surveillance of multiple electrical resistance–type corrosion rate probes was also used to ensure that pipeline corrosion would always be below an acceptable level, even under AC interference.

In another AC interference control practice, an AC interference risk ranking was established on an existing transmission pipeline of 10,300 km in length operated by Pacific Gas and Electric Company in the U.S., attempting to identify the "pipeline segments that are at greatest risk due to steady-state AC interference, to prioritize them based on the severity of risk, and to determine what further action is required."[32] For this purpose, historical AC voltage and soil resistivity measurements, as well as the calculated AC current densities (based on local soil resistivity), were used as the most effective indicators for the pipelines, which were located within 30 m of the HVAC transmission power lines during steady-state operation. Although AC corrosion of pipelines can be affected by many factors, the relationships between most factors and the AC corrosion rate could not be quantified. The AC current density has so far been the primary parameter used to evaluate the likelihood of AC corrosion on pipelines.[21] Generally, the AC corrosion rates increase with increased AC current density, especially when the AC current densities are greater than 100 A/m$^2$. For risk rating of safety and AC corrosion, a steady-state touch voltage of 15 V or more relative to local earth at above-grade or exposed sections of the pipeline is considered to cause a shock hazard.[3]. The AC current densities of 20 A/m$^2$, 50 A/m$^2$, and 100 A/m$^2$ categorize different corrosion rate ranges. The criteria for safety and AC corrosion risk ranking on pipelines are shown in Table 9-1.

| Risk ranking | AC voltages | AC current densities |
|---|---|---|
| Low | < 5 V | < 20 A/$m^2$ |
| Moderate | 5 - 15 V | 20 - 50 A/$m^2$ |
| High | - 20 V | 50 - 100 A/$^{m}$2 |
| Severe | > 20 V | > 100 A/$m^2$ |

**Table 9-1. Risk ranking of safety and AC corrosion by AC voltages and AC current densities.**[32]

The study program included three phases: data consolidation, date normalization, and pipeline risk ranking. As a result, the risks of safety and AC corrosion at various test stations were ranked, and the risk ranking was then prioritized by pipeline. For example, each pipeline was given a priority rating from 1 to 5, with rating 1 being the highest priority:

- Priority 1: one or more "severe" ratings
- Priority 2: "high" rating for both the AC voltage and the AC corrosion risk, or multiple high ratings
- Priority 3: "high" rating for the AC voltage or the AC corrosion risk, or multiple moderate ratings
- Priority 4: "moderate" ratings for either the AC voltage or the AC corrosion risk
- Priority 5: only "low" ratings

From the risk ranking and the priority rating, recommendations were given for each pipeline. All measurements and the resulting rankings and ratings were time sensitive to reflect the variability in power line load conditions and other affecting factors such as soil resistivity. According to periodic measurements and inspection, as well as data normalization, each pipeline would be evaluated so that the AC safety and corrosion risk could be ranked appropriately, and real-time measures could be taken.

# References


1. Gummow, R.A., *A/C Interference Guideline Final Report,* Canadian Energy Pipeline Association, Calgary, Alberta, 2014.
2. Matcor, *AC Mitigation: 4 Approaches,* Chalfont, PA, 2018.
3. NACE, *Mitigation of Alternating Current and Lightning Effects on Metallic Structures and Corrosion Control Systems,* SP0177-2014, Houston, TX, 2014.
4. Brelsford, C., "Mitigating AC corrosion on cathodically protected pipelines," *Pipeline Gas Journal* 242, 10: 71–74, 2015.
5. Finneran, S., Krebs, B., Bensman, L., *Criteria for Pipelines Co-Existing with Electric Power Lines,* Report 2015-04, Interstate Natural Gas Association of America, Washington, DC, 2015.

6. Markovic, D., Smith, V., Perera, S., "Evaluation of gradient control wire and insulating joints as methods of mitigating induced voltages in gas pipelines," Proc. Australasian Universities Power Engineering Conference, Hobart, Australia, 2001–2006, 2005.
7. Sirola, T., "Review of recent developments in induced AC , 2017current corrosion mitigation design, materials, installation and monitoring technologies," CORROSION/2017, paper no. 9711, NACE, Houston, TX.
8. Land, J., Yahner, T., Finneran, S., "Transformer use for mitigating induced alternating current effects on pipelines," CORROSION/2018, paper no. 11242, NACE, Houston, TX, 2018.
9. Finneran, S., Evans, K., Gossard, J., "Comparison of AC mitigation grounding materials through electrochemical analysis," CORROSION/2018, paper no. 10898, NACE, Houston, TX, 2018.
10. Croall, S.J., Hurley, W.G., "Case studies of prediction and mitigation of induced AC voltages on pipelines," *MP* 23: 30–33, 1984.
11. Tang, D., Lu, M., Du, Y., Gao, J., Wu, G., "Electrochemical studies on the performance of zinc used for alternating current mitigation," *Corrosion* 71: 795–806, 2015
12. Luk, M., Hebb, D., Gudino, E., Khattar, C., "Challenges in mitigating AC interference in remote areas," CORROSION/2017, paper no. 9090, NACE, Houston, TX, 2017.
13. Kouloumbi, N., Batis, G., Kioupis, N., Asteridis, P., "Study of the effect of AC interference on the cathodic protection of a gas pipeline," *Anti-Corrosion Methods and Materials* 49: 335–345, 2002.
14. Kioupis, N., Kouloumbi, N., Batis, G., Asteridis, P., "Study of the effect of AC-interference and AC-mitigation on the cathodic protection of a gas pipeline," CeoCor International Congress, European Committee for the Study of Corrosion and Protection of Pipes and Pipeline Systems, Brussels, 2003.
15. Pikas, J., "Looking back and understanding a case history of AC corrosion," CORROSION/2018, paper no. 10603, NACE, Houston, TX, 2018.
16. Brenna, A., Lazzari, L., Pedeferri, M., Ormellese, M., "Cathodic protection condition in the presence of AC interference," La Metallurgia Italiana, 6: 29–34, 2014
17. Ormellese, M., Goidanich, S., Lazzari, L., "Effect of AC interference on cathodic protection monitoring," *Corrosion Engineering, Science and Technology* 46: 618–623, 2011.
18. Xu, L.Y., Su, X., Cheng, Y.F., "Effect of alternating current on cathodic protection on pipelines," *Corrosion Science* 66: 263–268, 2013.
19. Büchler, M., Schöneich, H., "Investigation of alternating current corrosion of cathodically protected pipelines: Development of a detection method, mitigation measures, and a model for the mechanism," *Corrosion* 65: 578–586, 2009.
20. CSA (Canadian Standards Association), *Principles and Practices of Electrical Coordination Between Pipelines and Electric Supply Lines*, CAN/CSA-C22.3 No.6-M91, Ottawa, 2009.
21. NACE, *AC Corrosion State-of-the-Art: Corrosion Rate, Mechanism, and Mitigation Requirements*, Publication 35110, Houston, TX, 2010.

22. Wakelin, R.G., Sheldon, C., "Investigation and mitigation of AC corrosion on a 300 MM natural gas pipeline," CORROSION/2004, paper no. 04205, NACE, Houston, TX, 2004.
23. NACE, *Alternating Current Corrosion on Cathodically Protected Pipelines: Risk Assessment, Mitigation, and Monitoring*, SP21424-2018, Houston, TX, 2018.
24. INGAA (Interstate Natural Gas Association of America), *Criteria for Pipelines Co-Existing with Electric Power Lines*, Washington, DC, 2015.
25. Kajiyama, F., "Strategy for eliminating risks of corrosion and overprotection for buried modern pipelines," International Gas Union Research Conference, Seoul, 2011.
26. CEN (European Committee for Standardization), *Evaluation of AC Corrosion Likelihood of Buried Pipelines applicable to Cathodically Protected Pipelines*, CEN/TS-15280, Brussels, 2013.
27. ISO (International Organization for Standardization), *Corrosion of Metals and Alloys–Determination of AC Corrosion–Protection Criteria*, ISO 18086, Geneva, 2015.
28. Simon, P.D., Lopez-Garrity, M., Wagner, D., Garrity, K., "Identification and prioritization of potential HVAC interference locations on a 6,000 mile crude and product transmission pipeline system," CORROSION/2017, paper no. 9260, NACE, Houston, TX, 2017.
29. Adedeji, K.B., Ponnle, A.A., Abe, B.T., Jimoh, A.A., Abu-Mahfouz, A.M., Hamam, Y., "GUI-based AC induced corrosion monitoring for buried pipelines near HVTLs," *Engineering Letters* 26: 489–497, 2018.
30. Fieltsch, W., Shahinas, F. (2015), "CP considerations for pipelines subjected to AC interference," NACE Northern Area West Conference, Calgary, Alberta, 2015.
31. Lindemuth, D., Crabtree, D., Kustova, V., Lisovoy, S., "Challenging AC corrosion mitigation system for 100-mile long pipeline," CORROSION/2017, paper no. 9174, NACE, Houston, TX, 2017.Fieltsch, W., Silva, D., "AC interference risk ranking," CORROSION/2016, paper no. 7393, NACE, Houston, TX, 2016.
32. Fieltsch, W., Silva, D., "AC interference risk ranking," CORROSION/2016, paper no. 7393, NACE, Houston, TX, 2016.

10

# Corrosion of Pipelines under DC Interference

When buried pipelines parallel or cross DC sources including HVDC power transmission lines, DC-electrified railways, and DC rectifiers, the pipe steel at coating defects would be subject to DC corrosion. DC-induced corrosion is caused by electrochemical reactions associated with DC interference. DC corrosion has been recognized as one of the major threats to the integrity of buried pipelines. DC interference can accelerate pipeline corrosion, alkalize the soil environment, facilitate coating disbondment, and enhance hydrogen evolution and hydrogen permeation into steel.[1, 2] The increased reports on DC-induced pipeline failures are associated with the increasing frequency of parallelism between the buried pipelines and HVDC power lines. This is especially the case where HVDC power lines share the same ROW as energy pipelines over a long distance.[3, 4]

Buried pipelines can pick up DC stray current leaked from a DC source and discharge the current to the soil and then back to the source. Cathodic reaction (e.g., oxygen reduction, hydrogen evolution) occurs when the DC current enters the pipeline, and anodic reaction (e.g., steel oxidation) occurs when the current returns to the original path through the soil. Consequently, corrosion occurs at the anode, with no corrosion attack at the cathode.[5] During this process, the DC corrosion kinetics obey Faraday's law, where 1 A of DC current discharging from the pipeline consumes 9.1 kg of steel within one year.[6] Generally, DC corrosion occurs more rapidly than AC corrosion on pipelines.[5]

The first reported HVDC power transmission system was in operation in 1954. In the past several decades, conversion methods have been improved, allowing for higher voltages to be transmitted over long distances. Public needs and industrial development require substantial amounts of electric energy, and power lines provide a crucial mode of electrical transportation from source generators to end users. With the development of electric systems, such as HVDC power lines, and their adjacency to oil and gas pipelines, the problems resulting from DC interference on pipeline integrity are becoming an increasing concern.[7] For example, it was reported that the DC interference caused by Quebec-New England Intertie[8] and the interference induced by British Columbia Hydro's HVDC system on a crude oil pipeline could result in a 214 mV shift of pipe-to-soil potential under the DC current of 1,200 A that entered the earth.[9] Nowadays, compared with AC corrosion events, there has been limited research conducted on DC corrosion, making our understanding about the phenomenon limited.

## 10.1. Sources of DC interference

DC interference can be generated on buried pipelines or adjacent facilities when DC power equipment has multiple ground points within the working circuit.[10] Common devices that can induce DC interference include:

- DC-electrified railways, where the rail is the current conductor
- Overhead tram lines, where multiple wires are connected to the negative pole of the power supply station, or the current returns through the rail
- HVDC transmission power lines
- Electrolysis devices
- Internal and external ports, especially the DC operating equipment and DC welding equipment in the shipyard
- DC telephone network and traffic light equipment
- CP devices, which constitute the forced current device to discharge stray current

The interference can be stationary or nonstationary.[11] Stationary interference occurs on buried pipelines when they are under a stationary electric field, such as that generated by an external CP system. Nonstationary interference takes place when the electric field varies with time: one typical case is the stray current dispersed by DC traction systems, for which the interference takes place or becomes stronger when the train passes over the buried pipeline. Although the duration is short, usually several minutes, the effect may be severe owing to a high circulating current.[12]

### 10.1.1. DC device

CP is a common technique used to protect metallic structures such as buried pipelines from corrosion attack. However, the stray current flowing from the CP system can cause extensive damage to nearby unprotected pipelines. For example, current leakage from auxiliary anodes associated with the CP system can enter a nearby unprotected structure unintentionally and leave from the structure surface, where severe corrosion is caused. When the stray current enters a buried pipeline from the unintended potential difference of two objects, the current passes from the soil to the pipeline at one point (i.e., cathode) and returns to the soil from another point (i.e., anode). Corrosion occurs at the anode. Generally, the effect of the DC stray current from nearby DC devices on the buried pipelines depends on many factors. For example, when two pipelines cross, the DC protective current from the CP system of one pipeline can be picked up by another pipeline. Four main situations potentially affect the pipeline integrity:[13]

- The CP-applied pipeline is well coated, but the other pipeline is coated with an aged coating: DC interference occurs on the pipeline with the aged coating.
- The CP-applied pipeline has an aged coating, and the other pipeline is well coated: An interference effect occurs only the well-coated pipeline only.
- The two pipelines are well coated, and the CP system is too close to both lines: An interference effect occurs on both pipelines.
- The two pipelines are well coated, and the CP system is remote from both pipelines: There is no interference.

For ICCP systems, auxiliary anodes are connected to the DC power source, which generates a certain amount of current, depending on the pipe size, the coating performance, and the anode distribution. The anodes are commonly distributed in ground beds or deep vertical holes. By provision of cathodic current to the protected structure, the corrosion reaction is suppressed. However, for the adjacent unprotected pipelines, localized corrosion may occur at the stray current discharge site. Hu and Li reported that DC interference occurred on buried gas pipelines that were close to the installed anode bed (about 10 m away).[14] Both the field pipe-to-earth potential measurements and the calculations of the earth potential increase confirmed the presence of DC interference on the pipelines.

For SACP systems, an electrical connection is made from the sacrificial anodes to the pipeline being protected. Electrons are supplied to the steel pipe via the electrical connection. Similarly, the stray current generated during anodic reaction of the sacrificial anodes may flow into the unprotected pipelines, generating the DC interference.

In 1997, research was conducted by the Kansas Department of Transportation in the United States to investigate stray current corrosion due to the utility's CP.[15] It was found that the underground metallic structures, such as pipelines, that were adjacent to CP systems suffered from corrosion at certain positions of the structures relative to the ground beds of the CP systems. A change in potential of the underground pipeline was produced unless the pipeline was completely shielded by coating. However, the pipeline coatings usually contained various defects such as pinholes and breakages. The change of potential caused a current flow within the pipelines, and corrosion was proportional to the current flow leaving the pipe to return to the electrolyte (i.e., soil). Frequently, no available method can fully protect the structures when they are under the influence of the ground beds. It was thus suggested that the ground beds be moved if pipelines were located within the ROW. At the same time, all pipelines close to cathodically protected utilities should be assessed for stray current interference. Additional inspections should be conducted along the pipelines where DC interference is possible and could compromise the pipelines' structural integrity.

### 10.1.2. HVDC power lines

Another main source of DC interference is HVDC power transmission lines. It has been observed that, compared with HVAC power lines, HVDC lines are more efficient, more economical and more environmentally friendly over long distances.[4, 16] Energy transportation to end users through HVDC methods has grown rapidly in recent years. The requirements for line tower construction are simpler and more convenient than for HVAC transmission lines, and there are lower per-unit costs, including costs per kilometer of line and per megawatt of transmitted power.[17] Moreover, the costs for cables of the same transfer capacity in HVDC lines are significantly lower than the costs of HVAC lines. Furthermore, AC transmission must have the same synchronized frequency to connect with another system; this is not required for DC transmission because the current flows in one direction. This allows renewable energy, such as solar and wind energy, to be connected over long distances. Finally, it is possible for independent power flows and frequency regulation in power systems to be connected via HVDC power lines. It is also possible that the power transfer can continue via one pole of a bipolar line even when the second pole is shut down during an emergency. It is thus predicted that DC transmission will have a more prominent presence in the future.[4]

HVDC transmission lines are usually bipolar systems. However, they may operate either continuously during normal operation or periodically in monopolar operation. In bipolar mode, one wire is positive, and the other is negative. Current flowing through the earth is the unbalanced current between the two wires, which is usually less than 1% of the wire-carried current.[18] In the case of a fault or equipment repair, the bipolar mode can turn into the monopolar mode, where DC follows an earth path to flow between the system terminals. Operations as a monopole cannot be avoided in reality.[16] A monopole (single conductor) system consists of one terminal from the rectifier connected to the ground, while the other high-voltage terminal is connected to the transmission lines. In comparison, bipolar systems use two conductors: one from the source to the load and the other from the load to the source. Bipolar configuration systems are equipped with a ground electrode for any imbalanced current. Owing to differences in current flow between the two poles (i.e., a dissymmetry of the bipolar scheme), a prolonged current that passes through the ground is estimated to be 1% to 3% of the carried current value.[19]

A portion of the DC current flowing in the circuit system can be captured by the pipelines and other grounding systems located nearby. The buried pipelines may collect, conduct, or discharge the DC current, depending on the location of the pipeline relative to the HVDC lines and the polarity of the HVDC ground electrode. For example, when the electrode near the pipeline is operating in a positive mode, the current from the electrode is collected by the pipeline adjacent to the electrode, and the current is then discharged from the pipeline at locations remote from the electrode, resulting in corrosion at the discharging points.[20] Conversely, when the HVDC electrode near the pipeline operates in a negative mode, the current is collected in a location far from the electrode and discharged from the pipeline in the vicinity of the electrode. Again, the locations on the pipeline at which the DC current is discharged suffer from corrosion.

At the same time, external coatings on buried pipelines would experience degradation, such as breakage or disbondment, over long-term service. Coating disbondment can be further fostered when the pipeline is polarized negatively to exceed the recommended potential range, resulting in the so-called cathodic disbonding. The DC voltage can be sufficiently negative to cause coating disbondment at the current collection sites. Figure 10-1 illustrates the adverse impact on buried pipelines resulting from DC interference due to adjacent HVDC power lines.

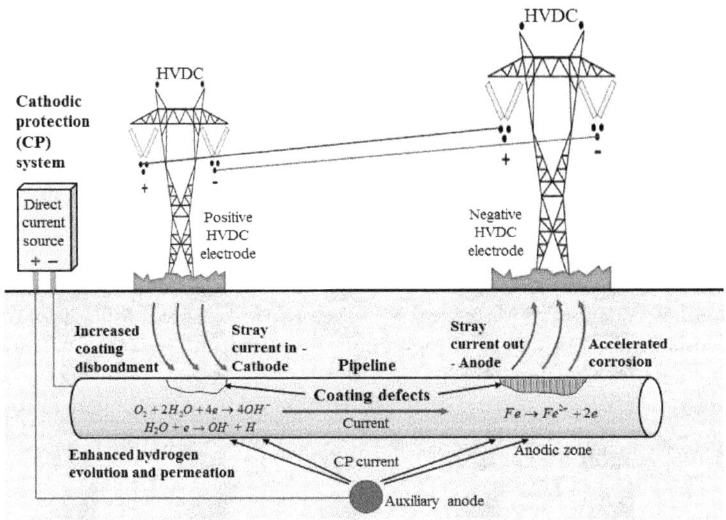


**Figure 10-1.** Schematic diagram illustrating the mechanism of the adverse impact of DC interference on buried pipelines.[21]

Like HVAC power lines, HVDC transmission lines can couple with parallel conductive structures such as pipelines by three mechanisms: mutual capacitive, mutual inductive, and direct conductive couplings.[22] Capacitive coupling refers to the interaction of the electric field of HVDC power lines with a parallel pipeline that is not grounded. This is particularly experienced by pipelines during construction near overhead transmission lines or by aboveground pipelines routed near the HVDC transmission lines. For buried pipelines, the electric field is eliminated by grounding with the earth. Inductive coupling occurs when the buried pipeline is influenced by the electromagnetic field generated by the HVDC power lines. Conductive coupling results from the DC stray current flowing into the earth near a buried pipeline from the adjacent HVDC power lines. Where the fault current enters the soil, a voltage gradient is created in the affected area, accompanying the current flowing through the earth. Owing to the high resistance of the external coating, the buried pipeline is coupled with a risen local earth potential.

During the steady-state operation of the HVDC power lines, negligible voltage is induced on the adjacent pipelines. Thus, there will be minimal risk to human safety and DC corrosion during the steady-state condition of the HVDC transmission lines. Aside from construction issues, extreme weather (i.e., lightning) is the main cause of HVDC line faults. Generally, there are two types of lightning-related flashover modes.[22]

- **Backflashover.** Any strike to the tower top or shield wires can cause a backflashover on the insulator if the voltage difference across the insulator exceeds its critical flashover value. The pipeline is at its normal potential relative to earth, whereas the tower top voltage is raised to a high value owing to the lightning discharge.
- **Shielding failure.** Normally, the shield wires are used to intercept lightning strikes to the power lines. However, low-intensity lightning strikes can pass over the shield wires and terminate on the phase or pole conductors. As a result, the voltage rise causes insulator flashover. The shielding failure strike current can lead to voltage stresses, and the pole conductor would then flash over. In the shielding failure mode, most of the fault current goes to the earth with thousands of amperes of current. The high amplitude of current can cause significant inductive coupling. High currents flowing in the pole conductor and shield wires owing to faults will induce voltages on the parallel pipelines. The interference touch voltage can cause step and touch voltage hazards for facilities along the pipeline. Thus, shielding failure faults are the worst case for the affected pipelines.[23]

### 10.1.3. DC transit systems

When a current is sent through a motor circuit, a torque proportional to the magnitude of the current is generated between the field and the rotating coils. A high current is produced when a voltage is applied. The high current can enable heavy objects like trains to start.[24] DC traction systems are one of the main sources of electrical interference to buried pipelines. Figure 10-2 shows how the stray DC flows from a DC-electrified railway to an adjacent buried pipeline. The current flows along the overhead catenary wire, through the traction motors, and goes to the traction power station along the running rails. The rail-to-ground resistance is not sufficiently high, especially when the soil contains moisture or water. A portion of current enters the soil to generate the so-called stray DC current. The DC current flows into the adjacent pipeline through the soil and returns to the rails, forming a closed circuit. The extent of DC interference is a function of pipeline steel properties, coating properties, soil environment, amount of DC current, and time.

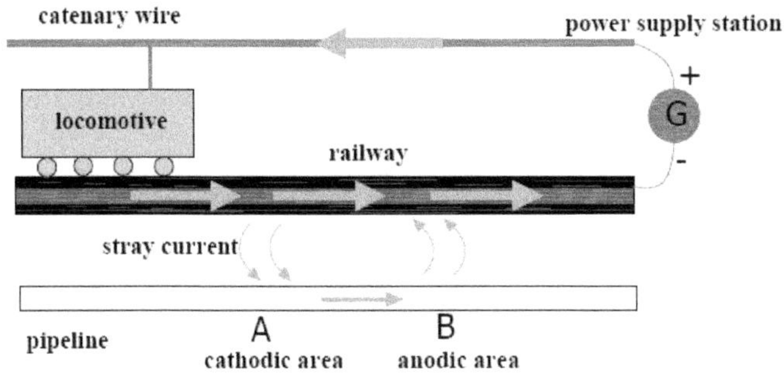


**Figure 10-2.** Schematic diagram showing the flow of stray DC current from rails to a buried pipeline.

The stray current generated by rail transit systems is usually dynamic in nature, which is attributable to the frequent change of track-to-earth potential in the railway. Mostly, the potential measured in the field varies with time in response to the acceleration and deceleration of trains, the number of vehicles passing through the area, and the seasonal alteration of the soil conductivity.

Kustova et al. reported that a natural gas pipeline operating with adjacency to a nearby DC-powered light rail transit system that started service in the 1990s suffered from DC interference.[25] While electrical isolation measures were applied on the transit tracks in design and construction to control stray current at the source, the measures degraded and stray current leakage increased over time. It was found that the track-to-earth potentials varied by approximately ±10 V, directly causing varied pipeline potentials, which could be either cathodically or anodically polarized. In some locations, the variation of the track-to-earth potential by 10 V could cause shifts of the pipe potential of 200 mV. Moreover, pipe-to-soil potentials and DC current densities were monitored on buried coupons, and the results indicated the effect of DC stray current on the pipeline. The measured coupon current density was cathodic, flowing from the soil to the coupon and to the pipe; the area that the stray current discharged out of the pipeline experienced corrosion. Furthermore, owing to the dynamic nature of rail transit systems, monitoring of the pipe-to-soil potentials for short periods of time cannot effectively identify the influences resulting from the transit system. Long-term monitoring on a buried pipeline collocated with Salt Lake City's light rail transit system, which was powered by a 750 V DC-electrified system, showed that the pipe-to-soil potential was more positive than the −0.850 V (CSE) criterion by 23.9% in the first 15 days of December 2016 and by 44.7% in the next 15 days.[26] The pipe segment in that location was thus categorized as being at high risk of DC interference.

### 10.1.4. Identification of DC interference

Generally, there is no universal criterion that can be used for identification of stray current. The pipe-to-soil potential and the potential gradient in the soil are usually the main parameters used to evaluate the existence and strength of stray DC current.[27] According to British Standard BS EN 50162:2004, DC stray current exists when the pipe-to-soil potential shifts positively over 20 mV.[28] However, in Germany, the criterion is that the pipe-to-soil potential positively shifts more than 100 mV. In China, according to Chinese Standard GB/T 19285-2014, DC corrosion can be considered possible when the positive shift of the pipe-to-soil potential is greater than 20 mV or when the potential gradient in the soil exceeds 0.5 mV/m.[29] When the pipe-to-soil potential shifts positively over 100 mV or the potential gradient of the soil is more than 2.5 mV/m, electrical drainage or other protective measures must be applied to protect the buried pipelines from DC corrosion.[30]

## 10.2. Features and parametric effects on DC corrosion of pipelines

### 10.2.1. Recognition of DC corrosion of pipelines

There are generally two categories of DC corrosion: structure current corrosion and earth current corrosion.[31] For the former, the DC current flows into the metal and causes potential gradients. However, for earth current corrosion, the current can flow into the electrolyte or environmental medium, inducing potential differences in the electrolyte or medium. As a result, metallic structures such as pipelines buried near to earth current sources can suffer from corrosion.

In reality, the DC current collected in the buried pipelines can be up to tens or even hundreds of amperes, causing significant corrosion.[32] DC corrosion is highly localized, concentrating at the current-discharging locations. Thus, DC corrosion can have dramatic consequences on pipelines. For example, when the DC current flows out of the pipeline through a small area, the local corrosion rate of the pipe steels is so fast that corrosion perforation can happen within a few months for a steel pipe with a wall thickness of approximately 7–8 mm.[33]

The main characteristics of DC corrosion include:

- **High corrosion rate.** In the absence of DC, corrosion may occur on buried pipelines in soil environments. The generated corrosion current is at most on the order of several tens of milliamps. However, DC corrosion of buried pipelines can generate a maximum corrosion current up to hundreds of amperes.[34] Tests were conducted at DC current densities more than 1,000 $A/m^2$, and an astonishingly high corrosion rate of up to 1,223 mm/y was reported in a simulated solution.[35]When the large current flows through the pipeline steel and goes into the soil, a significant dissolution of the steel would be produced.

The amount of steel corrosion is proportional to the DC current density. For example, as the DC current density increased from 0 A/m$^2$ to 1,200 A/m$^2$, the corrosion rate of an X70 steel increased linearly in a soil solution.[35]

- **Localized corrosion.** Generally, stray DC current flows out of the pipeline to the soil at sites where the pipe-to-soil resistance is small. Thus, DC corrosion is mostly concentrated at those local areas. As buried pipelines are protected by external coatings, DC corrosion usually occurs at sites where the coatings contain defects and where there is therefore low resistance locally. Therefore, pitting corrosion often initiates at the coating defects under DC interference, especially at high DC current densities.[35]

### 10.2.2. Parametric effects on DC corrosion of pipelines

Numerous factors affect the DC corrosion of buried pipelines, and most of them are associated with the physical–chemical properties of the soil, the pipeline burial conditions, and the properties of the pipe steels. Particularly, the intensity of the stray DC current depends largely on the current source and the resistance of the soil. Determination of the DC current density is thus critical in investigating the corrosion rate of pipelines.[36] Generally, the stray DC current density decreases with increasing soil resistivity when the buried depth and the pipe-to-soil potential are fixed. Thus, an increase in the soil resistivity of the buried area can decrease the DC corrosion of pipelines. A model was developed to simulate the electric field generated when DC current flew into the soil and the stray DC current density at coating defects, where it was assumed that the current flew between the soil and pipelines owing to their potential difference. The larger the potential difference between them, the greater the DC current density. If the soil is a uniform medium, the distribution of stray current is relatively uniform. The stray DC current density between the pipeline and the soil can be determined by:[37]

$$\frac{j_0}{j} = \frac{4\delta\rho}{D\rho_0} \tag{10-1}$$

where $j_0$ is the stray DC current density in pipelines, $j$ is the stray DC current density in soil, $\delta$ is the pipe wall thickness, $D$ is the inner diameter of the pipeline, $\rho$ is soil resistivity, and $\rho_0$ is the resistivity of the pipe steel. Generally, the resistance of soil is much larger than that of the pipe steels. Thus, the stray DC current flows primarily in the pipeline, instead of flowing through the soil.

In addition to soil resistivity, soil chemistry plays an important role in the DC corrosion of buried pipelines. It was demonstrated that, when the soil was contaminated by chloride ions, even at levels too low to initiate pitting corrosion, stray DC currents may be affected remarkably.[5] Moreover, a reduction of the separating distance between DC facilities and the buried pipelines and an increase in the DC power transmission voltage can cause an increase in the risk of DC corrosion on the pipelines. It has been confirmed that the buried depth of the pipelines and the coating performance affected the DC current density distribution. For example, it was shown that an intact external coating could reduce the amount of DC stray current to be picked up by the pipelines.[38]

# 10.3. DC corrosion measurements

## 10.3.1. Testing setup enabling DC corrosion measurements

Designing and building a setup to conduct DC corrosion measurements is not easy. It becomes even more complicated when the steel specimen under testing is under CP application. The complexity is mainly due to the mutual effect of the multiple current-flowing circuits involved: the electrochemical corrosion measurement circuit, the DC circuit applied on the testing specimen, and/or the CP circuit. Moreover, electrical signals from extraneous circuits could flow into the corrosion measurement equipment, shifting the recorded information and even damaging the facility. Thus, building a reasonable, reliable testing facility is the first step in conducting DC corrosion measurements.

Figure 10-3 shows an experimental setup specifically designed for DC corrosion testing. There are two cells–an anodic cell and a cathodic cell–to simulate the anodic and cathodic zones during DC corrosion of pipelines. Two electric circuits–the DC stray current circuit and the CP circuit–are included in the setup. In the DC current circuit, two carbon rods, which are used as counter electrodes in the anodic and cathodic cells, are connected to the negative and positive poles, respectively, of a DC power supply. The steel specimens, which are used as working electrodes in both cells, are connected through a slide rheostat to form a closed stray DC circuit. The DC current density flowing between the working electrode (WE in the figure) and the counter electrode (CE in the figure) is varied by adjusting the voltage output of the power supply. The DC output can be controlled galvanostatically so that the DC current density is maintained constantly in order to investigate the effect of DC current density on pipeline corrosion. In the CP circuit, the steel specimen and a carbon rod are connected to the negative and positive poles, respectively, of another DC power supply. An SCE is used as reference electrode (RE in the figure), which is immersed into a Luggin capillary. The distance from the tip of the capillary to the surface of the steel electrode is 2 mm to eliminate the ohmic drop. The potential of the steel cannot be maintained at the applied CP potential when the DC circuit is on. In other words, the potential would deviate from the applied CP potential under the DC effect (i.e., the potential of the steel does not automatically adjust itself to the applied CP potential). The potential of the steel electrode under various DC current densities is measured through a potentiostat. The recorded potential of the steel includes both the applied CP potential and the potential shift under the DC stray current.

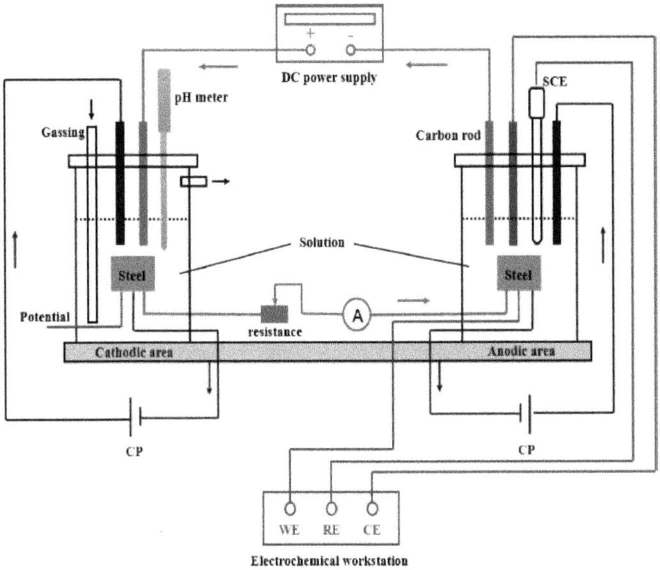


**Figure 10-3.** Schematic diagram of the experimental setup for DC corrosion testing on a pipeline steel specimen in a solution, where WE, RE, and CE refer to working electrode, reference electrode, and counter electrode, respectively, and SCE is the saturated calomel electrode.[39]

## 10.3.2. Effect of DC current density on steel potential

Figure 10-4 shows the time dependence of the potential of an X52 pipeline steel in the anodic zone at various DC current densities in a simulated soil solution.[39] In the recorded curves, three potential regions are observed: from 0 s to 1,200 s, no DC is applied, and the recorded potential is the corrosion potential of the steel in the solution; from 1,200 s to 2,400 s, the DC is applied, and the DC-ON potential is recorded; after 2,400 s, the DC stops, and the DC-OFF potential is recorded. It can be seen that, upon application of DC, the potential of the steel in the anodic zone shifts less negatively (i.e., toward the positive direction). The potential becomes less negative with increasing DC current density. At each DC current density, the potential is maintained at a relatively stable value. When the DC is stopped at the 2,400th second, the potential drops rapidly, and then gradually reaches a steady-state value, which is almost identical to the corrosion potential recorded prior to the DC application (i.e., the potential from 0 s to 1,200 s). At higher DC current densities, the potential recovery to the steady value is a bit slower.

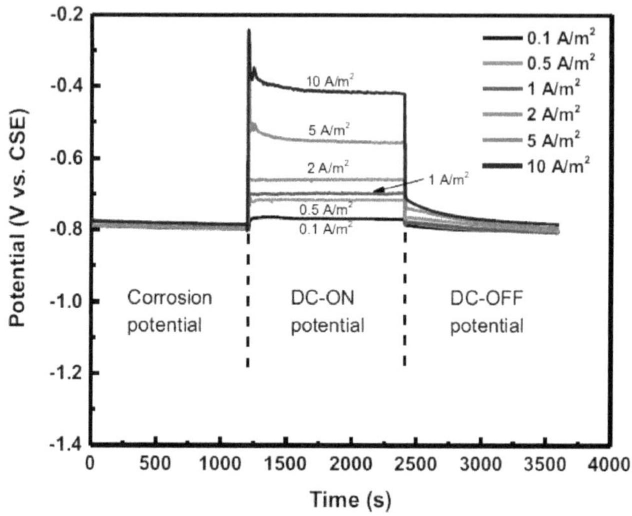


**Figure 10-4.** Time dependence of the potential of an X52 pipeline steel in the anodic zone at various DC current densities in a simulated soil solution.[39]

Figure 10-5 shows the time dependence of the potential of the X52 steel electrode under various DC current densities in the soil solution in the cathodic zone. Similar to Figure 10-4, before the DC is applied, the steel is at a corrosion potential of about –800 mV (CSE). Upon application of DC, the potential of the steel shifts negatively and remains at a constant value. With increasing DC current density, the potential tends to be more negative. When the DC is stopped at the 2,400th second, the DC potential shifts toward the positive direction but does not return to the corrosion potential recorded before the DC was applied. The new potential is generally less negative than the initial corrosion potential in the period from 0 s to 1,200 s.

Figure 10-6 shows the steady-state DC-ON potentials of the X52 steel electrode in anodic and cathodic zones in the soil solution as a function of DC current density. It can be seen that, in the anodic zone, the potentials become less negative with increasing DC current density, while in the cathodic zone, the potentials are more negative at higher DC current densities.

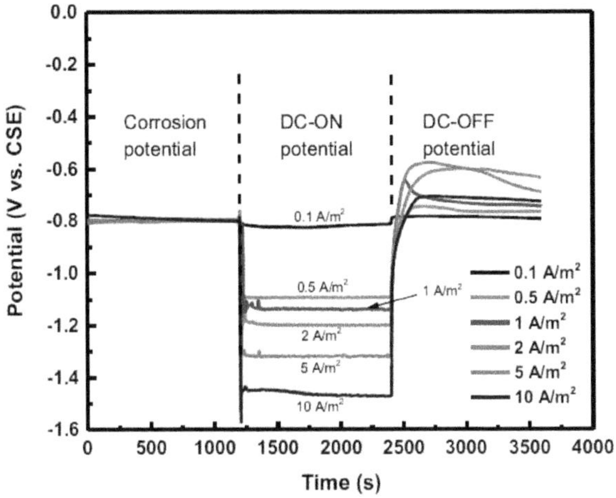


**Figure 10-5.** Time dependence of the potential of X52 pipeline steel in the cathodic zone at various DC current densities in a simulated soil solution.[39]

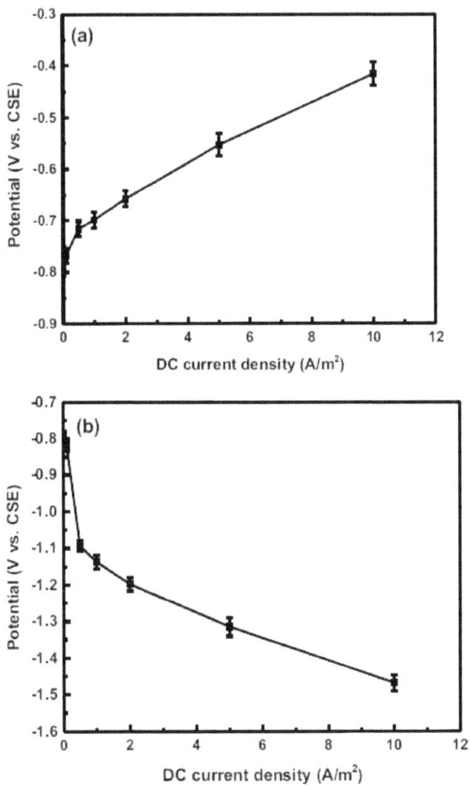


**Figure 10-6.** Steady-state DC-ON potentials of an X52 steel electrode in the (a) anodic zone and (b) cathodic zone as a function of DC current density in the soil solution.[39]

### 10.3.3. Effect of DC current density on solution pH in the anodic and cathodic zones

Figure 10-7 shows the solution pH in the anodic and cathodic zones as a function of DC current density. It can be seen that, in the anodic zone, the solution pH drops (tends to be more acidic) with increasing DC current density (Figure 10-7a). However, the solution pH in the cathodic zone increases (tends to be more alkaline) with the increase of the DC current density (Figure 10-7b).

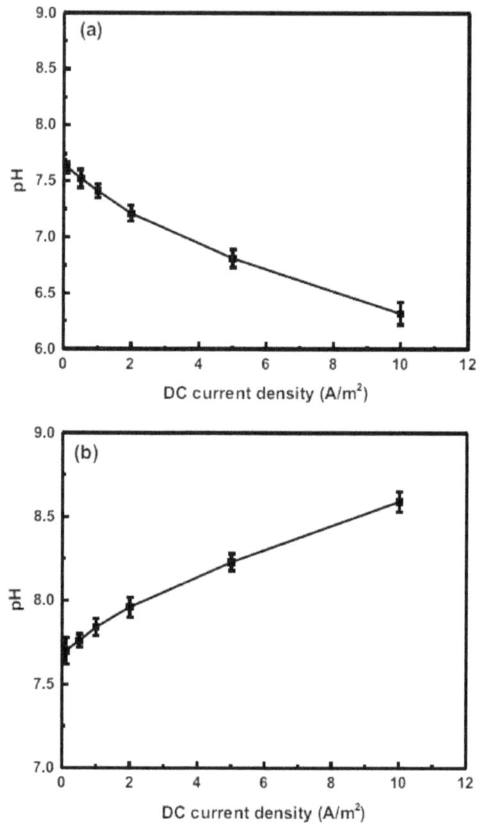


**Figure 10-7.** Solution pH in the (a) anodic zone and (b) cathodic zone as a function of DC current density.[39]

## 10.3.4. Corrosion rate and morphology of steel under various DC current densities

Figure 10-8 shows the corrosion rate determined by weight loss testing of X52 pipeline steel after 48 h of immersion in the soil solution as a function of DC current density. It can be seen that the corrosion rate of the steel increases with the DC current density. For example, in the absence of DC, the steel corrodes in the aerated soil solution at a corrosion rate of about 0.4 mm/y. At the DC current densities of 5 A/m$^2$ and 10 A/m$^2$, the corrosion rates increase to 6.5 mm/y and 12.5 mm/y, respectively.

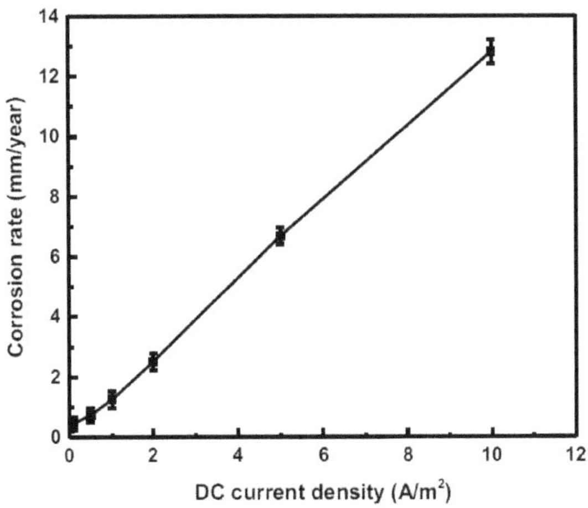


**Figure 10-8.** Corrosion rates determined by weight loss testing of X52 steel after 48 h of immersion in the soil solution as a function of DC current density.[39]

Figure 10-9 shows the morphology of X52 steel electrode after 48 h of testing in the soil solution under various DC current densities. In the absence of DC, the steel shows a bright, clear view, and corrosion products are not visible on the steel surface. At the small DC current density of 0.1 A/m$^2$, the morphology of the steel electrode is similar to that in the absence of DC. With the increase of DC current density, the steel surface becomes dark, and corrosion products can be seen on the steel surface. Generally, the DC application induces uniform corrosion on the steel.

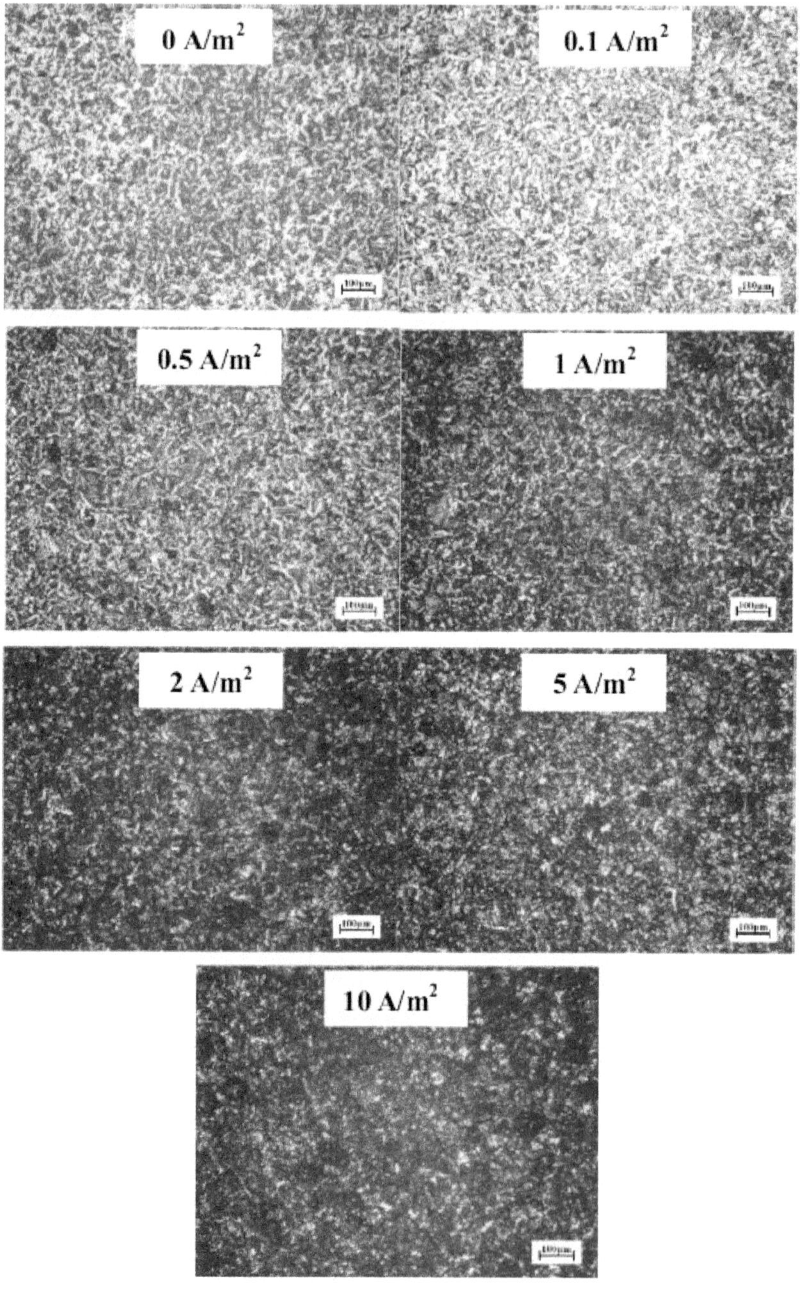


**Figure 10-9.** Optical views of the morphology of X52 steel electrode after 48 h of testing in the soil solution under various DC current densities.[39]

## 10.4. Mechanism of DC corrosion

DC corrosion of buried pipelines is typical of classic corrosion behavior: current flows from the anode to the cathode through the electrolyte by ionic conductivity and flows from the cathode to the anode through the metal by electric conductivity.[40] At the anode, the metal is oxidized; at the cathode, hydrogen evolution or oxygen reduction occurs. Anodic and cathodic zones can overlap on the metal surface, resulting in uniform corrosion. A certain area can also always be anodic, producing localized corrosion.

During electrochemical corrosion of pipeline steels, the primary anodic reaction is iron oxidation:

$$Fe \rightarrow Fe^{2+} + 2e \tag{10-2}$$

The cathodic reactions can be reduction of dissolved oxygen (or water) or hydrogen evolution, depending on the presence of oxygen in the electrolyte and the solution pH:

$$O_2 + 2H_2O + 4e \rightarrow 4OH^- \tag{10-3}$$

$$2H_2O + 2e \rightarrow H_2 + 2OH^- \tag{10-4}$$

$$2H^+ + 2e \rightarrow H_2 \tag{10-5}$$

The relationship between the amount of corroded steel and the current density obeys Faraday's law:

$$m = \frac{Mit}{nF} \tag{10-6}$$

where $m$ is the amount of the metal loss due to corrosion, $M$ is the atomic weight of steel, $i$ is the corrosion current density, $t$ is time, $F$ is Faraday's constant, and $n$ is the number of electrons exchanged in corrosion reactions. The corrosion rate, $V$, of the steel can be determined by:

$$V = \frac{dm}{dt} = \frac{M}{nF}i \tag{10-7}$$

Figure 10-10 illustrates the mechanism of DC corrosion reactions occurring at the anode and the cathode. Upon DC application, the potentials of the steel in the anodic and cathodic zones shift remarkably in the positive and negative directions, respectively (Figure 10-6). For example, at the DC current density of 10 A/m$^2$, the DC-ON potentials at the anode and the cathode can be up to about −400 mV (CSE) and −1,500 mV (CSE), respectively, in the soil solution, as compared with the corrosion potential of X52 steel of approximately −800 mV (CSE). Thus, the DC greatly polarizes, either anodically or cathodically, the steel at the anodic and cathodic zones, respectively. As a result, both the steel dissolution (corrosion) and the cathodic reduction of dissolved oxygen are accelerated in the presence of DC interference. Moreover, new steady-state potentials are established at the DC-ON levels, indicating that the steel corrodes at a new state with a high corrosion rate, as confirmed by the weight loss measurements and corrosion morphology observations in Figures 10-8 and 10-9.

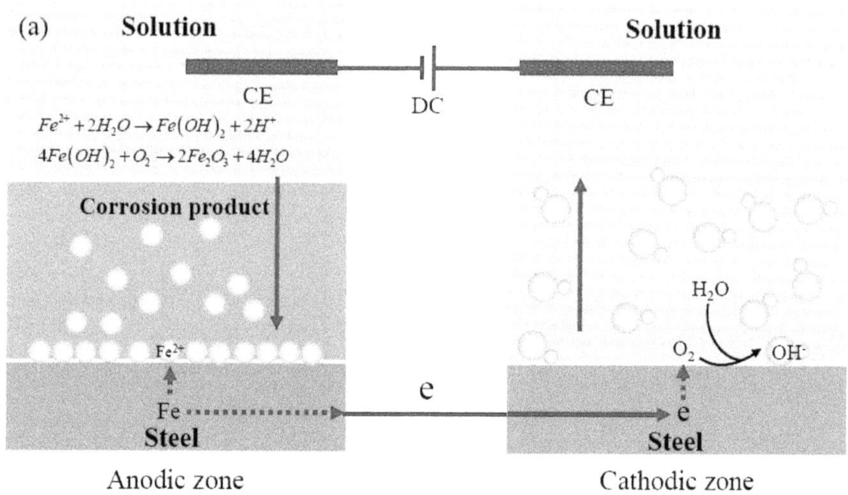


**Figure 10-10.** Schematic diagram of the mechanistic model to illustrate the electrochemical reactions occurring in the anodic and cathodic zones under DC interference.[39]

When DC is removed, the potential gradually recovers back to the steady-state value, which is generally identical to the initial corrosion potential prior to DC application in the anodic zone. This indicates that DC-accelerated corrosion is reversible under the present testing condition. Although the DC application can cause the drop of solution pH in the anodic zone (Figure 10-7a), the pH values of the solution in the anodic zone before and after application of the highest DC current density of 10 $A/m^2$ are still in the near-neutral pH range (changing from 7.7 to 6.4). The potential of the steel does not change greatly in the neutral pH range. If the DC current density is sufficiently large and the anodic solution pH drops greatly to make the solution acidic, it is expected that the steady-state DC-OFF potential of the steel would be different from the original corrosion potential.

However, after the DC is stopped, the potential at the cathodic zone does not return to the original corrosion potential. Generally, the steel is at a new steady state with a less negative potential in the cathodic zone. Thus, the DC effect in the cathodic zone is irreversible under the testing condition. The DC drives the cathodic reduction of dissolved oxygen to produce hydroxyl ions, elevating the solution pH. For example, the cathodic solution pH increases from 7.7 to 8.6 upon application of the DC current density of 10 $A/m^2$. The steel is expected to be in a relatively more stable state in the new alkaline environment, even after the DC is removed. Thus, the DC corrosion of pipelines would cause local alkalization of the solution through production of $OH^-$ by cathodic reaction. The degree of alkalization depends on the value of DC current density.

## 10.5. Effect of DC on coating properties and performance

When coated pipelines are under DC interference, it is expected that the external coatings will suffer from degradation in both their molecular structure and properties, and thus in their corrosion-resistant performance. Water uptake and impedance behavior are the primary coating properties that affect the coating's blocking ability.

### 10.5.1. Water permeability

To investigate the effect of DC interference on water uptake and permeability of pipeline coatings, an appropriate test rig should be designed and built for the purpose. Figure 10-11 shows the gravimetric cup method used to measure the water uptake of an FBE coating membrane. The FBE coating is mounted on two glass containers, in which a simulated soil solution is pre-filled. The two coated containers, which simulate the cathodic and anodic zones during the DC corrosion of pipelines are bonded on a conductive steel net. The assembly is put in an oven, kept at 65°C and a relative humidity of nearly zero, which are under continuous monitoring. Various DC voltages (0, 2, 5, 10, and 20 V) are applied between the carbon rods installed in the two containers. The containers with the sealed FBE coating membranes are weighed at intervals of 48 h. The weight change is used to calculate the water uptake.

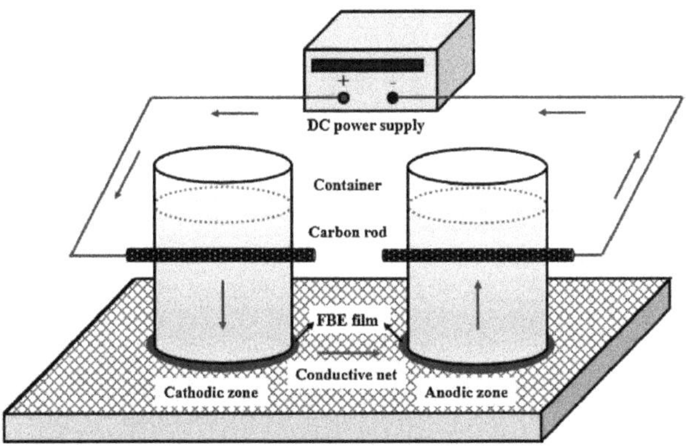


**Figure 10-11. Schematic diagram of the experimental setup for measurement of water uptake by FBE coating membrane.**[21]

Figure 10-12 shows the time dependence of the weight of the FBE-sealed test container in the cathodic and anodic zones under various DC voltages.[21] It can be seen that the weight of the container decreases with time in both the cathodic and anodic zones under DC interference, indicating that water permeates through the FBE membrane and leaves the containers in the absence and presence of DC interference. The weight loss increases with increasing DC voltage, suggesting that the DC application accelerates the water permeation through the FBE coating. Under high DC voltages, such as 20 V, water reduction and iron oxidation would occur at the cathode and the anode, respectively. The reaction rates and the amount of water to be reacted depend on the conductivity of the system. Calculations show that the water used for the electrochemical reactions is approximately $4.8 \times 10^{-4}$ g after 30 days of testing. This is much smaller than the measured weight loss. Thus, it is believed that the weight loss is primarily dominated by the water permeation through the coating membrane, while the amount of water consumed by anodic and cathodic reactions is negligible.

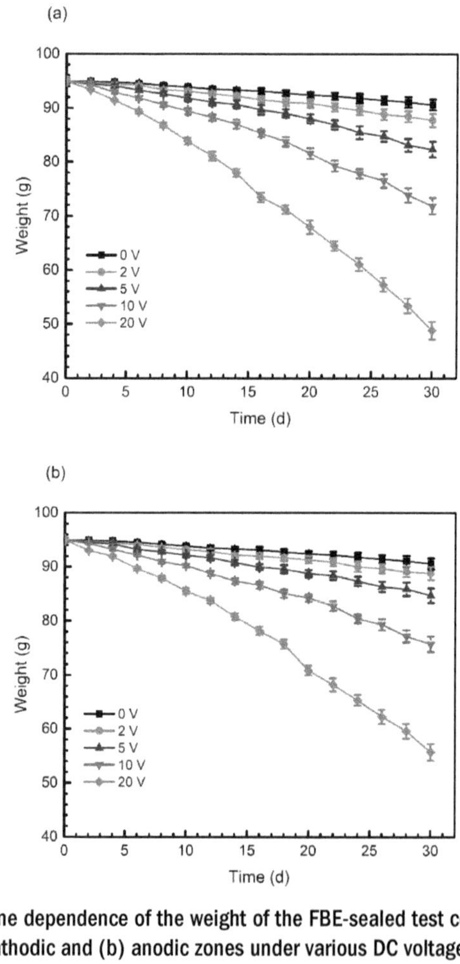


**Figure 10-12.** Time dependence of the weight of the FBE-sealed test containers in the (a) cathodic and (b) anodic zones under various DC voltages.[21]

The WTR is used to evaluate quantitatively the water permeability of polymeric coatings:[41]

$$WTR = \frac{\Delta G}{tA} \tag{10-8}$$

where $\Delta G$ is the weight loss (g), $t$ is time (days), and $A$ is the exposed area of the coating ($m^2$). The WTR of the FBE coating in the cathodic and anodic zones under various DC voltages is plotted in Figure 10-13. It can be seen that the WTR increases with the DC voltage. There is a larger WTR in the cathodic zone than in the anodic zone under specific DC voltages.

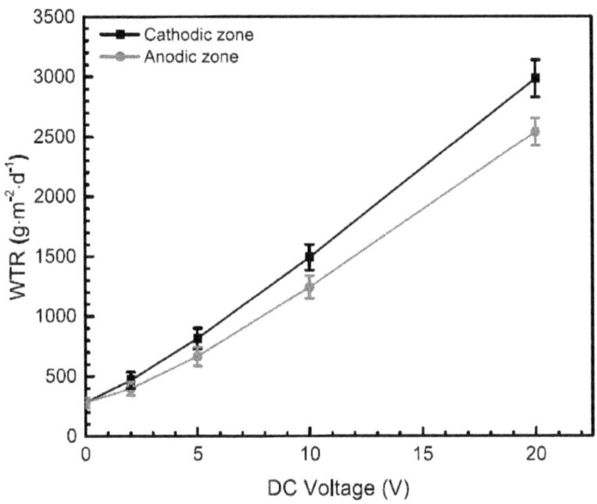


**Figure 10-13.** Water transmission rates of the FBE coating in cathodic and anodic zones under various DC voltages after 30 days of immersion in a simulated soil solution.[39]

## 10.5.2. Electrochemical impedance spectroscopy

Figure 10-14 shows the Bode plots (i.e., the phase angle and impedance modulus plots) of FBE-coated steel in cathodic and anodic zones after 30 days of immersion in a simulated soil solution at various DC voltages at 65°C. It can be seen that, in the absence of DC interference, typical capacitive impedance features—including a high phase angle over 80° for the majority of the measuring frequency range, a linear relationship between the impedance modulus and the measuring frequency, and a low-frequency impedance in excess of $10^{10}$ $\Omega \cdot cm^2$—are observed in both the cathodic and anodic zones. The coating stays intact to protect the steel from corrosion attack. Upon DC application, the feature of the Bode plots is changed. As the DC voltage increases, the frequency range of the high phase angle decreases. An impedance modulus plateau is observed in the low-frequency range. Moreover, the low-frequency impedance value reduces with increasing DC voltage. These features apply for the impedance plots measured in both the cathodic and anodic zones. The results show that the DC interference decreases the coating impedance and thus the ability of the coating to protect against corrosion. The coating changes from purely capacitive behavior to both capacitive behavior (in the high-frequency range) and resistive behavior (in the low-frequency range).

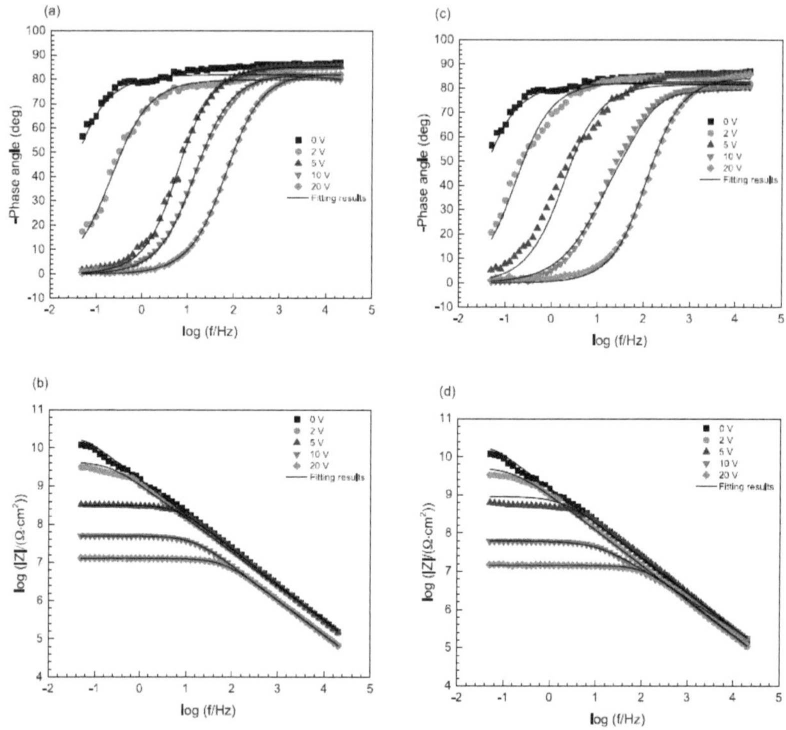


**Figure 10-14.** Bode plots: (a, c) phase angle and (b, d) impedance modulus of FBE-coated steel in (a, b) cathodic and (c, d) anodic zones after 30 days of immersion in the simulated soil solution at various DC voltages at 65°C.[21]

The electrochemical impedance parameters of the coated steel electrode under DC interference are derived. Table 10-1 lists the fitted impedance parameters, where $R_{ct}$, $R_c$, $C_{dl}$, $C_c$, $Q_{dl}$, and $Q_c$ are charge-transfer resistance, coating resistance, capacitance of the double-charge layer, capacitance of the coating, constant-phase element of the double-charge layer, and constant-phase element of the coating, respectively.[21] It can be seen that $R_c$ decreases and $C_c$ increases with increasing DC voltage in both the cathodic and anodic zones, showing the reduced resistance of the coating under DC. The $R_c$ is smaller and the $C_c$ is larger in the cathodic zone than in the anodic zone, showing that the coating in the cathodic zone suffers from more degradation than that in the anodic zone under specific DC voltages. Moreover, the $R_{ct}$ fluctuates around $10^7$ Ω·cm² under the DC voltages of 10 V and 20 V in both the cathodic and anodic zones, compared with the low-frequencyimpedance over $10^{11}$ Ω·cm² in the absence of DC. The results indicate that corrosion occurs on the steel under DC interference.

| | DC voltage (V, SCE) | $R_c$ (Ω cm²) | $Q_c$ | | $C_c$ (F/cm²) | $R_{ct}$ (Ω cm²) | $Q_{dl}$ | | $C_{dl}$ (F/cm²) |
|---|---|---|---|---|---|---|---|---|---|
| | | | Y ($\Omega^{-1}$ cm$^{-2}$ s$^n$) | n | | | Y ($\Omega^{-1}$ cm$^{-2}$ s$^n$) | n | |
| Cathodic zone | 0 | $3.77 \pm 0.27 \times 10^{10}$ | $1.29 \pm 0.09 \times 10^{-10}$ | $0.84 \pm 0.01$ | $1.74 \pm 0.16 \times 10^{-10}$ | | | | |
| | 2 | $4.32 \pm 0.32 \times 10^{9}$ | $1.88 \pm 0.16 \times 10^{-10}$ | $0.85 \pm 0.01$ | $1.81 \pm 0.19 \times 10^{-10}$ | | | | |
| | 5 | $3.17 \pm 0.07 \times 10^{8}$ | $2.06 \pm 0.42 \times 10^{-9}$ | $0.83 \pm 0.04$ | $1.89 \pm 0.47 \times 10^{-9}$ | | | | |
| | 10 | $3.99 \pm 0.02 \times 10^{7}$ | $8.72 \pm 0.54 \times 10^{-8}$ | $0.89 \pm 0.06$ | $1.02 \pm 0.20 \times 10^{-7}$ | $1.33 \pm 0.89 \times 10^{7}$ | $1.35 \pm 0.35 \times 10^{-8}$ | $0.87 \pm 0.02$ | $1.04 \pm 0.43 \times 10^{-8}$ |
| | 20 | $1.18 \pm 0.01 \times 10^{7}$ | $4.81 \pm 0.23 \times 10^{-7}$ | $0.91 \pm 0.01$ | $5.71 \pm 0.19 \times 10^{-7}$ | $1.37 \pm 0.01 \times 10^{7}$ | $1.98 \pm 0.02 \times 10^{-8}$ | $0.89 \pm 0.01$ | $1.69 \pm 0.23 \times 10^{-8}$ |
| Anodic zone | 0 | $3.77 \pm 0.27 \times 10^{10}$ | $1.29 \pm 0.09 \times 10^{-10}$ | $0.84 \pm 0.01$ | $1.74 \pm 0.16 \times 10^{-10}$ | | | | |
| | 2 | $5.12 \pm 0.37 \times 10^{9}$ | $1.89 \pm 0.09 \times 10^{-10}$ | $0.84 \pm 0.01$ | $1.88 \pm 0.12 \times 10^{-10}$ | | | | |
| | 5 | $8.89 \pm 0.86 \times 10^{8}$ | $1.24 \pm 0.19 \times 10^{-9}$ | $0.90 \pm 0.02$ | $1.25 \pm 0.23 \times 10^{-9}$ | | | | |
| | 10 | $6.03 \pm 0.26 \times 10^{7}$ | $5.11 \pm 0.29 \times 10^{-8}$ | $0.86 \pm 0.02$ | $6.14 \pm 0.73 \times 10^{-8}$ | $1.21 \pm 0.14 \times 10^{7}$ | $8.65 \pm 0.34 \times 10^{-9}$ | $0.86 \pm 0.02$ | $5.99 \pm 0.86 \times 10^{-9}$ |
| | 20 | $1.47 \pm 0.42 \times 10^{7}$ | $2.65 \pm 0.27 \times 10^{-7}$ | $0.88 \pm 0.01$ | $3.19 \pm 0.64 \times 10^{-7}$ | $1.05 \pm 0.01 \times 10^{7}$ | $1.46 \pm 0.01 \times 10^{-8}$ | $0.85 \pm 0.01$ | $1.05 \pm 0.01 \times 10^{-8}$ |

**Table 10-1.** Electrochemical impedance parameters fitted from the recorded impedance data in Figure 10-14.[21]

### 10.5.3. Molecular structure of coatings

Figures 10-15 and 10-16 show the SEM images of the cross-sectional morphology of an FBE coating under various DC voltages in the cathodic and anodic zones, respectively, after 30 days of immersion in a simulated soil solution.[21] It can be seen that there are some micropores in the coating prior to the DC application (Figures 10-15a and 10-16a). When the DC voltages of 2 V and 5 V are applied, bigger pores are observed, as shown in Figures 10-15b, 10-15c, 10-16b, and 10-16c. With the applied DC voltage increasing to 10 V and 20 V, some deep microvoids and microcracks are observed, as marked in Figures 10-15d, 10-15e, 10-16d, and 10-16e. The morphological features of the coating show that the presence of DC is associated with the integrity degradation of the coating. Further testing and evidence are required to verify the findings. Furthermore, there is no obvious difference in the coating integrity between the cathodic and anodic zones.

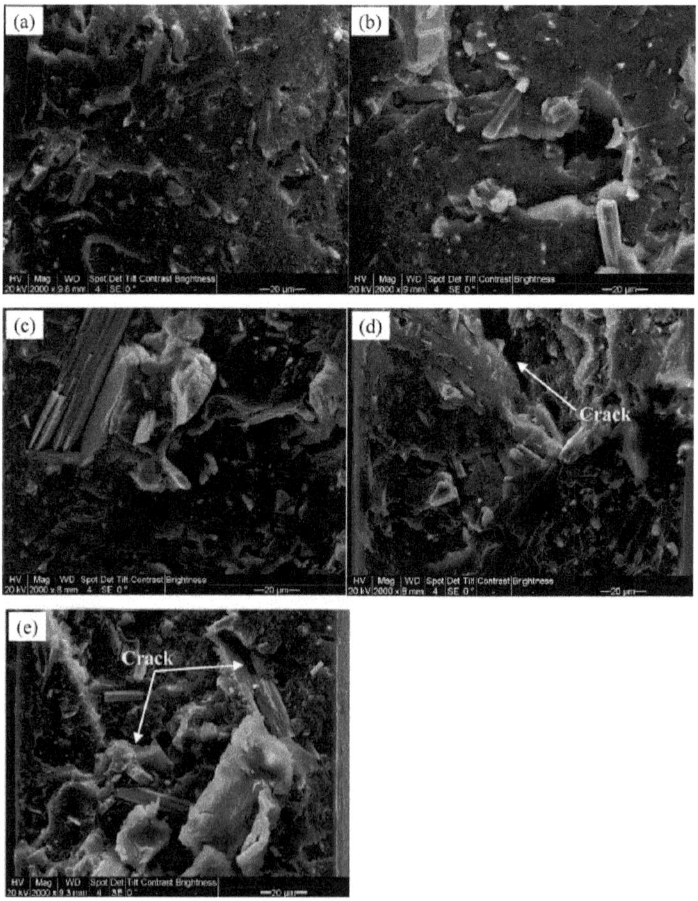


**Figure 10-15.** SEM images of the cross-sectional morphology of an FBE coating in the cathodic zone under the DC voltages of (a) 0 V, (b) 2 V, (c) 5 V, (d) 10 V, and (e) 20 V after 30 days of immersion in a simulated soil solution.[21]

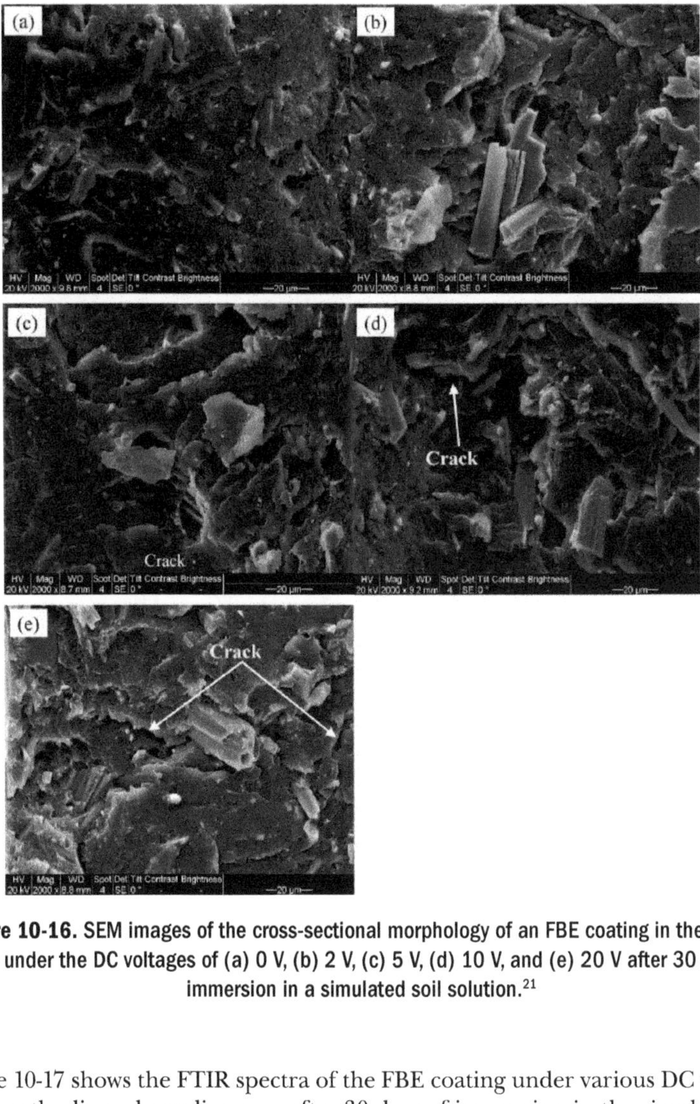


**Figure 10-16. SEM images of the cross-sectional morphology of an FBE coating in the anodic zone under the DC voltages of (a) 0 V, (b) 2 V, (c) 5 V, (d) 10 V, and (e) 20 V after 30 days of immersion in a simulated soil solution.[21]**

Figure 10-17 shows the FTIR spectra of the FBE coating under various DC voltages in the cathodic and anodic zones after 30 days of immersion in the simulated soil solution. The identified characteristic functional groups are listed in Table 10-2. Three peaks at 840 $cm^{-1}$, 1,056 $cm^{-1}$, and 1,260 $cm^{-1}$ are attributed to the C-O-C stretching vibrations of the oxirane group, aliphatic ethers, and aromatic ethers, respectively.[42, 43, 44] The peaks at 915 $cm^{-1}$, 2,980 $cm^{-1}$, and 3,580 $cm^{-1}$ are associated with the C-O of the oxirane group,[45] the C-H tension of the methylene group in the epoxy ring,[46, 47, 48] and the O-H functional group from the epoxy coating,[49, 50, 51] respectively. The C-O-C stretching vibrations become weaker in both the cathodic and anodic zones as the DC voltage increases. The C-O adsorption peak is observed only when the coating is under the DC voltages of 10 V and 20 V. These are attributed to the hydrolysis of C-O-C functional groups contained in the FBE molecules under DC interference in both the cathodic and anodic zones. Moreover, the adsorption band of C-H stretching becomes stronger, and the adsorption peak of the stretching O-H is increased, with increasing DC voltage. These results further confirm that the presence of DC causes the hydrolysis of C-O-C functional groups in FBE molecules.

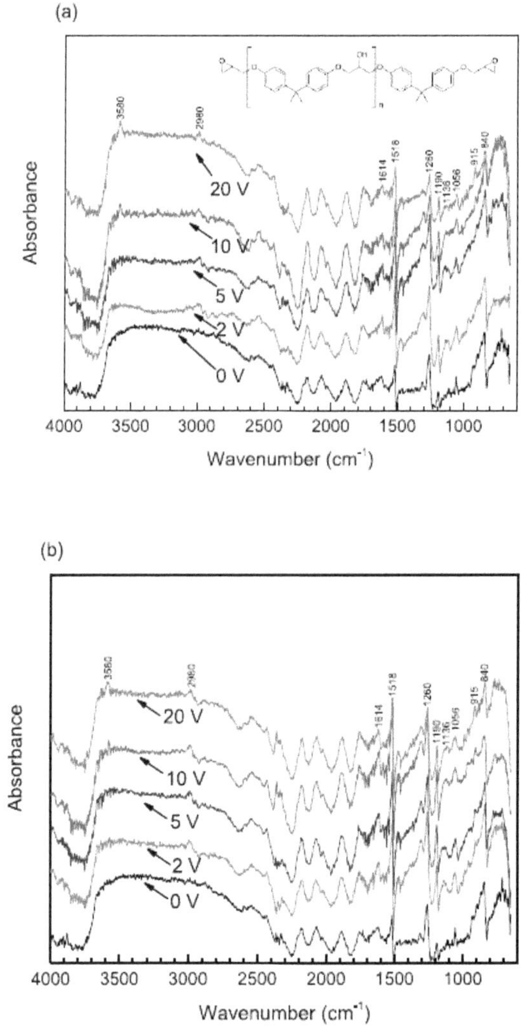


**Figure 10-17.** FTIR spectra of FBE coating under various DC voltages in the (a) cathodic and (b) anodic zones after 30 days of immersion in a simulated soil solution.[21]

| Bond ($cm^{-1}$) | Assignment |
|---|---|
| 3580 | Stretching O-H of alcohols |
| 2980 | Stretching C-H of methylene |
| 1614 | Stretching C=C of aromatic rings |
| 1518 | Semicircle stretching C-C of phenylene |
| 1260 | Stretching C-O-C of aromatic ethers |
| 1190 | Stretching C-C of phenyl group |
| 1136 | Stretching C-C of aliphatic chain |
| 1056 | Stretching C-O-C of aliphatic ethers |
| 915 | Stretching C-O of oxirane group |
| 840 | Stretching C-O-C of oxirane group |

**Table 10-2.** Characteristic bonds of FBE coating identified in the FTIR spectra in Figure 10-17.[21]

## 10.5.4. Mechanistic aspect of the role of DC interference on coating degradation

DC interference can contribute to increased water permeation through the FBE coating. The effect applies to the coating in both the anodic and cathodic zones. Generally, the coating capacitance, $C_c$, can be used to estimate the water content contained in a coating:[52]

$$w = \frac{K\log\frac{C_c}{C_0}}{\log\varepsilon_w} \qquad (10\text{-}9)$$

where $w$ is the water content, $K$ is a constant, $C_0$ is the capacitance of a "dry" coating, and $\varepsilon_w$ is the dielectric constant of water. According to the impedance results in Table 10-1, the $C_c$ increases with the DC voltage in both the cathodic and anodic zones. Thus, the DC increases the water content in the coating when exposed to an aqueous solution. The WTR increases obviously under the DC interference, where the WTRs in the cathodic and anodic zones are, respectively, 10.3 and 8.7 times more than those in the absence of DC.

It has been acknowledged that water permeation through a coating is predominated by the type of penetrant, the polarity of the coating, and its mechanical integrity.[53] The FBE coating is a highly polar molecule owing to the presence of hydroxyl groups. This kind of coating is usually a poor barrier to polar molecules such as water. Moreover, many micropores exist in the FBE coating, as shown in Figures 10-15a and 10-16a. Therefore, water can permeate into the FBE coating even in the absence of DC. In the presence of DC, in addition to increased water permeation, the electric field induced by DC accelerates the diffusion of chemical ions in the coating. This is confirmed by the decreased $R_c$ with the increasing DC voltage (Table 10-1). The permeation and inclusion of chemical ions in the coating can even make it change from nonconductive to semiconductive or conductive.[54] As a result, the ability of the coating to protect against corrosion is reduced. When water and chemical ions penetrate the coating, the existing micropores are aggrandized to form big pores.[55] Additionally, the interaction of water and chemical ions along with the pores can impact the functional groups and change the molecular structure of the polymeric coating.[56, 57] The result in Figure 10-17 shows that the C-O-C functional groups of the FBE coating become hydrolyzed with increased DC voltage.

The presence of cathodic and anodic zones is a typical feature of the DC corrosion of coated pipelines. In the presence of DC interference, opposite electric fields are generated in the cathodic and anodic zones: the electric field extends from the solution to the steel in the cathodic zone, and from the steel to the solution in the anodic zone. Cations such as $Na^+$, $K^+$, $Mg^{2+}$, and $Ca^{2+}$ and anions such as $NO_3^-$, $SO_4^{2-}$ and $HCO_3^-$ tend to be attracted and accumulate on the coating surface in the cathodic and anodic zones, respectively. In addition to the electric field intensity, the size of the ions is important for their diffusion in the micropores and voids existing in the cross-linked structure of the FBE coating. Generally, the smaller the ions, the faster the diffusion. As the cations are usually smaller than the anions, the diffusion of the cations is faster than the anions under the identical electric field intensity. The ion diffusion can cause accelerated water permeation into the coating membrane. Therefore, the WTR in the cathodic zone is bigger than that in the anodic zone, as can be seen in Figure 10-13.

The analysis above can be illustrated by a conceptual model, which was developed to illustrate the mechanistic aspects of DC-induced water permeation, degradation of the FBE coating, and steel corrosion, as shown in Figure 10-18. When water permeates the coating and reaches the steel surface in both the cathodic and anodic zones, the DC electric circuit is closed under the DC interference, causing water reduction in the cathodic zone and iron oxidation in the anodic zone. In the cathodic zone, hydroxyl ions are generated in the presence of DC interference, which elevates the solution pH to accelerate the coating degradation.[58, 59, 60] In the anodic zone, the steel corrosion happens when the DC electric circuit becomes closed.

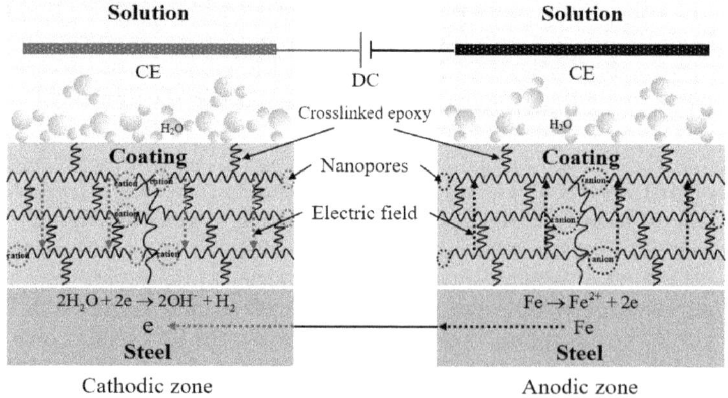


**Figure 10-18.** A conceptual model to illustrate the FBE coating degradation in the presence of DC interference.[21]

## 10.6. Adverse effect of DC interference on CP performance

As stated above, DC corrosion measurements should be conducted on an experimental test rig where mutual interferences of multiple electrical circuits are avoided or minimized. The setup shown in Figure 10-3 was used to investigate the effect of DC interference on CP performance applied on an X52 pipeline steel in a simulated soil solution. Figure 10-19 shows the time dependence of the potential of the X52 steel electrode under CP potentials of –850 mV (CSE) and –1,000 mV (CSE) in the anodic zone at various DC densities in the soil solution.[39] Prior to DC application, the steel is at the applied CP potential. Upon DC application, the potential cannot be maintained at the applied value and shifts toward the positive direction. The steady-state DC-ON potential (i.e., the potential of the steel under both CP and DC) is less negative with increasing DC current density. When the DC is stopped at the 2,400th second, the potential drops but does not recover to the initial CP potential values. Generally, the new steady-state DC-OFF potential is more negative than the applied CP potential, especially at high DC current densities.

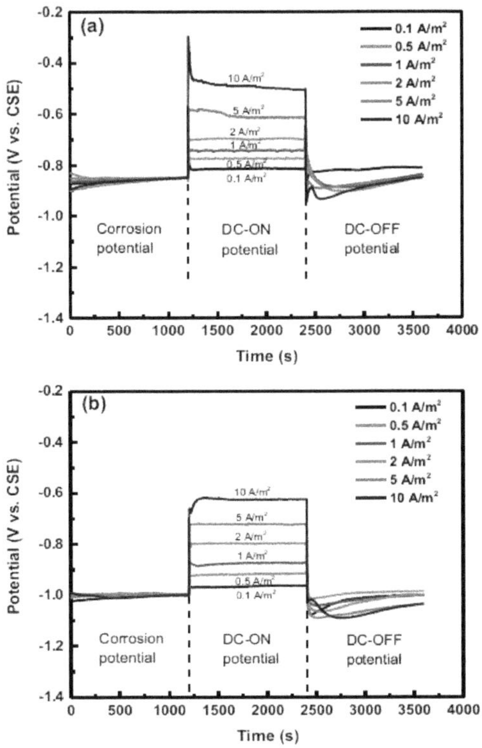


**Figure 10-19.** Time dependence of the potential of X52 steel electrode at CP potentials of (a) -850 mV (CSE) and (b) -1,000 mV (CSE) in the anodic zone at various DC current densities in a simulated soil solution.[39]

Figure 10-20 shows the time dependence of the potential of the X52 steel electrode at CP potentials of –850 mV (CSE) and –1,000 mV (CSE) in the cathodic zone at various DC densities in the soil solution.[39] It is seen that the potential of the steel is shifted negatively upon the DC application in the cathodic zone. The DC-ON potential becomes more negative with increasing DC current density. When the DC is stopped, the potential of the steel increases rapidly and gradually reaches a relatively stable value, which is close to the initially applied CP potential.

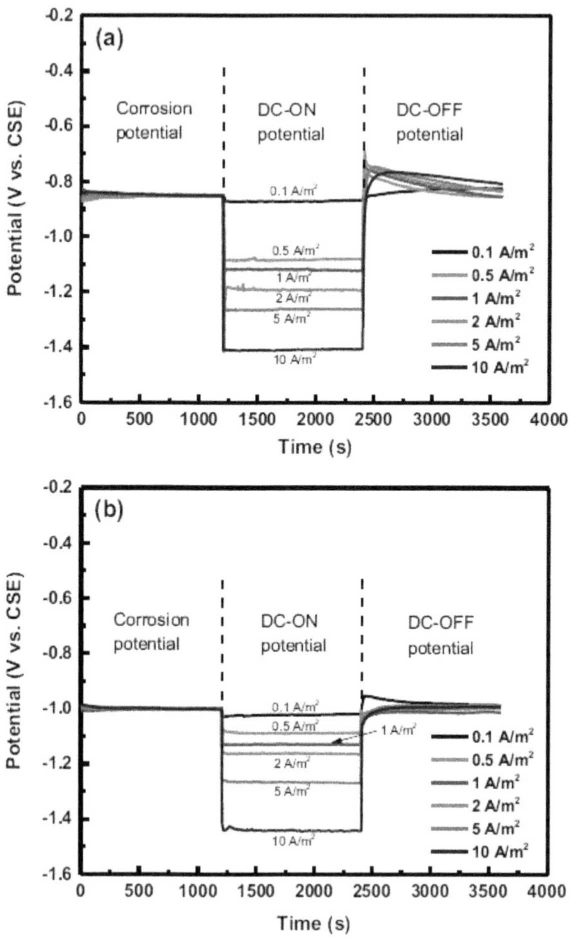


**Figure 10-20.** Time dependence of the potential of X52 steel electrode at CP potentials of (a) -850 mV (CSE) and (b) -1,000 mV (CSE) in the cathodic zone at various DC current densities in a simulated soil solution.[39]

Figure 10-21 shows the potential of X52 steel at corrosion potential and varied CP potentials in the soil solution as a function of the DC current density. It can be seen that, in the anodic zone, the DC application shifts the CP potential positively. At specific DC current densities, the steel without the CP application has the least negative DC-ON potential, and the steel applied with −1,000 mV (CSE) has the most negative potential. In the cathodic zone, the DC shifts the applied CP potential more negatively as the DC current density increases. At individual DC current densities, the steel without CP application has the most negative potential, and the steel applied with −1,000 mV (CSE) of CP has the least negative potential.

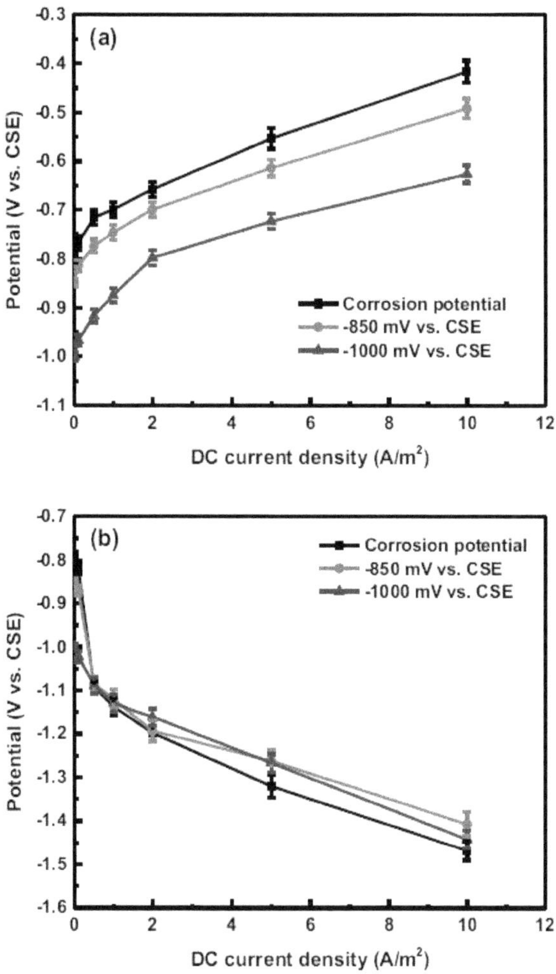


**Figure 10-21.** Potential of X52 steel at corrosion potential and varied CP potentials as a function of the DC current density in a soil solution: (a) anodic zone, (b) cathodic zone.[39]

Figure 10-22 shows the corrosion rate of X52 steel at corrosion potential and the applied CP potentials of –850 mV (CSE) and –1,000 mV (CSE), determined by weight loss testing after 48 h of immersion in the soil solution as a function of the DC current density.[21] Generally, the corrosion rate of the steel increases with the DC current density. At specific DC current densities, the corrosion rate decreases if CP is applied. Moreover, the more negative CP potential (i.e., –1,000 mV, CSE) causes a reduced corrosion rate. At a small DC current density of 1 A/m$^2$, the corrosion rate of the steel exceeds 1 mm/y, even when the CP of –850 mV (CSE) is applied. The industry-acceptable corrosion rate of pipelines is usually less than 0.1 mm/y.[61] Only when the CP potential shifts negatively to –1,000 mV (CSE) is the corrosion rate negligible at 1 A/m$^2$ of DC current density.

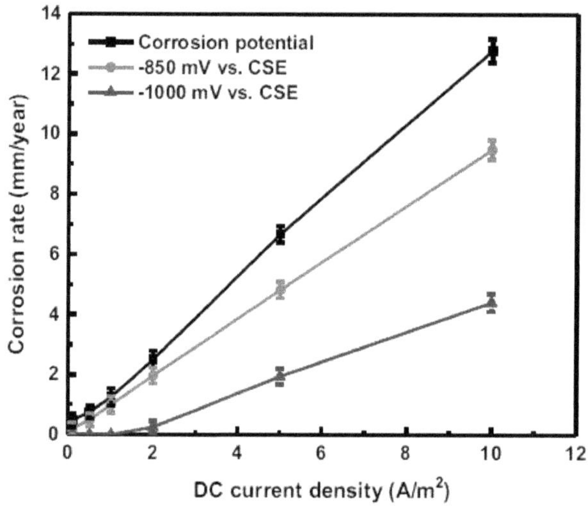


**Figure 10-22.** Corrosion rate of X52 steel at corrosion potential and the applied CP potentials of -850 mV (CSE) and -1,000 mV (CSE), determined by weight loss testing as a function of the DC current density after 48 h of immersion in a soil solution.[39]

Figure 10-23 shows the optical views of the morphology of the steel at corrosion potential and the CP potentials of –850 mV (CSE) and –1,000 mV (CSE) after 48 h of testing in the soil solution at various DC current densities. It can be seen that the steel experiences uniform corrosion under DC interference. Moreover, the CP potential of –850 mV (CSE) is not sufficient to protect the steel from corrosion even at a small DC current density of 0.1 A/m$^2$. When the CP potential is as negative as –1,000 mV (CSE), the steel is protected from corrosion at DC current densities up to 1 A/m$^2$. However, when the DC current density is increased to 2 A/m$^2$, this CP level is not able to offer full corrosion protection. Obviously, in the presence of DC interference, the CP effectiveness for corrosion protection is reduced.

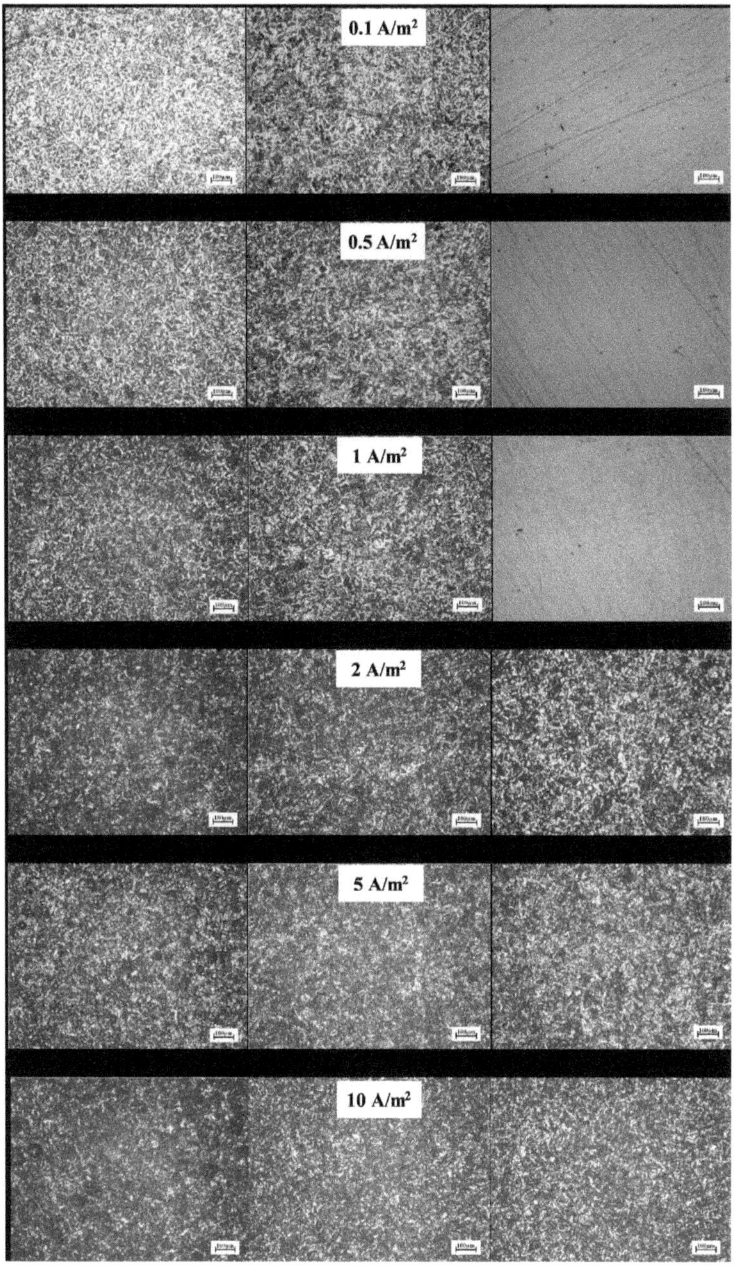


**Figure 10-23.** Optical views of morphology of steel at corrosion potential [left], -850 mV (CSE) [center] and -1,000 mV (CSE) [right] after 48 h of testing in a soil solution at various DC current densities.[39]

The adverse effect of DC on CP performance is illustrated mechanistically in Figure 10-24. When the applied CP potential, such as –850 mV (CSE), is not sufficient to offset the DC-induced anodic polarization, corrosion would occur in the anodic zone. Unless the CP potential is sufficiently negative for the steel to become cathodically polarized even in the presence of DC interference (i.e., –1,000 mV [CSE]), the steel can be protected from corrosion. The effect of DC on the potential in the cathodic zone results in a negative shift of the steel potential with increasing DC current density, which accelerates the cathodic reaction to produce hydroxyl ions and elevate the solution pH. The application of CP can further enhance the solution alkalinization.

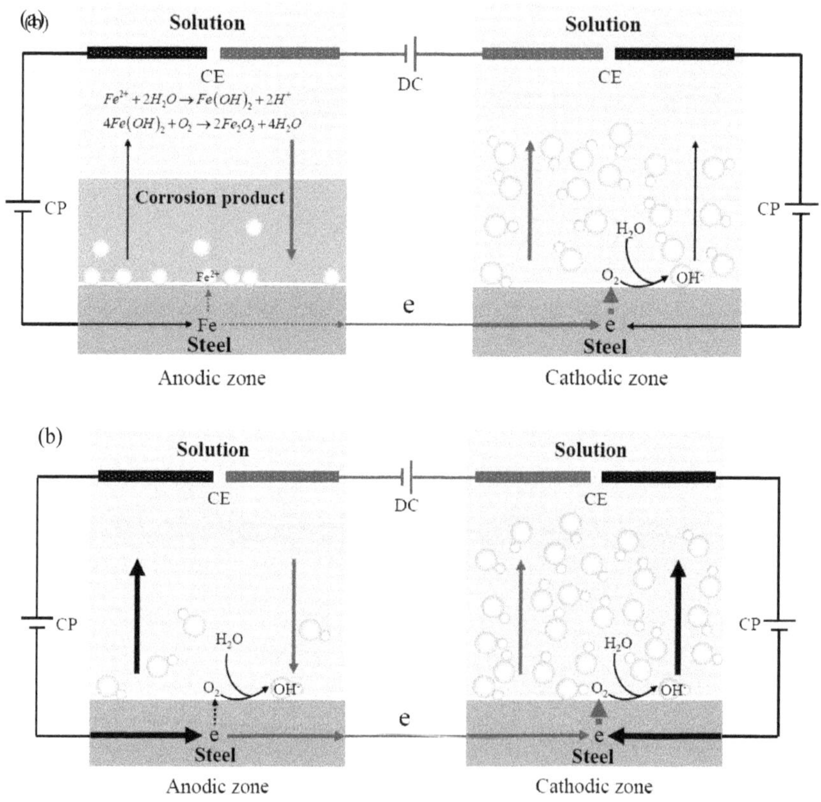


**Figure 10-24.** Mechanistic model to illustrate the electrochemical reactions occurring in the anodic and cathodic zones under DC interference at CP potentials of (a) -850 mV (CSE) and (b) -1,000 mV (CSE) when the DC current density is smaller than 1 A/m$^2$.[39]

In the presence of DC interference, the pipeline integrity can be adversely affected by three aspects. First, accelerating corrosion would occur in the anodic zone, which is usually associated with coating defects, where the DC stray current flows out of the pipeline into the conducting soil. Even at small DC current densities, such as 5 $A/m^2$ and 10 $A/m^2$, the local corrosion rate of the steel can be 16 or 31 times that in the absence of DC interference. The accelerated corrosion at the anode can result in rapid perforation of the pipelines at the coating defect. Second, the presence of DC stray current causes alkalinization of the electrolyte remaining at the cathodic zone (i.e., the coating defect where the DC current flows into the pipeline steel). Although corrosion would not occur locally, the highly alkaline electrolyte can degrade the polymeric coating, causing disbondment of the coating from the steel substrate (i.e., cathodic disbonding). This problem is even worse on CP-applied pipelines because, in addition to the DC stray current, the CP could further elevate the solution pH. Finally, the DC can shift the applied CP potential, reducing the CP effectiveness for corrosion protection. Thus, in the presence of DC interference, the CP criterion should be modified to provide full protection to the pipeline steel from corrosion attack.

## 10.7. Management of DC corrosion of pipelines

Generally, there are four main methods of measuring and evaluating DC interference and DC current density: voltage measurements and recording, current span-line drop, current clamp-on meters, and electromagnetic stray current mapping. The voltage measurements are usually performed in an array over time to monitor the effect of DC interference. However, this technique does not give current or current density, only the change in voltage at each measurement spot.[62] The current span-line drop or line current survey testing is to measure the resistance of the pipeline and determine the current flowing through the pipeline via the change in voltage across a length of the pipe according to Ohm's law.[63] However, the access points for taking measurements along the pipeline are often difficult to access. Moreover, any changes in the diameter of the pipe would affect the calculations. The current clamp-on meters use a pickup coil to measure the magnetic field around the pipeline. They can measure the combined AC–DC stray currents. Electromagnetic stray current mapping is a new technique that measures the electromagnetic field of the stray current interference. Measurements by an array of sensors can provide the size, shape, intensity, and depth of an electromagnetic field to accurately calculate the magnitude and direction of stray current.[62] In addition to detecting stray currents, the technique is able to determine the resistivity of the soil as the current flows through underground metallic structures (i.e., pipelines). Since it is difficult, if not impossible, to eliminate the presence of DC stray current in environments where the infrastructure has already been in place for some time, measurements of DC interference can enable effective management of the adverse effect of the electrical interference.

### 10.7.1. Detection of DC voltage and stray current for evaluation of DC interference

Generally, the most accepted method for evaluation of DC interference is to measure the polarization potential of the affected pipeline through the coupon approach.[64] However, the coupon approach is associated with measurement error owing to the ohmic drop caused by the stray current, which cannot be interrupted during field testing. The measured coupon potential can be affected by various parameters, such as the polarization retention period, delay in sampling the instant-OFF potential, coupon burial depth, and the proximity of ground beds.[65] It was suggested that a reasonable coupon potential monitoring period should exceed 3 h and that for accurate measurements, monitoring should be conducted for more than 24 h. Quick-sampling data loggers with a sampling interval less than 100 ms should be used. The instant-OFF potential sampling delay should be avoided. Depending on the coupon surface, 200 ms and 300 ms were suggested for coupons of 6.5 $cm^2$ and 15 $cm^2$ in area, respectively. The coupon buried depth should not be less than 50 cm. Moreover, the instant-OFF potential measurements could be affected remarkably by ground bed or adjacent buried metal structures. Attention should be paid to the proximity of the coupon to foreign buried structures.[65]

Detection of DC interference can also be based on potential measurements on buried pipelines by using a reference electrode (i.e., CSE). When a metallic structure is influenced by DC, the potential of the structure shifts in the positive or negative direction when the current leaves or enters the metal, as shown in Figures 10-4 and 10-5. The pipe potential and its variation are used for corrosion assessment. In a field measurement, it was found that the pipe-to-soil potential was shifted up to –304 V by the adjacent HVDC power lines.[66] However, this finding is associated with many uncertainties owing to the system complexity. Moreover, the interpretation of the results is sometimes ambiguous.[67] Therefore, there are a number of issues to be addressed in the detection of pipeline DC corrosion.[68]

The best approach to assessing DC interference is to determine the change in stray current density. This is complex owing to frequent uncertainties of current flow paths and the influencing factors such as soil resistivity, coating properties, and coating performance condition. Hosokawa and Kajiyama installed coated steel coupons adjacent to a buried pipeline that was close to a DC-powered rail transit system, and the DC coupon current density was measured.[69] It is believed that the ON potential measurements in a conventional CP survey are not able to assess the DC corrosion of pipelines because large errors due to ohmic potential drops are included. Instead, the ohmic-free potential (i.e., the OFF potential) was measured on the coupons continuously, although the CP level fluctuated remarkably owing to the adjacency to the rail transit system. Nevertheless, the DC coupon current density was measured without the errors induced by ohmic drops. Moreover, the DC current density measurement was able to directly demonstrate the local tendency of DC interference and the performance of CP in the assessment of the DC corrosion risk of the pipeline. Thus, it was proposed that the DC coupon current density provided advantages over the pipe-to-soil potential for the assessment of DC corrosion under DC interference, as well as the effectiveness of CP design criteria for corrosion prevention.

Furthermore, for pipelines collocated with electric transit systems in congested urban and suburban areas, it is usually difficult to accurately evaluate the interference from multiple sources. Hilleary et al. reported the use of a web-based monitoring system with remote data-logging capability to acquire data and develop profiles of the interference on pipelines resulting from electric transit systems.[70] The target was to accomplish effective synchronization of time-based events occurring across multiple sites and to correlate the data to the intermittent transit system events causing the interference. Moreover, other foreign current sources that were not directly associated with the transit system but could potentially affect the interference profile could also be identified. For confirmation of the effectiveness of the system, pipelines collocated with electric rail transit systems in two metropolitan areas were selected for field testing. For the affected pipelines, the DC pipe-to-soil potential, as well as AC and DC current densities, were measured at 1 min intervals, and all data were logged to the device memory. The units were accessed periodically through the internet to download the stored measurements. The results demonstrated that the remote monitoring systems were effective in monitoring and compiling the measured data, and a detailed analysis was conducted to understand the effects of periodic bursts of DC interference at electric transit crossings. Remote monitors provided the flexibility to customize the data recordings at any site.

In addition to the field measurements, numerical simulation has been developed using business software to simulate the interference of HVDC on buried pipelines. Software based on the boundary element method used to solve Maxwell's equations, which were claimed to possess greater accuracy than traditional circuit methods, was used to assess the interference of a 500 kV HVDC power line on a gas pipeline in southern China.[18] It was found that a large DC stray current entered the earth in monopolar mode, causing a rapid increase of the soil potential near the earth electrode, while the pipeline potential raised only slightly. The HVDC interference potential was thus generated. Moreover, a good external coating resulted in high HVDC interference on the pipeline. The modeling results were verified by the field measurement data. Furthermore, the numerical modeling technology was also used to study the coupling interaction between the railway and the pipeline network, as well as the corrosion risk of the pipelines.[71] The modeling results showed that the corrosion rate of the pipeline was higher at the rail crossings than at the tracks.

## 10.7.2. Mitigation of DC corrosion

To date, there has been a general agreement regarding the criteria for corrosion mitigation of pipelines in the presence of DC interference from sources such as HVDC power lines, a third-party CP system, or electric traction systems.[72, 73] An effective approach for the control of DC corrosion of buried pipelines is the reduction of the DC current density flowing through the pipe steel. The relationships between DC current density, relevant resistances, and DC voltage are shown as:

$$I_{DC} = V_{DC}/(R_{pipe} + R_{soil} + R_{leak\text{-}loading} + R_{leak\text{-}substation}) \tag{10-10}$$

$$R_2 = \rho \frac{l}{A} \tag{10-11}$$

where $I_{DC}$ is DC current; $V_{DC}$ is DC voltage; $R_{pipe}$, $R_{soil}$ (aka $R_2$), $R_{leak\text{-}loading}$, and $R_{leak\text{-}substation}$ are pipeline resistance, lateral resistance of the soil, leakage resistance between the current-loading side and the soil, and leakage resistance between the substation and the soil, respectively; $\rho$ is soil resistivity; $l$ is the distance between the load terminal and the substation; and $A$ is the lateral cross-sectional area of the soil. Since the value of $A$ tends to be infinity, $R_{soil}$ is approximately zero, and Equation (10-10) can be written as:

$$I_{DC} = V_{DC}/(R_{pipe} + R_{leak\text{-}loading} + R_{leak\text{-}substation}) \tag{10-12}$$

Thus, DC current decreases with the increase of $R_{leak\text{-}loading}$ and $R_{leak\text{-}substation}$, as well as the reduction of $V_{DC}$. For buried pipelines, the main methods to control DC corrosion are decreasing the voltage of leakage DC and increasing the leakage resistance.

Generally, techniques or methods that can be used to reduce or eliminate DC interference include the use of electrical bonds, galvanic anodes, coatings, and electrical shields. Grounding systems for an HVDC control station should be situated some distance from the station to eliminate the possibility of corrosion of the underground components.[16] The grounding installation must exclude the possibility of dangerous "step voltages" generating in the grounding electrodes zone. Measures should be applied to prevent the ground from losing its properties as a conductor.

Coatings and CP are primary techniques for controlling the external corrosion of pipelines. The CP criterion to control DC interference on freely corroding metallic structures is that the positive potential shift (excluding ohmic drop) is lower than 20 mV for steels. For the pipelines under CP, the anodic interference is considered unacceptable if the ohmic-free potential is outside the potential range (i.e., if the ohmic-free potential of the pipe steel does not match the –0.850 V [CSE] criterion).[74, 75] The CP is usually operated by potentiostatic mode, where continuous measurements of the pipe-to-soil potential are conducted relative to a reference electrode (e.g., CSE) to drive an automatic DC feeder. The feeder adjusts the applied current from the external power supply (i.e., rectifier) to maintain the measured pipeline potential at the desired protection value. The measured pipe-to-soil potential (i.e., the ON potential) consists of the ohmic-free potential of the pipeline and the ohmic drop due to the current flowing in the soil. Since the ohmic drop greatly affects the potential records, especially in dry soils or in the presence of stray currents, various techniques have been used to eliminate the ohmic drop contribution. These include reduction of the distance between the reference electrode and the pipeline or interruption of the current pathway to measure the instant-OFF potential. Under DC interference, the ON potential of the buried pipelines measured through the reference electrode would keep changing owing to the unsteady stray currents. Since the stray currents usually last for a short time (e.g., a few minutes when a train passes by), there is not sufficient time for the pipe steel to change the polarization status. The ohmic-free potential of the pipeline may be assumed to be constant.[76, 77] As a result, the measured ON potential that controls the DC feeder based on the constant potential varies in response to the change in ohmic drop in the soil. Thus, the CP feeder does not recognize the true potential of the pipeline. Instead, the changes of the ohmic drop in the soil are indicative of the varying cathodic current in the soil. To address this problem, a new approach is developed to measure and maintain the pipe-to-soil potential constant using a potential probe, which consists of a carbon steel coupon and an internal reference electrode, to contrast stray current.[12] The potential probe is effective at controlling the stray current interference, and the current, rather than the potential, is used to control the DC feeder, which can adjust the current according to the variation measured by the probe in order to maintain the pipe-to-soil potential at the appropriate value.[73]

### 10.7.3. Modeling and calculation of DC interference on pipelines

As with HVAC power lines, the interference induced by HVDC power lines on collocated pipelines can be modeled to analyze the electric field distribution around the ground electrode in the soil and the effect of DC interference on the pipelines. Consider a coated pipeline with defects (pinholes) present on the coating, the pipe steel is exposed to the soil directly at the defects. As the coating resistivity is much higher than the resistivity of the soil, the current of the ground-coating loop circuit at the defects is negligible. The resistance of the coating defect is equivalent to the sum of the contact resistance between the steel and the soil, and the resistance of the soil inside the defect. Compared with the pipe size, the diameter of the defect is usually small, so the curvature of the surface of the defect can be ignored. Thus, the surface of the defect is approximated as a plane. An equivalent circuit of the coated pipeline containing defects is then developed on the basis of these assumptions. According to the transmission line theory,[78] the pipeline without defects is subdivided infinitely into small sections. The potential difference generated by the grounding current of HVDC is equivalent to a voltage source in each section, and the stray current and the pipe-to-soil potential are then calculated. For the coating containing defects at individual locations, the local admittance is calculated, and the formulas for the calculation of stray current and pipe-to-soil potential are adjusted correspondingly.

Yu et al. modeled and in situ measured the stray current and pipe-to-soil potential of an oil pipeline that was adjacent to 800 kV ultrahigh voltage DC power lines by considering the influences of potentiostat output, coating defects, and the insulation joint.[79] The stray current and the pipe-to-soil potential were calculated. The correctness of the algorithm was checked by comparison of the calculation results and the measured data. It was found that the DC stray current affected the pipe-to-soil potential remarkably, with the potential offset beyond the allowable value according to the industry standard. The potentiostat could not provide an effective protection for the pipeline when the HVDC was operated in single-pole mode with a transmission load up to 2,500 MW. Actually, the protective effect of the output current of the potentiostat decreased with the increase in stray current. The modeling results were coincident with the measured data. It was suggested that the DC interference modeling provides a meaningful basis for the development of a comprehensive risk assessment system for buried pipelines in the presence of DC interference.

Zhang et al. reported the use of COMSOL Multiphysics software to simulate, for buried pipelines coated with epoxy, the corrosion dynamics induced by DC stray current in a soil with a resistivity of 50 Ω·m.[80] The software was used to calculate the distribution and intensity changes of electrolyte potential in the cathodically protected system by solving Laplace's three-dimensional equation. The modeling results showed that an increased DC voltage caused positive shifts of the pipe-to-soil potential, accelerating the steel corrosion. When the pipeline was protected with a sacrificial anode, DC corrosion was reduced. Thus, CP was able to mitigate the DC corrosion.

## References


1. Schwalm, L., Sandor, J., "Stray current: The major cause of underground plant corrosion," *MP* 6: 31–36, 1969.
2. Martin, B., "A history of stray current corrosion," *Corrosions & Materials* 31: 12–14, 2006.
3. Henderson, M., Gagnon, J., Bertagnolli, D., Hosie, B., DeShazo, G., Silverstein, B., "Building a plan for HVDC," *IEEE Power Energy Magazine* 5: 52–60, 2007.
4. Figueroa-Acevedo, A.L., Czahor, M.S., Jahn, D.E., "A comparison of the technological, economic, public policy, and environmental factors of HVDC and HVAC interregional transmission," *AIMS Energy* 3: 1–1 8, 2015.
5. Bertolini, L., Carsana, M., Pedeferri, P., "Corrosion behaviour of steel in concrete in the presence of stray current," *Corrosion Science* 49: 1056–1068, 2007.
6. Andrade, C., Martinez, I., Castellote, M., "Feasibility of determining corrosion rates by means of stray current-induced polarisation," *Journal of Applied Electrochemistry* 38: 1467–1476, 2008.
7. Zhu, Q., Cao, A., Wang, Z., Song, J., Chen, S., "Stray current corrosion in buried pipeline," *Anti-Corrosion Methods and Materials* 58: 234–237, 2011.
8. Nicholson, P., "High voltage direct current interference with underground/underwater pipelines," CORROSION/2010, paper no. 10102, NACE, Houston, TX, 2010.
9. Verhiel, A.J., "The effects of high-voltage DC power transmission systems on buried metallic pipelines," *IEEE Transactions on Industry and General Applications* 3: 403–415, 1971.
10. Langelund, E.S., "Pipeline stray current (DC) interference," CORROSION/2018, paper no. 10689, NACE, Houston, TX, 2018.
11. Lazzari, L., Pedeferri, P., *Cathodic Protection*, Polipress, Milan, 2006.

12. Brenna, A., Lazzari, L., Ormellese, M., "Stray current control by a new approach based on current monitoring on a potential probe," *Corrosion Engineering, Science and Technology* 52: 359–364, 2017.
13. Didas, J.L., "Cathodic protection interference: Interference testing and mitigation of interference issues," CORROSION/2018, paper no. 10656, NACE, Houston, TX, 2018.
14. Hu, Y., Li, J., "DC interference of impressed anodes on metallic structures buried in an adjacent gas station," CORROSION/2013, paper no. 5423, NACE, Houston, TX, 2015.
15. Shipley, R.W., Darwin, D., Locke, C.E., *Stray Current Corrosion due to Utility Cathodic Protection*, Kansas Department of Transportation, Topeka, 1997.
16. Koshcheev, L.A., "Environmental characteristics of HVDC overhead transmission lines," 3rd Workshop on Power Grid Interconnection in Northeast Asia, Vladivostok, 2003.
17. Rudervall, R., Charpentier, J.P., Sharma, R., "High voltage direct current (HVDC) transmission systems technology review paper," presented at Energy Week, Washington, DC, 2000.
18. Jiang, Z., Gu, Q., Cao, G., Wang, X., Ge, C., Du, Y., "Study on the mechanism and influencing factors of HVDC interference," CORROSION/2017, paper no. 9293, NACE, Houston, TX, 2017.
19. Molburg, J.C., Kavicky, J.A., Picel, K.C., *The Design, Construction, and Operation of Long-Distance High-Voltage Electricity Transmission Technologies*, Argonne National Laboratory, IL, 2007.
20. Gong, Y., Xue, C., Yuan, Z., Li, Y., Dawalibi, F.P., "Advanced analysis of HVDC electrodes interference on neighboring pipelines," *Journal of Power and Energy Engineering* 3: 332–341, 2015.
21. Qian, S., Cheng, Y.F., "Degradation of fusion bonded epoxy pipeline coatings in the presence of direct current interference," *Progress in Organic Coatings* 120: 79–87, 2018
22. Nejad, B.M., Blumhagen, L., "Inductive interference from HVDC transmission lines on close proximity pipeline segments," CORROSION/2018, paper no. 10913, NACE, Houston, TX, 2018.
23. CAPP (Canadian Association of Petroleum Producers), *Influence of High Voltage DC Power Lines on Metallic Pipelines*, Publication 2014-0034, Calgary, Alberta, 2014.
24. Prabhu, S., "Stray current corrosion and preventive measures," Corrosion-Pedia, www.corrosionpedia.com/stray-current-corrosion-and-preventive-measures/2/1623, 2017.
25. Kustova, V., Lindemuth, D., Kuykendall, C., "DC transit stray current corrosion control: Everyone wins when you work together," CORROSION/2018, paper no. 11270, NACE, Houston, TX, 2018.
26. Llewellyn, E., Sharon, Z., "Long-term monitoring of influence from DC transit stray currents on underground pipelines," CORROSION/2018, paper no. 11215, NACE, Houston, TX, 2018.
27. Chen, Z., Qin, C., Tang, J., Zhou, Y., "Experiment research of dynamic stray current interference on buried gas pipeline from urban rail transit," *Journal of Natural Gas Science & Engineering* 15: 76–81, 2013.

28. British Standards Institution, *Protection against Corrosion by Stray Current from Direct Current Systems*, BS EN 50162:2004, London, 2005.
29. Chinese Standard, *Inspection of Corrosion Protection for Buried Steel Pipelines*, GB/T 19285-2014, Beijing, 2014.
30. Yang, Y., He, R., Li, G., Han, Y., Li, Z., "Measurement and reduction of stray current on buried steel pipeline," *Corrosion & Protection* 33: 324–327, 2012.
31. Bortels, L., *Study and Evaluation of Stray Current Influences on Cathodic Protection Systems of Buried Pipelines*, Elsyca, Kranenberg, Belgium, 2000.
32. Song, Y.W., Wang, X.H., He, R.Y., He, C.F., Wang, X.Y., "Status in research on stray-current corrosion of buried steel pipelines," *Corrosion & Protection* 30: 515–525, 2009.
33. Revie, W., *Corrosion and Corrosion Control: An Introduction to Corrosion Science and Engineering*, 4th ed., Wiley, Hoboken, NJ, 2008.
34. Barlo, T.J., Zdunek, A.D., *Stray Current Corrosion in Electrified Rail Systems: Final Report*, Northwestern University, Evanston, IL, 1995.
35. Wang, J., Li, Z., Cui, G., Liu, J., Kong, C., Wang, L., Gao, G., Guo, J., "Corrosion behaviors of X70 steel under direct current interference," *Anti-Corrosion Methods and Materials* 66: 307–316, 2019.
36. Ogunsola, A., Mariscotti, A., Sandrolini, L., "Estimation of stray current from a DC-electrified railway and impressed potential on a buried pipe," *IEEE Transactions on Power Delivery* 27: 2238–2246, 2012.
37. Cao, A., Zhu, Q., Hou, B., Zhang, S., "Stray current corrosion on oil and gas pipeline and its protection," *Gas & Heat* 29: 6–9, 2009.
38. Brichau, F., Deconinck, J., Driesens, T., "Modeling of underground cathodic protection stray currents," *Corrosion* 52: 480–488, 1996.
39. Qian, S., Cheng, Y.F., "Accelerated corrosion of pipeline steel and reduced cathodic protection effectiveness under direct current interference," *Construction and Building Materials,* 148: 675–685, 2017.
40. Brichau, F., Deconinck, J., "A numerical model for cathodic protection of buried pipes," *Corrosion* 50: 39–49, 1994.
41. Barbucci, A., Delucchi, M., Cerisola, G., "Organic coatings for concrete protection: Liquid water and water vapour permeabilities," *Progress in Organic Coatings* 30: 293–297, 1997.
42. Mertzel, E., Koenig, J.L., "Application of FT-IR and NMR to epoxy resins," in *Epoxy Resins and Composites II*, Advances in Polymer Science, vol 75 (K. Dušek, ed.), Springer, Berlin, 73–112, 1986.
43. MijoviĐ, J., AndjeliĐ, S., "A study of reaction kinetics by near-infrared spectroscopy. 1. Comprehensive analysis of a model epoxy/amine system," *Macromolecules* 28: 2787–2796, 1995.
44. Cao, P., Gu, R., Tian, Z., "Electrochemical and surface-enhanced Raman spectroscopy studies on inhibition of iron corrosion by benzotriazole," *Langmuir* 18: 7609–7615, 2002.
45. González, M.G., Baselga, J., Cabanelas, J.C., *Applications of FTIR on Epoxy Resins Identification, Monitoring the Curing Process, Phase Separation and Water Uptake*, IntechOpen, New York, 2012.
46. Chike, K., Myrick, M., Lyon, R., Angel, S., "Raman and near-infrared studies of an epoxy resin," *Applied Spectroscopy* 47: 1631–1635, 1993.

47. Gupta, G., Birbilis, N., Cook, A.B., Khanna, A.S., "Polyaniline-lignosulfonate/ epoxy coating for corrosion protection of AA2024-T3," *Corrosion Science* 67: 256–267, 2013.
48. Al-Borno, A., Chen, X., Kewaldas Dhoke, S., "Effect of high temperature sodium hydroxide immersion on fusion bond epoxy coating," *International Journal of Corrosion* 2015: 1–7, 2015.
49. Musto, P., Mascia, L., Ragosta, G., Scarinzi, G., Villano, P., "The transport of water in a tetrafunctional epoxy resin by near-infrared Fourier transform spectroscopy," *Polymer* 41: 565–574, 2000.
50. Ji, M.G., Hu, J.M., Liu, L., Zhang, J.Q., Cao, C.N., "Water uptake of epoxy coatings modified with Đ-APS silane monomer," *Progress in Organic Coatings* 57: 439–443, 2006.
51. Ramezanzadeh, B., Attar, M.M., "Studying the corrosion resistance and hydrolytic degradation of an epoxy coating containing ZnO nanoparticles," *Materials Chemistry and Physics* 130: 1208–1219, 2011.
52. Brasher, D., Kingsbury, A., "Electrical measurements in the study of immersed paint coatings on metal. I. Comparison between capacitance and gravimetric methods of estimating water-uptake," *Journal of Chemical Technology and Biotechnology* 4: 62-72, 1954
53. Thomas, N.L., "The barrier properties of paint coatings," *Progress in Organic Coatings* 19: 101–121, 1991.
54. Buchanan, R., "A critical review of industry codes and standards as they relate to electrically resistive coatings and what that means to shielding," CORROSION/2013, NACE, Houston, TX, 2013.
55. Wei, Y.H., Zhang, L.X., Ke, W., "Comparison of the degradation behaviour of fusion bonded epoxy powder coating systems under flowing and static immersion," *Corrosion Science* 48: 1449–1461, 2006.
56. Xu, Y., Yan, C., Ding, J., Gao, Y., Cao, C., "Permeation behavior of water vapour in organic coatings of alkyd and polyurethane," *Corrosion Science and Protection Technology* 13: 493–496, 2001.
57. Xu, Y., Yan, C., Ding, J., Gao, Y., Cao, C., "Diffusion and existing form of water vapour in organic coatings," *Acta Physicochimica Sinica* 18: 649–652, 2002
58. Nazarov, A.P., Thierry, D., "Scanning Kelvin probe study of metal/polymer interfaces," *Electrochimica Acta* 49: 2955–2964, 2004.
59. Kuang, D., Cheng, Y.F., "AC corrosion at coating defect on pipelines," *Corrosion* 71: 267–276, 2015.
60. Wang, X., Wang, Z., Chen, Y., Song, X., Yamg, Y., "Effect of a DC stray current on the corrosion of X80 pipeline steel and the cathodic disbondment behavior of the protective 3PE coating in 3.5% NaCl solution," *Coatings* 9: 1–1 2, 2019.
61. Davis, J.R., *Corrosion: Understanding the Basics*, ASM International, Materials Park, OH, 2000.
62. Gord, P., Walton, J., "New methods ease multiline stray-current measurement," *Oil and Gas Journal* 105, 36, 2007.
63. Bahadori, A., *Cathodic Corrosion Protection Systems*, Gulf Professional Publishing, Sydney, 2014.
64. Peelen, W.H.A., Neeft, E.A.C., Leegwater, G., Kanten-Roos, W.V., Courage, W.M.G., "Monitoring DC stray current interference of steel sheet pile structures in railway environment," *HERON* 56: 107–122, 2011.

65. Tang, D.Z., Du, Y.X., Liu, J., Lu, M.X., "Some considerations on the use of coupon under stray direct traction current interference," CORROSION/2018, paper no. 11028, NACE, Houston, TX, 2018.
66. Qin, R., Du, Y., Peng, G., Lu, M., Jiang, Z., "High voltage direct current interference on buried pipelines: case study and mitigation design," CORROSION/2017, paper no. 9049, NACE, Houston, TX, 2017.
67. Zakowski, K., Darowicki, K., "Potential changes in an electric field and electrolytic corrosion," *Anti-Corrosion Methods and Materials* 50: 25–33. 2003.
68. Freiman, L.I., "Stray-current corrosion criteria for underground steel pipelines," *Protection of Metals* 39: 172–176, 2003.
69. Hosokawa, Y., Kajiyama, F., "Case studies on the assessment of AC and DC interference using steel coupons with respect to current density CP criteria," CORROSION/2006, paper no. 06161, NACE, Houston, TX, 2006.
70. Hilleary, J., DeWitt, J., Krissa, L., "Using remote data-logging techniques to evaluate transit system interference," CORROSION/2015, paper no. 5918, NACE, Houston, TX, 2015.
71. Baeté, C., Bortels, L., "Investigating railway corrosion caused by cathodic protection systems," CORROSION/2015, paper no. 6015, NACE, Houston, TX, 2015.
72. CEI (Comitato Elettrotecnico Italiano), *Protection against Corrosion by Stray Current from Direct Current Systems*, CEI EN 50162, Rome, 2005.
73. NACE, *Control of External Corrosion on Underground or Submerged Metallic Piping Systems*, SP0169-2013, Houston, TX, 2013.
74. ISO (International Organization for Standardization), *Petroleum, Petrochemical and Natural Gas Industries–Cathodic Protection of Pipeline Systems–Part 1: On-Land Pipelines*, ISO 15589-1, Geneva, 2015.
75. Angst, U., Büchler, M., Martin, B., Schöneich, H.G., Haynes, G., Leeds, S., Kajiyama, F., "Cathodic protection of soil buried steel pipelines: A critical discussion of protection criteria and threshold values," *Materials and Corrosion* 67: 1135–1142, 2016.
76. W. Von Baeckmann, W., Prinz, W., *Electrochemical Measuring Techniques of Stray Current Influenced Pipelines*, Dechema Monographs, Vol. 101, VCH Verlagsgesellschaft: 65–177, 1986.
77. Bazzoni, B., Lazzari, L., "The lateral gradient technique for potential measurements in presence of stray current," CORROSION/96, paper no. 96202, NACE, Houston, TX, 1996.
78. Lagace, P.J., Houle, J.L., Greiss, H., Mukhedkar, D., "Computer aided evaluation of pipeline current near toroidal HVDC ground electrodes," *IEEE Transactions on Power Delivery* 4, 216–222, 1989.
79. Yu, Z., Liu, L., Wang, Z., Li, M., Wang, X., "Evaluation of the interference effects of HVDC grounding current on a buried pipeline," *IEEE Transactions on Applied Superconductivity* 29: 45–54, 2019.
80. Zhang, Y., Feng, Q., Xue, H., Yu, L., Wu, C.M., Ng, S.P., Tang, X., "Numerical modelling of buried pipelines under DC stray current corrosion," *Journal of Electrochemical Science and Engineering* 9: 125–134, 2019.

# Index


## A

AC accelerated pipeline corrosion, 43–48, 44*f*, 45*f*, 46*f*, 47*f*, 48*f*, 49*t*
AC corrosion of pipelines, 5–6, 8–9, 25–26. *see also* pipelines
  AC interference management program for pipelines, 214–219
  active dissolution systems, 43–52
  characteristics of, 27–28, 27*f*
  at coating defects, 60–74
  control of, 201–219
  DC decoupling technique, 211–212
  electrolyte alkalization mechanism during, 30–31, 31*f*
  general principles for mitigation and control of AC interference, 201–203
  gradient control facilities, 212–213
  industry standards and practical recommendations, 213–214
  inspection and monitoring techniques for, 163–177
  measurements, experimental setup for, 41–42, 42*f*
  mechanisms, 28–34
  mitigation of, 201–219
  overvoltage arresters, 208–210
  parametric effects, 34–41
  passive systems, 52–60
  schematic diagram illustrating process of, 30*f*
  surface and deep grounding, 203–208
  transformer use for AC interference mitigation, 210
AC corrosion of steels
  double-charge layer effect, 52
  evolution of hydrogen and oxygen, 52
  vibrating effect, 52
AC current density, 40–41
  and AC corrosion, 20, 36–37
  and coating defect, 34, 38–39
  and CP-applied pipelines, 151–152
  and DC potential, 46, 47*f*
  measurement, 166
  modeling on pipelines, 192–194, 193*f*, 194*f*
  ratio to DC current density, 166–167
AC frequency, 36, 52, 190
AC harmonic interference, 190–191, 191*f*
AC-induced passivity degradation, 53–56, 53*f*, 54*f*, 54*t*, 55*f*, 56*f*, 57*f*
AC-induced pitting corrosion. *see also* corrosion
  conditions for initiation of, 81–87, 82*f*, 83*f*, 84*f*, 86*f*

high pH aqueous environments, 90–93, 90*f*, 92*f*, 93*f*
mechanisms for, 94–97, 95*f*, 96*f*
near-neutral pH aqueous environment, 87–89, 88*f*, 89*f*
threshold conditions for initiation of, 94
AC interference mitigation
copper materials, 207
planned, 218
purposes of, 201
steel materials, 207
transformer use for, 210, 210*f*
zinc materials, 206
and zinc ribbons, 193–194
AC-powered rail systems, 14, 14*f*
AC-powered rail transit systems, 6–7
active dissolution systems
AC accelerated pipeline corrosion, 43–48, 44*f*, 45*f*, 46*f*, 47*f*, 48*f*, 49*t*
AC corrosion of pipelines, 43–52
environments resulting in, 43
mechanistic analysis, 49–52
AC voltage, 36
CIS modeling on pipelines, 195
and coating resistance, 188–189, 189*f*
modeling on pipelines, 192–194, 193*f*, 194*f*
parametric effects, 36
pipeline diameter and, 189–190, 190*f*
and pipelines burial depth, 5–6
AC voltage modulation technique, 169–171, 172*f*
adhesion, and coatings, 100–101
Ahlborn, G.H., 29
alternating current (AC) interference, 8. *see also* AC corrosion of pipelines
associated risk, 20–22
coating structure and performance degradation under, 113–116, 113*f*, 114*f*, 115*f*, 116*t*
collecting and analyzing data, 202
conceptual model for coating degradation under, 117, 117*f*
corrosion rate of CP-applied pipeline steels under, 144–151
CP criteria for pipeline corrosion protection under, 153–158, 154*f*, 155*f*, 156*t*, 157*f*
CP potential shift under, 131–143
dependence of the water resistance of coatings on, 108–110, 109*f*, 110*f*
designing AC mitigation system, 202
effect on resistance of coatings, 100–105
from electrified rail systems, 22–23, 22f–23*f*
general principles for mitigation and control of, 201–203
by HVAC power transmission lines, 15–19, 16*f*, 17*f*, 19*f*
management program for pipelines, 214–219
measuring water uptake by coatings under, 108, 108*f*
microbial corrosion of buried pipelines under, 59–60
modeling and evaluation of, 181–191
pipeline corrosion in presence of, 25–74
pipeline failure under, 4–7
potentiostatic current density of CP-applied pipeline steels under, 144–151, 144*f*, 145*f*, 147*f*, 149*f*
principles for modeling, 181–182
reliability analysis of, 198
sources of, 13–15
testing and monitoring results, 202
testing for cathodic disbonding under, 101–102, 102*f*
testing for resistance of coatings under, 101–102, 102*f*
American Gas Association, 180
American Railway Engineering and Maintenance-of-Way Association, 181
anodic zones
effect of DC current density on solution pH in, 236, 236*f*
solution pH in, 236*f*, 240
time dependence of weight of FBE-sealed test containers in, 242*f*
Association of American Railroads, 180–181

## B

backfill materials, 207–208
backflashover, 228
Baeckmann, W.V., 37
blistering, 107
Brenna, A., 152, 157
British Columbia Hydro, 224
British Standard BS EN 15280, 37
British Standard BS EN 50162:2004, 230
buried coupon test station, 164–168, 165*f*
Butler–Volmer equation, 32

## C

Canadian Energy Pipeline Association (CEPA), 1, 3, 214
CAN/CSA-C22.3 No. 6-M91, 213
capacitive coupling, 8, 15, 22, 227
capacitive coupling effect, 15–16, 16*f*, 22
Carpentiers, P., 154
cathodic disbonding
  and cathodic protection, 101
  concept and principle, of coatings, 100–101
  of pipeline coatings under AC interference, 102–104, 105*f*
  testing under AC interference, 101–102, 102*f*
cathodic protection (CP), 4, 9. *see also* pipelines
  and buried pipelines, 119
  and cathodic disbonding, 101
  criteria for pipeline corrosion protection under AC interference, 153–158, 154*f*, 155*f*, 156*t*, 157*f*
  driven cathodic reactions, 53
  effectiveness for corrosion protection and AC, 151–153
  "external feedback," 122, 122*f*
  "internal feedback," 122, 122*f*
  and permeable coatings, 52
  potential, 101
  principle of impressed current, 120–122
cathodic zones
  effect of DC current density on solution pH in, 236, 236*f*
  solution pH in, 236*f*
  time dependence of weight of FBE-sealed test containers in, 242*f*
CEN/TS-15280 standard, 214
chemical-mechanical models, 97
Cheng, Y.F., 155
Chin, D.T., 59
Chinese Standard GB/T 19285-2014, 230
coating defects
  and AC corrosion, 38
  AC corrosion of pipelines at, 60–74, 62*f*
  and AC current density, 34, 38–39
  corrosion morphology, 65*f*–66*f*
  and DC potentials, 68f
  shape of, 71–74, 71*f*, 72*f*, 73*f*, 74*f*
  size of, 60–71
coating properties and performance. *see also* coatings
  effect of DC on, 241–250
  electrochemical impedance spectroscopy, 243–245, 244*f*, 245*t*
  molecular structure of coatings, 246–249, 246*f*, 247*f*, 248*f*, 249*t*
  role of DC interference on coating degradation, 249–250, 250*f*
  water permeability, 241–243, 241*f*, 242f, 243*f*
coating resistance, 188–189, 189f
coatings, 8
  and AC corrosion, 38
  characteristics of pipeline, 100
  concept and principle of cathodic disbonding of, 100–101
  effect of AC on electrochemical impedance of, 110–113, 111f, 112*f*
  essential properties, for pipeline integrity maintenance, 99–100
  "fail-safe," 52
  high-performance composite coating, 60
  molecular structure of, 246–249, 246*f*, 247*f*, 248*f*, 249*t*
  permeable, 52
  properties associated with water resistance, 106–107
  resistance testing under AC interference, 101–102, 102*f*

structure under AC interference, 113–116, 113*f*, 114*f*, 115*f*, 116*t*
testing for resistance of, 101–102
Code of Federal Regulations (CFR) of the U.S. Department of Transportation, 123
Cohn, P., 29
computational models
for assessment of AC interference on pipelines, 192–197
modeling of AC voltages and AC current densities on pipelines, 192–194, 193*f*, 194*f*
modeling of fault currents, 195–197, 196*f*
optimization of power line design, 192
COMSOL Multiphysics software, 262
conductive backfills, 208
conductive coupling, 180, 227
control of AC corrosion of pipelines, 201–219
copper materials, for grounding, 207
corrosion
AC (*see* AC corrosion of pipelines)
external, 4
internal, 3
localized, 231
and pipeline failure, 3–4
corrosion of pipelines under DC interference, 223–262
adverse effect of DC interference on CP performance, 251–257
DC corrosion measurements, 232–238
effect of DC on coating properties and performance, 241–250
features and parametric effects on, 230–231
management of, 257–262
mechanism of DC corrosion, 239–241
overview, 223–224
sources of DC interference, 224–230
corrosion or passivation protection, 208
corrosion potential, 46, 50
coupling effect
capacitive, 15–16, 16*f*, 22
inductive, 18–19, 19*f*, 23
resistive, 16–18, 17*f*, 22
coupon-to-soil potential, 166
CP current density, 40–41
CP monitor, 167–168
CP permeation
effect of AC on, 124–131, 125*f*, 126*f*, 128*f*, 129*f*, 131*f*
mechanistic analysis, 129–130, 131*f*
testing results, 124–127, 125*f*, 126*f*, 128*f*
CP potential, 40–41, 81–82
applied on more negative than NACE-recommended value, 136–138, 136*f*, 137*f*, 138*f*, 139*f*
NACE-recommended, 132–134, 133*f*, 134*f*, 135*f*
shift, mechanistic aspect of effect of AC on, 140–143, 142*f*
shift under AC interference, 131–143
CP shielding
described, 123
and water resistance, 123–124

## D

data acquisition (DAQ) system, 41–42
DC corrosion, 25
characteristics of, 230–231
effect on coating properties and performance, 241–250
high corrosion rate, 230–231
localized corrosion, 231
measurements, 232–238
mechanism of, 239–241
mitigation of, 259–261
DC corrosion measurements, 232–238
corrosion rate and morphology of steel under various DC current densities, 237–238, 237*f*, 238*f*
effect of DC current density on solution pH in anodic and cathodic zones, 236, 236*f*
effect of DC current density on steel potential, 233–235, 234*f*, 235*f*
testing setup enabling, 232–233, 233*f*
DC corrosion of pipelines
detection of DC voltage and stray current for evaluation of DC interference, 257–259

features and parametric effects on, 230–231
management of, 257-262
mitigation of DC corrosion, 259–261
modeling and calculation of DC interference on pipelines, 261-262
parametric effects on, 231
recognition of, 230–231
DC current density
coupon, 153
effect on solution pH in anodic zones, 236, 236*f*
effect on solution pH in cathodic zones, 236, 236*f*
effect on steel potential, 233–235, 234*f*, 235*f*
ratio of AC current density to, 166–167
DC decoupling technique, 211–212, 211*f*, 212*f*
DC device, 224–225
DC potential
and AC current density, 46, 47*f*
corrosion morphology, 65f–66*f*
coupon, 165
measured on coated steel electrodes containing coating defects, 69*f*
potentiodynamic polarization curves, 63, 64*f*
recorded on steel electrode at various AC current densities, 132–133, 133*f*
time dependence of, 60–61, 61*f*, 63*f*
DC transit systems, 228–229
deep drilled ground wells, 205
deep grounding, 205, 205*f*
depassivation-repassivation theory, 96
dielectric strength, 107
direct conductive coupling, 227
direct current (DC) interference, 9
adverse effect on CP performance, 251–257
corrosion of pipelines under, 223–262
DC device, 224–225
DC transit systems, 228–229
detection of DC voltage and stray current for evaluation of, 257–259
HVDC power lines, 226–228
identification of DC interference, 230
modeling and calculation of, 261–262
sources of, 224–230
double-charge layer effect, 52, 84, 84*f*

## E

Electric Power Research Institute, 180
electrified rail systems, 14
AC interference from, 22–23, 22f–23*f*
electrochemical impedance spectroscopy, 243–245, 244*f*, 245*t*
electrochemical impedance spectroscopy (EIS) scans, 206
electroendosmosis, 107
electrolyte alkalization
mechanism during AC corrosion, 30–31, 31*f*
two step and AC corrosion, 32
Electromagnetic and Conductive Coupling Analysis, 180
electromagnetic coupling. *see* inductive coupling effect
Enbridge, 2, 4
enhanced surface area, 208
European Committee for Standardization, 214
"external feedback," CP systems, 122, 122*f*

## F

"fail-safe" coatings, 52. *see also* coatings
Faraday's law, 54
fault currents, 195–197, 196*f*
FBE coating membrane, 241, 241*f*
Federal Energy Regulatory Commission (FERC), 3
flake-off mechanism, 152
Fourier transform infrared (FTIR) spectrum
characteristic bonds of epoxy coatings, 116*f*
of FBE coating, 114, 115*f*
of liquid epoxy coating, 115*f*
Fu, A.Q., 155

fusion bonded epoxy (FBE) coating, 241. *see also* coatings
cross-sectional view of, 114, 114*f*
degradation in the presence of DC interference, 250*f*
FTIR spectrum of, 115f, 248*f*
and North American pipelines, 60
SEM images of the cross-sectional morphology of, 246–247, 246*f*, 247*f*
water transmission rates of, 243, 243*f*

## G

German Standard DIN 50925, 37
Goidanich, S., 32
gradient control facilities, 212–213
graphical user interface-based AC corrosion monitoring technique, 176–177
grounding
deep, 205, 205*f*
materials, 206–207
point drain, 205
surface, 204, 204*f*
grounding materials, 206–207
copper materials, 207
new grounding electrode, 207
steel materials, 207
zinc materials, 206
Gummow, R.A., 28, 151

## H

Hamlin, A.W., 29
high-performance composite coating (HPCC), 60
high pH aqueous environments, 90–93, 90*f*, 92*f*, 93*f*
high pH carbonate-bicarbonate environments, and pitting corrosion, 85–87, 86*f*
high pH corrosion, 53
high-voltage alternating current (HVAC) power transmission lines, 4–5, 163, 179–180
and buried pipelines, 35
and collocation length of pipeline, 187, 188*f*
fault conditions of, 21–22
normal conditions of, 20–21
orientation between the pipeline and, 188
pipelines in collocation with, 5–6
and separation between pipeline, 186–187, 187*f*
as source of AC interference, 13–14, 14*f*
types and principles of AC interference by, 15–19, 16*f*, 17*f*, 19*f*
Hilleary, J., 259
HVAC sources, 182–184, 183*f*, 184*f*
HVDC power lines, 226–228
hydroscopicity, 208
hygroscopic backfills, 208

## I

ICCP systems, 207
identification of DC interference, 230
impressed current cathodic protection (ICCP), 120–121
in pipeline corrosion protection, 121–122, 122*f*
Indian Oil Corporation Limited, 6
inductive coupling, 18–19, 19*f*, 23, 23*f*, 188, 196, 227
inductive coupling effect, 18–19, 19*f*, 23
industry standards and practical recommendations
AC corrosion of pipelines, 213–214
INGAA Foundation, 214
in-line inspections (ILIs), 2, 3
"internal feedback," CP systems, 122, 122*f*
ISO 18086, 214

## J

Jiang, Z., 38
Johar, U.M., 188
Jones, D.A., 29

## K

Kajiyama, F., 153, 167
Kansas Department of Transportation, 225
Kioupis, N., 209
Kustova, V., 229

## L

Lalvani, S.B., 50
Land, J., 210
lightning-related flashover modes
  backflashover, 228
  shielding failure, 228
Lin, X., 50
local acidification theory, 96
localized corrosion, 231
Luk, M., 207

## M

management of DC corrosion of pipelines, 257–262
McCollum, B., 29
measurement techniques
  for AC corrosion of pipelines, 164–171
  AC voltage modulation technique, 169–171, 172*f*
  buried coupon test station, 164–168, 165*f*
  graphical user interface-based AC corrosion monitoring technique, 176–177
  online, for AC corrosion, 172–175, 173*f*, 174*f*, 175*f*, 176*f*
  ultrasonic-based coupon testing method, 168–169, 169*f*
mechanism of DC corrosion, 239–241
mechanistic analysis
  of AC accelerated corrosion of pipeline, 49–52
  CP permeation, 129–130, 131*f*
  passive systems, 57–60
mechanistic aspect of role of DC interference on coating degradation, 249–250, 250*f*
metallurgical microstructure of steels and AC corrosion, 41
microbial corrosion of buried pipelines under AC interference, 59–60
mitigation
  of AC corrosion of pipelines, 201–219
  of AC interference, general principles for, 201–203
  of DC corrosion, 259–261
modeling and calculation of DC interference on pipelines, 261–262
moisture vapor transfer rate, 106–107
molecular structure of coatings, 246–249, 246*f*, 247*f*, 248*f*, 249*t*
mutual capacitive coupling, 227
mutual coupling effect, 197, 197*f*
mutual inductive coupling, 227

## N

NACE
  SP0177, 213
  SP21424-2018, 214
Nakamura, Y., 153, 167
National Institute of Science and Technology, 25
National Instruments DAQ data-logging software package, 172
near-neutral pH aqueous environment, and AC-induced pitting corrosion, 87–89, 88*f*, 89*f*
near-neutral pH bicarbonate environments, and pitting corrosion, 81–84, 82*f*, 83*f*, 84*f*
new grounding electrode, 207
Nielsen, L.V., 29
Nyquist diagrams, 110–113, 111*f*, 112*f*

## O

Ohm's law, 257
online AC corrosion monitoring and assessment technique, 172–175, 173*f*, 174*f*, 175*f*, 176*f*
osmosis, 107
overvoltage arresters, 208–210
  alternating voltage arrester in continuous function, 209–210
  semiconducting, 208–209

## P

Pacific Gas and Electric Company, 218
parametric effects
  AC corrosion, 34–41
  AC current density, 36–37
  AC frequency, 36
  AC voltage, 36
  coatings and coating defects, 38

collocation of buried pipelines with HVAC power lines, 35
CP current density, 40–41
CP potential, 40–41
on DC corrosion of pipelines, 231
metallurgical microstructure of steels, 41
soil properties and chemistry, 38–40
passive systems
AC corrosion of pipelines, 52–60
AC-induced passivity degradation, 53–56, 53*f*, 54*f*, 54*t*, 55*f*, 56*f*, 57*f*
environments resulting in, 52–53
mechanistic analysis, 57–60
Pipeline and Hazardous Materials Safety Administration, 3
pipeline coatings. *see* coatings
pipeline corrosion. *see also* corrosion
AC accelerated (*see* AC accelerated pipeline corrosion)
in presence of AC interference, 25–74
pipeline corrosion under AC interference
European Committee for Standardization, 214
NACE International criteria, 213–214
pipeline failure
under alternating current interference, 4–7
and corrosion, 3–4
pipeline integrity management, 2–3
essential coating properties for, 99–100
pipelines. *see also* AC corrosion of pipelines
with adjacency to AC-powered rail transit systems, 6–7
burial depth and AC voltage, 5–6
collocation length and HVAC power transmission lines, 187, 188*f*
in collocation with HVAC power transmission lines, 5–6
corrosion of, under DC interference, 223–262
described, 1
development of AC interference management program for, 214–219
diameter and AC voltages, 189–190, 190*f*
economically viable for oil and gas transportation, 1
features and parametric effects on DC corrosion of, 230–231
growth, 1
management of DC corrosion of, 257–262
mitigation and control of AC corrosion of, 201–219
modeling and calculation of DC interference on, 261–262
orientation between HVAC power lines and, 188
recognition of AC corrosion of, 25–26
point defect models, 96
point drain grounding, 205
polar effect, 117
polarized potential, 165
potentiostatic current density
of CP-applied pipeline steels under AC interference, 144–151, 144*f*, 145*f*, 147*f*, 149*f*
primary mitigation, 203

**Q**

Quebec-New England Intertie, 224

**R**

railway electrification systems, 14
recognition of DC corrosion of pipelines, 230–231
resistive coupling effect, 16–18, 17*f*, 22
right-of-way (ROW), 4–5
road crossings, 163

**S**

sacrificial anode cathodic protection (SACP), 121
Safe Engineering Services & Technologies, 180
secondary mitigation, 203
Sheldon, C., 214
shielding failure, 228
Shwehdi, M.H., 188

soil properties
and AC corrosion, 38–30
soil resistivity, 184–186, 185*f*, 186*f*
solution pH
effect of DC current density on
in anodic and cathodic zones,
236, 236*f*
square wave polarization (SWP),
82, 82*f*
steel
corrosion rate and morphology
under various DC current
densities, 237–238, 237*f*, 238*f*
materials for grounding, 207
step potential, 20, 213
surface and deep grounding, 203–208
surface grounding, 204, 204*f*

## T

Tan, T.C., 59
Tang, D., 152
tertiary mitigation, 203
Thompson, N.G., 29
touch potential, 20
transformer use for AC interference
mitigation, 210
Transmission Line Model, 181
tubercles, 19

## U

ultrasonic-based coupon testing
method, 168–169, 169*f*
Union Gas, 5
U.S. Department of Energy, 1

## V

vibrating effect, 52, 117
voltage peak locations, 163

## W

Wakelin, R., 28, 214
Wan, H., 59
Warner, J., 168
water absorption, 106
water permeability, 241–243, 241*f*,
242*ff*, 243*f*
water resistance
coating properties associated with,
106–107
dependence, of coatings on AC
interference, 108–110, 109*f*, 110*f*
described, 106
water transmission rate (WTR), 109–110

## X

X52 pipeline steel, 233, 234*f*, 235*f*, 237,
251
X52 steel electrode, 234–235, 235*f*, 238*f*,
251–252, 252*f*
Xu, L.Y., 32

## Y

Yunovich, M., 29

## Z

Zhang, Y., 262
zinc grounding systems, 206
zinc materials, for grounding, 206

www.ingramcontent.com/pod-product-compliance
Ingram Content Group UK Ltd.
Pitfield, Milton Keynes, MK11 3LW, UK
UKHW020200290926
15471UKWH00005B/986